ZERO

ZERO
Worlds Beyond Carbon

Pete Rive

WORLD BENDERS
TĀMAKI MAKAURAU, AOTEAROA

Dedication

*To my darling daughters who make the world a
better place,*

Sophia and Lucia

And

to my World Maker,

Sarah

Also

to my parents who built my first World,

Bryan and Robin

Contents

Copyright

A Note on the Author

PETE RIVE has researched and written for journals, international conferences and contributed to books relating to design, innovation and creative collaboration. This is his second book. He currently lives in New Zealand, Aotearoa.

He has a PhD in digital design, and has worked in film, TV, and interactive design, including VR, AR, and virtual worlds for the past 30 years.

He is a consultant with S23M working with organisations on creative collaboration and improving knowledge flow through employee wellbeing and psychological safety.

You can follow him on Twitter #WorldBender. Join the Facebook group: World Bender. Continue reading and join the conversation on his blog: www.launchsite.co.nz/worldbender

Preface

On Tuesday, the 22nd October, 2019, my friend Jackie Grant generously hosted my book launch for, *Worldbending*, at her restaurant, Amano in downtown Auckland. As the sun set the blackened sky had a strange apocalyptic glow and an acrid smell hung in the air. Sirens wailed and the city was grid-locked as people tried to find any escape they could from the toxic fumes and clogged motorways. Claustrophobia blanketed the inhuman night. I had anticipated that many of my guests would not show because if ever there was a sign of the end of the world this was surely it. I was delighted that most of my friends and family did brave the poisonous air and there were at least one hundred guests at the event. The brick walls were graced with Cathy Carter's striking pieces of watery art; photos that showed a world bent by her creativity, hinting at other worlds that might be and probably were out there in the multiverse. Her talented literary husband, Paul Hewlett, interviewed me before the book signing, probing the dark recesses of the worlds within worlds.

Worldbending: a survivor's guide had been conceived when I was teaching creative collaboration at Auckland's university of technology, AUT. It had had a very academic gestation as it included a number of papers that I had presented at international conferences. I had finished it after I had left AUT and after I had reflected on what I perceived as the concerning and appalling state of tertiary education in New Zealand. My research had shown me that the parlous health of universities around the world is a result of the *'weaponisation of education'* and that students, teachers, and researchers have almost entirely lost all sense of agency and have become cowed by corporatised administrators – at least that's how I saw it. I mused, if the university was no longer *'the critic and conscience of society'* then who was going to provide us with wisdom? As someone who has researched and taught creativity I believe in the construction of reality – not just *'thinking by doing, but also doing by thinking'*. I believe that philosophy has played, and still plays, a pivotal role in building our worlds, and that as creative agents we all have a responsibility to question the worlds we build; then bend, and even break the worlds that are toxic, dangerous, and harmful to all things living and nonliving.

I have written *Zero: worlds beyond carbon,* because I continue to try and communicate how philosophy and ideas have a material and existential relevance to all worldbenders, breakers and makers. I have tried to untangle some of the justified beliefs, ideologies, and practices in a more accessible narrative about the way in which myths, religion, economics, politics and the arts have blossomed out of our creativity. Yet, I have concentrated on how they have also created a straight jacket for our imagination about our future. I have zeroed in on Aotearoa, New Zealand as an example of a country with a radical faith in markets, and therefore devotion to carbon pricing, to try and expose why it is that technocrats, educators, politicians and the public continue to experiment with solutions to climate change that are most likely to fail. Humanity is suffering from a massive mind meld, a life threatening case of groupthink, that has almost no sense of agency, but instead has finally capitulated to all those who say *'don't worry we've got this. It's the market stupid.'* The invisible hand, we are told, will magically fix the mess we are in, and we go along with it. Not because we believe those that tell us this, but because we have run out of ideas; don't believe we can do anything about it; and because we think to carry on business-as-usual is best because we think it won't upset our lifestyle. This book hopes to convince you that you can still do something about the economic system that is pushing us to the brink of oblivion. If you think there is zero you can do about it, read on – there are worlds beyond our obsession with carbon ignition. We all built this one, let's experiment with new ones beyond this carbonised world.

Introduction

This is the story of paradox, metaphor, creativity and hope. It is about the worlds we make up, the worlds we have lost but only half-forgotten, and the worlds we are yet to discover. While many might assume there is only one world I define a world as a world-view, and not one that is exclusive to humans, but includes all the things that are nonhuman, which is of course, the majority of things on this planet and in the universe. I am going to explore the worlds of *Zero Carbon* that builds on my previous book, *WorldBending: a survivor's guide*. This book is grounded in my local experience but cannot avoid the global context; both are strangely more real today because of the global pandemic that has stopped us hopping on planes and forced us into reflective isolation. I am filled with cautious optimism that *'the times they are a changing'*. I am hopeful that this narrative might resonate with you whether you live here in Aotearoa, or alternatively if you are unfamiliar with our little country you can still recognise our common dilemma – our planet is in crisis.

To many *Zero Carbon* means only one thing – the reduction of GHGs (greenhouse gases). There is so much more to unpack, to question, and to contemplate because this is no time to be relying on the default common sense approach. That has clearly failed us. The *Zero Carbon* slogan is a global call to action to reduce the anthropogenic or human-related damage caused by our addiction to growth, speed, energy, and consumerism. It is all made possible by turning fossilised energy from the Sun into an oil and gas explosion. The process takes the energy from the Sun that has been stored for millions of years and then, with our help, almost instantaneously converts it back into a gas. The carbon conversion cycle began at the dawn of time with an ancient nuclear explosion: eventually it morphed from a gas, using the energy from the Sun, into a biological form, forming microbes, plants and animals; then when those died, they were composted into rock, oil, and gas; then we set fire to them again in what amounts to a rapid conversion back into gas, and other carbon forms, with the *waste* given off as heat, light, and noise. Paradoxically, it is both destructive and creative – a part of the cycle of life.

In Aotearoa, New Zealand, where I live, as I write this we have a coalition government that consists of the Labour Party, the Greens, and New Zealand First – yes, I have just realised that sounds very Trumpian! The impetus to write this book started with a cross-party piece of legislation colloquially known as the Zero Carbon Bill, or the Climate Change Response (Emissions Trading Reform) Amendment Bill. New Zealand is only one of a few countries in the world who has enshrined climate mitigation in the law. This is potentially the most important legislation to pass in this country, or it could turn out to become a symbol of lost opportunities and the death knell of humanity.

The Labour party had previously tried to declare a climate emergency, however, this was voted down by the National party.[1] The debate in the New Zealand Parliament did not result in any party or parliamentarian denying that global heating is a very serious issue, however, political gamesmanship had the National and ACT opposition parties voting against the climate emergency declaration saying that *actions speak louder than words*. A country is ruled by the actions of voters and National voted in opposition to the climate emergency declaration. National's Stuart Smith called the climate emergency declaration a *stunt* and attempted to worry voters that big government would deprive them from a *life worth living*, that the government and its bureaucracy were trying to tell them how to live their life and what car they could drive. What Smith wanted was for New Zealand to meet our Paris Agreement targets with actions that were *affordable and effective* without changing our lives. However, Kiritapu Allan, the Minister of Conservation, replied in her speech, that Stuart Smith may say we want to continue to drive the cars that we want to drive, but actually, we can no longer just maintain the status quo because: "Papatūānuku, [the Earth Mother], is demanding something more of us.'[2]

According to the Facebook page of the Pacific Island Climate Action Network: "the question of who will pay for climate loss and damages in the world's most vulnerable

---

1. Following the 2020 election, on the back of a substantial victory, the Labour government successfully declared New Zealand is facing a climate emergency in November 2020. This declaration was passed into law on the 2nd December 2020 with 76 in favour and 43 opposed.

2. See Hansard, 2nd December 2020, https://www.parliament.nz/en/pb/hansard-debates/rhr/combined/HansDeb_20201202_20201202_08

nations – countries least responsible for their causes – remains. Ralph Regenvanu, a prominent Ni-Vanuatu politician, has a simple answer. He's behind the push to hold large countries accountable." The governments of the Pacific Islands did want to hear New Zealand recognise that there is a climate emergency. Words do have consequences and politicians will be held to account by the public record. Aupito William Sio, the Minister for Pacific Peoples, said that words matter and repeated the Pacifica saying, *'rocks will turn to dust but words will remain'*. He went on to say : "From a Pacific perspective, climate change is a consequence of the failure of the market. They know that, since the Industrial Revolution, the rich and wealthy have gotten richer, more wealthy, while the poor become poorer. And, since the Industrial Revolution, they have seen how the costs of pollution have been socialised and the benefits have been privatised."[3] According to the Minister of Climate Change, James Shaw, due to sea level rise it is already costing the Cook Islands 25% of their annual budget. We no longer have any choice – either we dramatically change our lifestyle or it will become our deathstyle. Decarbonisation is not some minor adjustment to the way we live. If we are going to keep below a temperature rise of 1.5°C then we will require a radical change to the way we eat, drink, and be merry. Our occupations, our houses, our transportation, land use and our food must all change radically.

The challenge will be for the Labour government to use their majority in the Parliament to prove that the previous political compromise with New Zealand First and National was no longer a hindrance to climate action. The

---

3.   See Hansard, 2nd December 2020, ibid.

declaration of a climate emergency will be applauded by many who want to see change but those who are resisting any wrinkles to their lifestyle will continue to try and force political compromise as they did with agricultural emissions. While National was in power it signed the Paris Agreement, however, it may seem disingenuous when they continue to try and slow climate action while in opposition.

As a political compromise the previous Labour government agreed to exempt methane and nitrous oxide emissions from the Zero Carbon legislation until 2025 as long as farmers make progress in reducing agricultural GHGs, and measuring them. After 2025 agriculture will be included in the ETS with a 95 percent discount. That is despite the fact that agriculture is our biggest carbon emitter, and that methane and nitrous oxide are many times more powerful at heating our planet than carbon dioxide. Quite simply, agriculture has been more about profit and not sustaining life by providing healthy nutritious food and avoiding ecological harm. We have been warned that we only have until 2030 to attempt to reduce GHGs or we face the prospect of a tipping point with runaway heating. In the first nine years of its life methane from agriculture has 84 times the heating potential of carbon dioxide, and over the period of 100 years nitrous oxide has the GWP (global warming potential) of almost 300 times that of CO_2. The amount of nitrous oxide may be small but when we are running out of time we should be reducing it as soon as possible. The irony of any delay in reducing agricultural emissions is that we are in the last decade before it may be too late and methane will simply give it a push by speeding up global heating by 84 times more than CO_2 which has been the biggest contributor to date. According to the Environmental

Defence Fund: "Cutting methane emissions is the fastest opportunity we have to immediately slow the rate of global warming, even as we decarbonize our energy systems".[4]

From an energy point of view the arable land used for livestock would be far more efficient at feeding humanity if farmers were to switch to plant based food and it would be much less wasteful of resources such as fossil fuels, water and soil. Livestock and export dependency have been perennial problems for New Zealand; today our agricultural exports feeds 40 million people. 88 percent of those consumers live thousands of kilometres away increasing agriculture's already large carbon footprint. In a 2016 report by the World Resources Institute they wrote that: "an ambitious animal protein reduction' – focused on reducing overconsumption of animal based foods in regions where people devour more than 60 grams of protein and 2,500 calories per day – holds the greatest promise for ensuring a sustainable future for global food supply and the planet."[5] The UN FAO has pointed out that: "In all, livestock production accounts for 70 percent of all agricultural land and 30 percent of the land surface of the planet."[6]

Many people believe that they need far more protein than they actually do. As the WHO's research has shown we only need 10-15 percent of our daily calories from protein and that is something easily achieved from a plant based diet. According to the book, *Drawdown*: "agricultural land use and associated energy consumption to grow livestock

4. https://www.edf.org/climate/methane-crucial-opportunity-climate-fight

5. Hawken, P. (Ed.). (2017). Drawdown: The most comprehensive plan ever proposed to reverse global warming.

6. United Nations Food and Agriculture Organization. (November 29, 2006) Livestock's long shadow: environmental issues and options. http://www.fao.org/3/a-a0701e.pdf

feed produce carbon dioxide emissions, while manure and fertilizer emit nitrous oxide. If cattle were their own nation, they would be the world's third-largest emitter of greenhouse gases."[7] Food marketing makes the product both a fashion and cultural byproduct that is buffered by large multinationals and their big advertising budgets.

New Zealand faces an uncertain future as new markets open up to non-dairy, non-meat products fuelled by massive IPOs (share market listing or initial public offering) such as the oat milk company, Oatly, with an opening valuation of USD$10 billion.[8] All around the planet politicians are negotiating with each other in consultation and negotiation with the scientists to reduce their countries output of carbon dioxide equivalence based on the science of GWP, or Global Warming Potential. Unfortunately, although New Zealand is only one of a few countries to have zero emissions enshrined in law, the Climate Action Tracker (CAT) has given our government a fail mark for our efforts to bring down global warming and keep to our commitments under the Paris Climate Agreement. According to CAT: "The adoption of the Zero Carbon Act in 2019 was a step forward, but implementation is key and the methane exemption weakens the target considerably."[9]

As far back as the end of the 19th century. New Zealand has been concerned about our dependency on primary produce exports and various politicians, policy makers, and researchers have advocated that we diversify away from

7.   Hawken, P. (Ed.). (2017). ibid.

8.   May 20, 2021. Oatly shares soar 18% in company's public market debut on Nasdaq. CNBC.https://www.cnbc.com/2021/05/20/oatly-ipo-otly-starts-trading-on-nasdaq.html.

9.   See https://climateactiontracker.org/countries/new-zealand/

agriculture. The global trend of urbanisation continues to put pressure on limited arable land as fewer and fewer people are still in touch with local food production with dire social, cultural and environmental consequences. There is a growing tension between housing, food production and an economy dependent on exporting food and agriculture rather than self-sufficiency.[10]

Sixty years ago Aotearoa was awakened to environmentalism. The country was challenged by the threat posed by industrialisation and globalisation as our government colluded with a multinational aluminium corporation to extract hydro energy from Lake Manapouri. The engineers and their masters were determined to get the most profit out of the free exploitation of power by raising the level of the lake by up to 30 metres. If they had succeeded the iconic Fiordlands would have seen massive ecological and environmental damage with drowned trees left as ghostly corpses, their broken and twisted limbs left jutting out of the water. The campaign to preserve the natural beauty of Lake Manapouri may have been mostly backed by those who had aesthetic concerns but this environmental activism awoke what is now a perennial concern amongst New Zealanders; conservationism and the care and preservation of the lakes, land and seas. New Zealanders now commonly feel that they are the kaitiakitanga, or guardians of the land, water, and air and many want to restore nature to what it was before we began to colonise and farm this relatively untouched island

---

10. See *Our Land 2021*, "*Overseas markets are a significant driver of land use, and with global populations projected to reach 10.9 billion by 2100, market-based pressures on land are set to increase. Most of our agriculture and forestry products are exported, and these activities currently cover about half our land area, the report says.*" https://www.stats.govt.nz/news/new-report-shows-impact-of-demands-on-land-in-new-zealand

nation. 61 percent of the population would like a more ambitious emissions target, yet while 62 percent are optimistic that individuals can make a difference, and 72 percent believe individuals are responsible for climate change, the big emitters are largely responsible for our GHG emissions even if we participate. It is the supply chains and products of the number 1 emitter, Fonterra (20% of NZ emissions), and number 2, Z Energy (10% of NZ emissions) that are really responsible.[11]

The climatologist, Michael Mann accuses the big corporations for deflecting the blame and distracting us from their business-as-usual. He blames their secretive PR campaigns for why so little has been done about protecting our planet.[12] How did we get here? We have the exploitation of cheap hydroelectric power by Rio Tinto's[13] Tiwai Point aluminium smelter to thank. In the late 1960s, the National party, lead by Keith Holyoake, was confronted by a petition signed by around 250,000 citizens demanding that Lake Manapouri be left alone and the lake levels kept as they were. In 1972, National lost the election to Labour who listened to the concerns of a public that wanted to preserve the area from the ravages of hydroelectric dams built to generate power for multinational industrial profits from energy intensive aluminium smelting.[14]

This book explores the historical, philosophical, ontological, and scientific worlds of zero carbon and the

11. Rod Oram.(Aug. 2021) Z Energy needs to refuel its ambition.
     https://www.newsroom.co.nz/z-energy-needs-to-refuel-its-ambition

12. Mann, M.E. (2021) The New Climate War: the fight to take back our planet.
     Scribe Publications. New York.

13. Formerly known as Comalco.

14. Fox, A. P. (2001). The Power Game: the development of the Manapouri-
     Tiwai Point electro-industrial complex, 1904-1969 (Thesis, Doctor of
     Philosophy). University of Otago. Retrieved from http://hdl.handle.net/
     10523/335

interconnected ideas, philosophies, and sciences that are embedded in this slogan. From the well-meaning and inspirational lobbying and marketing of Generation Zero, 'the movement of 38,397 young kiwis working to cut carbon pollution', to 150-year-old agricultural lobbies, and coalitions of fossil fuel deniers – here I outline my findings in the hope that we can critique the past in order to creatively imagine future worlds, and thereby co-inhabit the Earth with the nonhuman majority of animals, and things, such as the atmosphere, biosphere, hydrosphere and lithosphere. Our ability to adapt to the dangerous worlds that lie ahead will be determined by our speculative creativity and creative collaboration within our species and between humans and nonhumans. Disturbingly our ability to think and act creatively is confined by the physical temperature range of the planet. We will literally be unable to think clearly as Earth faces hotter temperatures, more carbon dioxide and less oxygen.

According to the psychologist Antonio Damasio who is the director of the Brain and Creativity Institute at the University of Southern California: "...it is known that as ambient temperatures rise, not only do we need to adjust our internal physiology to losses of water and electrolytes, but we also function less well cognitively. That poor adjustment of internal physiology spells disease and death is no surprise." [15] Surprisingly, the editors of *Drawdown* identified the phasing out of refrigerants, chloroflurocarbons (CFCs) and hyrdocholorofluorocarbons (HCFCs) – the culprits behind the holes in the ozone layer, and their replacement hydroflurocarbons (HFCs) as the number 1 way that we could reduce global heating. New

15.   The Strange Order of Things: Life, Feeling, and the Making of Cultures by Antonio Damasio

Zealand is a signatory to the mandatory Montreal Protocol on Substances That Deplete the Ozone Layer and has begun phasing them out. However, as the planet gets hotter New Zealand will continue to buy air conditioning units that will ironically contribute to global heating. In 2018 it was estimated that New Zealanders have, on average, added another 100,000 new air conditioning units per year over the previous 10 years.[16] In 2016 more than 170 countries including New Zealand signed the Kigali amendment that will also start phasing out HFCs. The former Secretary of State, and President Biden's climate change advisor, John Kerry said HFC reduction was "the biggest thing we can do [to mitigate climate change] in one giant swoop." However, Lawrence Berkeley National Laboratory estimates that 700 million air-conditioning units will come online by 2030.[17] While these refrigerants cause emissions throughout their life, ninety percent of those emissions happen at the end of the appliance's life. Between now and 2050: "87 percent of refrigerants likely to be released could avoid emissions equivalent to 89.7 gigatons of carbon dioxide, and an additional equivalence of 25 to 78 gigatons of carbon dioxide due to the Kigali accord. To properly dispose of these refrigerants so they don't leak will be costly at a projected net cost of USD$903 billion by 2050.[18] This is a technological problem in which our complex world is facing societal collapse as global heating threatens our lives and our sanity.

Strange as it might sound I am suggesting here that we

---

16.   See https://www.mfe.govt.nz/sites/default/files/media/ Climate%20Change/hfc-consumption-in-nz.pdf

17.   Hawken, P. (Ed.). (2017). Drawdown: The most comprehensive plan ever proposed to reverse global warming.

18.   Hawken, P. (Ed.). (2017). ibid.

reacquaint ourselves with our feelings. In a world where technological solutions are increasingly based on AI and machine learning, it is important to critically assess the semantics and origins of these words. We should be aware of the etymology of words such as *'artificial'*, *'intelligence'*, *'machine'* and *'learning'*. We are constantly talking about moving into a *'digital'* world but is the word digital a misnomer. *Digital* originally referred to the digits on our hands, base 10. However, most information technology is actually binary based on the yes or no mathematical answer of a base 2 number system, that is zero or one. This black and white world is very different from the biological reality our brains inhabit, in other words, our minds are more analogue as opposed to the binary digits used in computation. A grossly simplified model of the human brain often depicts it as a digital or binary computer, but to forget that this is a model is to mistake the map for the territory. Biology is founded on the graduated changes in electrical potential and chemical concentrations that smoothly transition from one state to another as opposed to the binary on/off switch of a digital semiconductor. The colonisation of words and phrases is something that can be exposed by returning to the origin or etymology of words. The so-called digitisation of the planet suggests that we are not only imagining the Earth as some sort of binary machine but a reminder that even a one to one scale map is not a comprehensive replication of the planet by any stretch of the imagination.

I explore this in my chapter, *Binary Worlds*, and consider the risks of ill-conceived metaphors and why marketing slogans like *Zero Carbon* confound our attempts to imagine better worlds. As the ecologist Daniel Christian Wahl has pointed out in his book, *Designing Regenerative Cultures* it is more important that we stop, think, and listen to what others think the deep questions are? Our fast paced culture that has been accelerating towards the technological singularity is far too obsessed by design answers rather than design questions. We suffer from digital ADHD, skittishly zipping from one screen to the next, too busy and too panicked about the end of the world to slow down and consider philosophical questions. We need to slow down and imagine what parts of our worlds do we wish to retain, and envisage speculative and utopian worlds we would prefer?

Creative collaboration is grounded in our biological reality and not some imagined digital and binary science fiction substrate. Antonio Damasio explains how our creative imagination evolved during the billions of years of symbiotic collaboration between microbial organisms and their organelles eventually emerging as feelings and emotions that inform our rational minds. The coordination of life-giving systems that desire the nutrients and sustenance of life are the foundations and precursors of feelings, emotions and human creativity. The scientific imagination has unlocked biology from its individualistic past and exposed collective new worlds inhabited by microbial organisms that are as much a part of us as they are dependent on others. The understanding of plant and animal life as a symbiotic network decentres humanity and forces us to reconsider the nonhuman majority of *things* who help build our worlds, and yet are still ultimately mysterious, alien, and unknowable to themselves and us. As the mycologist, Merlin Sheldrake, explains: "Learning more about these associations changes our experience of our own bodies and the places we inhabit. 'We' are ecosystems that span boundaries and transgress categories. Our selves emerge from a complex tangle of relationships only now becoming known."[19]

This acknowledgement of the primal prerequisites of how we think corrects previous theories of evolution and the theories that have led us down a path that is biased against creativity. Rather than relegating feelings and emotions behind the almighty rational logic of our scientific and technological world, Damasio points out that feelings are the foundation of our survival. It is by sensing our environment outside and inside our bodies that we navigate the dangers and joys of our creative presence here on Earth. It all began with the primaeval electro/chemical reactions that gave those early organisms ways to sense what is both bad and good for life, not in some binary way but in a biological way smelling and tasting the concentration of chemicals flowing across their porous outer membranes. This was how the prototypes of feelings and emotions evolved.

It was also how proto-minds and primitive creative intelligence began to design simple worlds. Billions of years before IDEO franchised '*design thinking*' simple single-cell prokaryotic microbes were '*design questioning*' their environment.[20] These early designers lacked the neuronal networks that eventually appeared in multicellular organisms, but even without the very basic means of storing memories they were able to sniff, taste, and feel their environments. We should recognise this alien behaviour as intelligence rather than pretend we have the only viable form of working out our environment. Of

19. See Merlin Sheldrake. (2020) Entangled Life: How Fungi Make Our Worlds, Change Our Minds and Shape Our Futures.

20. Liz Jackson, Honouring the Friction of Disability. https://www.youtube.com/watch?v=cZhiu-jGbdE&feature=youtu.be

course, those sensory probes were nothing like what we would recognise as sensory organs in humans but the principles of how they explored and questioned their surroundings and their internal metabolism are strikingly familiar to us when we analyse our own way that we sense our world. I call them designers because in effect they achieved a prehistoric design methodology similar to our earliest 16th-century definitions of design. They achieved a proto-mental plan of a sensed environment, and they 'sketched' a very simple world by a process of homeostasis that 'imagined' a future in which it consumed a nutrient that it had sensed in the present. As Sheldrake has pointed out: "Complex information-processing is evidently not restricted to the inner workings of brains. Some use the term 'swarm intelligence' to describe the problem-solving behaviour of brainless systems. Others suggest that the behaviour of these network-based life forms can be thought of as arising from 'minimal' or 'basal' cognition, and argue that the question we should ask is not whether an organism has cognition or not. Rather, we should assess the degree to which an organism might be cognisant. In all these views, intelligent behaviours can arise without brains. A dynamic and responsive network is all that's needed."[21]

The means by which these proto-designers sustained and reproduced themselves give us hints about how we should return to our evolutionary womb in order to imagine how we might build, bend, and break new worlds beyond our obsessive compulsion to ignite carbon bombs that will immolate us in the fiery future of our own making. By understanding where creative intelligence comes from and how it evolved we can reset our imagination and

---

21.   Sheldrake, Merlin. Entangled Life (Kindle Locations 1295-1300). Random House. Kindle Edition.

envisage biological models that predate the past three hundred years before the mechanical models came to blinker our brains. Damasio points out that: "Creative intelligence was the means by which mental images and behaviors were intentionally combined to provide novel solutions for the problems that humans diagnosed and to construct new worlds for the opportunities humans envisioned." [22]

It is precisely this creative intelligence that we need to apply to our world building, bending, and breaking today. *Zero Carbon* is a deliberate and urgent call to action but it is also a frightening prospect to many as it suggests the void, the dark abyss, an empty space that haunts us, a precursor to chaos, a phrase that suggests death, a capitalist heresy which recalls a dark age with some vague promise of progress. While zero was once a number to be feared, zero is essential for the successful running of our modern world and paradoxically responsible for its current demise and the existential threat we face as a species. To cite Robert Kaplan in his book about the number zero, "If you look at zero you see nothing; but look through it and you will see the world." [23]

We are in a perilous situation that now demands urgent action to avoid a very real existential threat brought on by our own technological prowess made possible by mathematics and science, and yet we need time to think and question how we think? The art of mathematics and the mindset of science have become specialities that the majority of us do not share. The phrase *Zero Carbon* hides more than it reveals. Fear, reluctance, and denial are all

---

22.   The Strange Order of Things: Life, Feeling, and the Making of Cultures by Antonio Damasio

23.   Kaplan, R. (1999). The Nothing that Is: A Natural History of Zero

feelings that have an ancient association with the emptiness of the void. If we explore the etymology and context of these words I hope to remove their spell and uncover hidden and subconscious fears that are blocking our ability to imagine novel futures. Charles Seif explains that "Simply being able to count was considered a talent as mystical and arcane as casting spells and calling the gods by name. In the Egyptian *Book of the Dead*, when a dead soul is challenged by Aqen, the ferryman who conveys departed spirits across a river in the netherworld, Aqen refuses to allow anyone aboard "who does not know the number of his fingers." The soul must then recite a counting rhyme to tally his fingers, satisfying the ferryman. (The Greek ferryman, on the other hand, wanted money, which was stowed under the dead person's tongue.)"[24] Counting was a mystical component of ancient cosmology in which non-written cultures, such as the Dogon and Māori enumerated the stages of cosmic evolution beginning before the number one, or as Māori knew it Te Kore, *'the void'* or *'great nothingness'*.[25] What we have forgotten about the void, before we count the material reality around us, is that zero did not just symbolise death, but more importantly the beginnings of creation and life. According to Tregear, Te Kore refers to: "the primal power of the Cosmos, the Void or negation, yet containing the potentiality of all things afterwards to come."[26] Māori do not fear Te Kore but venerate *them* as the beginning of their whakapapa, or their lineage from IO, the primal force, connecting each

24.   Seife, Charles, (2000). Zero: the biography of a dangerous idea. Penguin Books.

25.   Scranton, L. (2018) Decoding Māori Cosmology: the ancient origins of New Zealand's Indigenous Culture. Inner Traditions. Vermont. p.37

26.   Scranton, L. (2018) ibid. p.35.

individual to the cosmos and to all things in the universe, their ancestors. Western pākehā culture no longer relates to the void as a creative force and is alienated from the nonhuman life force.

Yet, while it is healthy to call out the origins of our nightmares, this philosophical analysis is also essential if we are going to reimagine our future and employ our creative intelligence in a non-binary way. We must become more comfortable with the metaphors of paradox and poetic design, engaging nonhuman designers in the worlds that we will co-design with them – we must reconnect with our kith and kin.

Chapter Outline

This book is a polemic exposing why after sixty years of knowing better we have carried on with business-as-usual, and done little to prevent the harm that is predicted to result from a geological epoch that has been largely caused by humanity, known as the Anthropocene. It is an attempt to show how our common sense has imprisoned us in a cognitive construct that needs urgent attention, deconstruction, and a total rebuild. Our philosophies, beliefs, theories, religions and even science build, bend, and break worlds, but more importantly we only see these worlds, and not the multiplicity of other worlds inhabited by things we do not know.

Chapter 1, *The Story of Zero*, begins to examine what many people might imagine is purely a mathematical number but is revealed to be a powerful philosophical, and conceptual invention that is steeped in a mythological and religious past. We could not live in our current world without zero, and yet it has a dark past, that is associated with death, darkness, the void, the abyss, hell, and evil. Its archetypal cosmology is why we both fear zero and worship it in the form of computing, and digital algorithms.

If zero represents the dark void then, in chapter 2, *The Story of Carbon*, we will trace how carbon and zero have reversed their good and evil roles. It is carbon that is now associated with death, and zero that is identified with life. This paradox is not only confusing because of its reversal, but also because humanity is capable of holding contradictory assumptions at the same time. However, because of our new found computerised religion we hold fast to the law of non-contradiction[27] because algorithms are incapable of paradox, metaphor, and therefore true creativity. Carbon represents our cultural, evolutionary, and economic past that are inexorably entwined together. As we recall that life is our most important project we must find ways to find co-existence with carbon, and not demonise it, or zero for that matter.

Carbon, is essential for life as we know it. In chapter 3, *Zero Humans*, I look at the Earth's paleontological past, before humans, and what we might anticipate if we don't manage to reduce our GHGs emissions. We have had CO_2 levels as high as this before but it may have not been conducive to human populations. Of course, we had not evolved at that point. I consider how so-called *civilisation* has been frightened of *nature* and the *wilderness* in the past, and how we imagined our primitive existence to be a battle of survival in which wild nature wanted to kill us. Authoritarian regimes have long used plagues to command and control the people through fear, and this is no different today. The threat of extermination and the extinction of

---

27.   See Timothy Morton's book Dark Ecology.

humanity is an old trick that is still being applied. Today, it is not just the Church and the State that are peddling fear, but corporations who claim that the end of fossil fuels, and capitalism will also cause the end of humanity. According to these corporations the status quo must prevail; it is the argument of those in power. And so, around the world the industrial agri lobby has convinced our governments that nothing should change, or at least nothing fundamental. Meanwhile, population growth is problematic, but when we understand that the environmental impact and energy consumption of one child in the U.S. is ten times higher per child compared with India, the population growth is politically contentious.

In many ways we are addicted not to our lifestyle, but rather our *deathstyle*, in Chapter 4, *Zero Growth*, I examine one of the biggest contradictions to our attempts to reduce GHG emissions while continuing to grow the global economy. In 2015 the UN announced its 17 Sustainable Development Goals; number 8 is '*Decent Work and Economic Growth*' and yet year on year economic growth is impossible without environmental damage and there are hard limits imposed on us by the depletion of resources. The metrics by which economists, policy makers, capitalists, and politicians claim success are reduced down to one number GDP; a simplistic number that hides real harm to the planet, people, nonhumans, and creativity. Economics constrains the imagination and prevents us from building worlds that put aside capital accumulation and the financialisation of sustainable alternatives. The *market* is an illusion that will rob the earth of its fertility and only leave behind poverty and misery, yet that was the promise of colonisation, capitalism, and the growth of GHGs. In Aotearoa, growth was all about exporting the nutrients sucked from the land and the water, leaving behind waste and toxicity. Agriculture was only primarily about profit and not food and by those metrics it was successful. It would be impossible to suddenly stop fossil fuels, and synthetic fertilisers and to continue to grow the economy, or even the plants and animals that sustain our deathstyle. Capitalism is not just our drug but our religion.

Quite simply, if capitalism grew from the liberal philosophy of the Enlightenment, and capitalism and economics continue to preach ideas that will cause the end of the world as we know it, then we have to change our philosophy of life. It is time to start speculating about reality and reimagining futures that are not bound by a wilful ignorance of the destructive consequences of our outmoded ideas.

Chapter 5,000 *Zero Worlds,* introduces Speculative Realism, and Object Oriented Ontology, a philosophy I am hoping will be more accessible if it is called, *Thing Related Reality.* The pandemic, COVID 19, is used to illustrate how a transdisciplinary enquiry into such a wicked thing can begin to change the way we look, and think about the worlds we are building around us. These worlds begin to look more like cognitive panopticons[28] rather than useful conceptual tools for adaptation and change. The virus begins to look like a *thing* that has emerged and evolved from our ignorance rather than our friend when it comes to our adaptation to a faster, hotter, more toxic world. Viruses have given us immunity to other diseases, COVID 19 could

---

28. "The panopticon is a type of institutional building and a system of control designed by the English philosopher and social theorist Jeremy Bentham in the 18th century. The concept of the design is to allow all prisoners of an institution to be observed by a single security guard, without the inmates being able to tell whether they are being watched."
https://en.wikipedia.org/wiki/Panopticon

be our canary in our coal mine that has killed large numbers of people but has forced us to change. We have to meditate on the ecological causes of the pandemic to learn its lessons or suffer the unintended consequences. Philosophy helps us to envisage new worlds.

Industrial agriculture should provide us all with, *Food for Thought*, (chapter 6). If we stop and think about what food is we might see that it is very subjective and depends on our daily dose. Even too much water, oxygen, and carbon can be, toxic depending on your species, and your tolerance. The qualities of the ecology in which food grows and is consumed will determine the health of the ecosystem that ingests it. We can not understand the impact on our local and global ecologies if we do not follow the life cycle of all the dependent components of the *zero carbon* policies. Methane is just one component that is brushed off as a short term emitter, yet it is 84 times more powerful at heating the planet than carbon dioxide. It does have a 9 year half life but this next decade is our most urgent according to scientists if we are to avoid a tipping point in which global heating becomes a run away problem. In 100 years methane still has the global warming potential of 23 times carbon dioxide and when it degrades methane turns into the long lived carbon dioxide which will take centuries before it has no effect. Our agricultural and land use practices must change if we have any hope of reducing global heating.

In chapter 7, I discuss *Binary Worlds*. Algorithms and our binary computerised technology have constricted our ability to think. For many of us we are living in a world in which the computer is the new hammer and everything looks like a nail. We no longer see the very material existence of the technology we use to look through, and yet the physical components of our tools are often toxic as e-waste continues to illegally accumulate in less developed countries. We have gone from fearing zero to worshipping it, as we only see 0/1, an in/out world of binary choices. The universe is not binary but has many non-computable things that we should consider, especially if we are looking for novel ideas. Unfortunately, Aotearoa has accepted the political compromise that puts its faith in market mechanisms and the capitalist wonders of the *invisible hand of the market*. The economists and policymakers believe that the Emissions Trading Scheme and carbon pricing will magically reduce our GHGs despite the fact that for over a decade it has simply encouraged more international GHG emissions, corruption, fraud and billions of dollars of cash to our biggest carbon emitters.

Chapter 8, *Ground Zero*, examines the nuclear history behind the term, and the close link between cartography, meteorology, colonialism, capitalism and climate change. Just as a nuclear bomb appears to wipe the ground free of all that came before, the history of the invaders starts again from when they say it does. Indigenous cultures are ignored or considered pre-historic and just like on a blank slate the invading culture imprints on those that came before. The shock and awe of invasion or a nuclear attack creates the psychological state of ground zero. Writing over the top of those who have lost their memories becomes easy. New Zealand's state of amnesia enabled the economic blitzkrieg of Rogernomics and Friedmanite neoliberalism to thrive at the expense of equity, neurodiversity, biodiversity, and resulted in our harmful contribution to the Anthropocene. 1984 was simply the next ground zero in which our history was rewritten. The New Zealand Treasury, the Business Roundtable, and a small group of politicians were indoctrinated by the 'Washington Consensus' with the help of economic pundits and local politicians – neoliberalism swept the globe and has stayed embedded in our economic policy that only trusts the markets.

Chapter 9, *Zero One Worlds*, looks to new worlds that are not so constrained as our current binary world. It looks at open worlds that explicitly publish their rules and algorithms, that expose the closed worlds that imprison our thoughts. These worlds have moved beyond money, carbon pricing, and the social inequities of capitalism. They are worlds created by the nonhuman majority, and those that were once disenfranchised who think, act, and create differently from the 'neurotypical' minds that some call 'normal'. It offers a prosocial approach to living together as communities that co-exist with the nonhuman majority and is sensitive to the ecological opportunities and limitations of unbridled desires that are fuelled by consumerism, advertising, and neoliberal rhetoric. This is intended as the jumping off point for creative world builders who can philosophically reshape their worlds; they are doing by thinking and thinking by doing. *Zero Carbon* may appeal as an advertising strapline, or a political slogan, but it hides more than it reveals. It is important to think about the words you use, their history, who brought them into popular use, and who do they serve? Behind every word is a story, and that story can hide the political objectives of those who tell it. Words, stories, algorithms, and code are technologies and as Ted Nelson once wrote:

"All human artefacts are technology. But beware of anybody who uses this term. Like "maturity" and "reality" and "progress", the word "technology" has an agenda for your behaviour: usually what is being referred to as "technology" is something that somebody wants you to submit to. "Technology" often implicitly refers to something you are expected to turn over to "the guys who understand it." This is actually almost always a political move." It may well be that *Zero Carbon* is just another algorithmic technology that vested interests want you to submit to in ways you don't fully comprehend?

1

The Story of Zero

On the surface of it a carbon emissions target of zero, set by countries who signed the Paris Climate Change Agreement, sounds eminently sensible. We are told that as a planet we need to achieve net zero carbon emissions by 2050 if we are going to have any hope of keeping global heating below 1.5°C hotter than pre-industrial temperatures. Despite regular meetings of the UN Climate Summit and a growing number of countries that have declared a *'Climate Emergency'* GHGs and temperatures continue to rise, and they have done despite the fact that we have known about this problem for 60 years. And yet, we would not want to achieve the literal objective of zero carbon emissions because, not only would that be impossible, it would literally signal the death of the entire planet, and our rock would be snap-frozen.

I will go into this in more detail in later chapters but for now, I want to give you a brief biography of the number zero because the history of ideas and their components really does matter, and the history of zero has shaped the way we have thought about our world for over two thousand years. Zero has come into focus because the world agrees we are facing an environmental catastrophe because of an abundance of GHG, (Green House Gas), emissions.

Zero is more than a number, it is a symbol, and a semiotic algorithm, a technological invention. Like every *thing* it is more than the sum of its parts and its past. Even if in the present these things appear partially hidden, buried beneath layer upon layer of events, resources, and agents – human and nonhuman things from the past shape the present and our future. They have created processes and caused events driving agents of desire and the creation of consumer goods, made from limited resources, and expelling growing amounts of waste. All of that activity also comprises our creative ability to survive and thrive.

The story of zero will help us to understand how we got to now and how we might creatively adapt and define new worlds worth inhabiting? The history of zero and infinity will reveal some unconscious fears and desires not discussed at the Paris Agreement on Climate Change. These unconscious memories lurk in the obsidian darkness of a *Zero Carbon* world that continues to frighten our psyche and paralyse our imagination. Our culture harbours deep psychological fears around the concept of zero that counter our conscious realisation that we need to dramatically reduce carbon emissions very very quickly. *Zero Carbon* as an advertising slogan, or a call to action, should never have ended up as a marketing campaign because it has a dark past that we have forgotten. Just as some people have an instinctual fear of spiders and snakes, Western culture inherits a deep subconscious phobia infused with zero, its past, and what it represents.

In his book about the history of zero, Charles Seife wrote that:

"Zero clashed with one of the central tenets of Western philosophy, a dictum whose roots were in the number-philosophy of Pythagoras and whose importance came from the paradoxes of Zeno. The whole Greek universe rested upon this pillar: there is no void. The Greek universe, created by Pythagoras, Aristotle, and Ptolemy, survived long after the collapse of Greek civilization. In that universe, there is no such thing as nothing. There is no zero. Because of this, the West could not accept zero for nearly two millennia. The consequences were dire. Zero's absence would stunt the growth of mathematics, stifle innovation in science, and, incidentally, make a mess of the calendar. Before they could accept zero, philosophers in the West would have to destroy their universe."[1]

The 2 Meanings of Zero

Zero has had two distinct meanings. The first meaning was as a placeholder with no numerical value. It came from the East, and not the West, and ironically it came from the oil region we now know as Iraq – the Fertile Crescent, the birthplace of agriculture which began in the warm and stable climes of the Holocene some 10,000 – 12,000 years ago. According to Scranton there are ancient cultures that are still very much alive today that embraced the concept of the void, the great nothingness, that existed even before zero was articulated. After extensive research into the

---

1.   Seife, Charles, (2000). Zero: the biography of a dangerous idea. Penguin Books.

linguistics and shared cosmology of other ancient cultures, Scranton has concluded that Māori share a primitive cosmology with Dogon, Egyptian, and Indian beliefs that pre-dates Buddhism, something like 10,000 BCE. In Māori cosmology the universe began with Te Kore, the void, before space and time.[2]

Zero is a major foundation of our modern world, but with an ancient legacy, and without it we would not be experiencing the current global heating – an unprecedented, non-analog epoch, the Anthropocene – the geological epoch that was nudged into existence by human population rise and its voracious resource consumption. It is a rift in the history of this planet and it: "stands for the notion that human beings have become the primary emergent geological force affecting the future of the Earth System."[3] When zero was first born in Babylonia almost two and a half thousand years ago, 300 BCE, it was first created to be a placeholder.[4] For example, the date 2020 has two zeros to signify the decimal places separating the twos and creating a unique number without ambiguity. However, as the mathematician and philosopher Alfred North Whitehead has pointed out, "The point about zero is that we do not need to use it in the operations of daily life. No one goes out to buy zero fish. It is in a way the most civilized of all the cardinals, and its use is only forced on us by the needs of cultivated modes of thought."[5]

Throughout the ancient regions of the world zero was physically denoted by a space on an abacus, a column where all the stones were at the bottom. According to Seife the

---

2. Scranton, L. (2018) ibid. p.35

3. Angus, I., 2016, Facing the Anthropocene. Kindle Location 81

4. Seife, C. ibid

5. Seife, C. ibid

origin of the words, calculate, calculus, and calcium all come from the Latin word for pebble: *calculus*. Pebbles were first used for counting and were then incorporated into the invention of the abacus. While the Egyptians did not have a zero the Babylonians used a system of numbering as if an abacus was inscribed on a clay tablet. The problem they had was that using their sexagesimal system, based on the number 60, they did not have any way of differentiating some numbers such as 1 and 60, that were both represented by a single wedge. Their solution was to invent zero using the symbol of two slanted wedges to solve the problem, a symbolic way to represent an empty column on the abacus and to tell which column a number was in. "Zero was born out of the need to give any given sequence of Babylonian digits a unique, permanent meaning."[6] However, zero had no numerical meaning and was simply designed to separate the columns as if on an abacus.

The second meaning of zero came around a thousand years later in 628CE. An Indian mathematician and astronomer, Brahmagupta first enumerated zero as a dot, a novel number in its own right and without it there would be no computers, lasers, spaceships, and most likely no global heating – the binary number system of 0 and 1 would be a lonely base 1. The mathematician Marcus du Sautoy wrote: "Some of these ideas that we take for granted had to be dreamt up. Numbers were there to count things, so if there is nothing then why would you need a number?" "The whole of modern technology is built on the idea of something and nothing."[7]

6.   Seife, C. ibid

7.   Revell, T. (2017) History of zero pushed back 500 years by ancient Indian text. Read more: https://www.newscientist.com/article/2147450-history-of-zero-pushed-back-500-years-by-ancient-indian-text/#ixzz6cz6el6cX

However empty the concept is, it was once considered heresy to even mention it by name – it was the work of the devil; the void and the hellish abyss, where there was nothing but dark pain and suffering. But this concept of zero eventually enabled the development of calculus and the foundation of electronics, modern warfare, and the computer age. People no longer feared it and mentioned it in casual conversations. The story of zero illustrates how culture and science have always been entwined and while it may appear that a subject such as pure mathematics stands alone, free of mythology and the foibles of culture, even maths grew from a past that is far more mysterious and still bound to ancient myths and legends.

While most of the population would not recognise how important it is to untangle this complex web, imagine you are presented with the most dangerous predicament you have ever encountered. It is not just your life but the life of all your loved ones and everyone else on the planet that is being threatened. You are given a large tool box full of the most amazing technological inventions ever designed. There are the usual things like a hammer and nails but there are other oddities you recognise but don't know a lot about. One is a thing that is both a tool and a dangerous weapon. You know how to use the hammer and you have used it many times before. The mysterious object is frightening because you don't understand it but you sense people have died using it in the past, and will most likely die if you misuse it now. That tool; that weapon is zero. Most of humanity feels safe because they know that the politicians will never take it out of its lead-lined box – its fine to talk about it, but just don't touch it! We know that people are terrified of the unknown but they can overcome their fear through gradual exposure and understanding.

If we can acknowledge that ideas are inextricably bound together and that words have incredible power and force then we can begin to see how ancient myths could and should be excavated to be examined for any archaeological wisdom that they may contain, especially when we are stuck in an endangered world running out of time and ideas.

According to Charles Seife in his book, *Zero: the biography of a dangerous idea*, the ancient cultures that shaped our contemporary worldview believed that zero and its sibling concepts, the void and the infinite could destroy the universe. They knew the idea of zero was dangerous so they fought the void and the infinite valiantly for two thousand years, but because they rejected them, Western culture was stuck with a paradox that once resolved would give us the insane technology of today. Ironically, oil and gas would be impossible without it. But don't we still fear zero? What if I said I will take everything away from you and you will be left with nothing?

The Paradox of Zeno

Born in 490BCE Zeno of Elea was known as one of the most annoying philosophers in the West. He used a simple running race to mathematically confound the Western world, and to add insult to injury he used infinity to really upset them. Aristotle and almost every philosopher, up until the Enlightenment in the 16th-17th century, denied there was such a thing as infinity. The Catholic church throughout the middle ages sided with Aristotle and believed physical space and the spheres in the heavens were limited. To talk about an infinite universe was quite simply, heresy. Aristotle and Zeno believed that the universe could not move because if it did it would leave behind empty space. This void, known as zero, was also considered an abomination for two millennia. And yet, both zero and the infinite lie at the heart of Zeno's paradox.

It was because the Greek's did not have zero at the time that they could not calculate a problem we now find absurd. Zeno's riddle went like this.

The speedy Achilles was racing the slow tortoise. The tortoise gets a head start and begins a foot ahead of Achilles. Achilles is running a foot a second and the tortoise only half a foot a second. In one second the speedy Achilles catches up to the tortoise. But by the time he reaches him the tortoise is already half a foot on. No problem Achilles is faster and makes up the half foot. But the tortoise is a quarter of a foot a head. As Seife explains it the faster Achilles runs and runs, it was in vain because the tortoise slowly inches ahead each time. What confounded the brightest minds in ancient Greece was that without zero it appeared that the race would go on forever and even if you divided the distance by half, then quarters ad infinitum there was no end point ever reached and so according to that logic Achilles never catches up and the tortoise is still in the lead. Seife noted: "The philosophers of his day were unable to refute the paradox. Even though they knew that the conclusion was wrong, they could never find a mistake in Zeno's mathematical proof. The philosophers' main weapon was logic, but logical deduction seemed useless against Zeno's argument. Each step along the way seemed airtight, and if all the steps are correct, how could the conclusion be wrong?"[8]

---

8.   Seife, C. ibid

While the Greeks were flummoxed by this logic and reason they knew that infinity was the cause of their problem. What they didn't know then is that it is possible to add infinite numbers together and to approach a finite number – in this case, zero. We also know that Achilles is running at a foot a second so he will reach 1 foot in 1 second and 2 feet in 2 seconds. Even if they are racing for infinity he will reach these milestones in a finite time and beat the tortoise getting there. Even if you infinitely keep dividing a number by half there is a limit to the division as you get closer and closer to zero and closer and closer to one foot.

To sum up, Seife wrote: "The Greeks couldn't do this neat little mathematical trick. They didn't have the concept of a limit because they didn't believe in zero. The terms in the infinite series didn't have a limit or a destination; they seemed to get smaller and smaller without any particular end in sight. As a result, the Greeks couldn't handle the infinite. They pondered the concept of the void but rejected zero as a number, and they toyed with the concept of the infinite but refused to allow infinity—numbers that are infinitely small and infinitely large—anywhere near the realm of numbers. This is the biggest failure in Greek mathematics, and it is the only thing that kept them from discovering calculus."[9]

Zeno is thought to have devised this and other paradoxes to support the doctrine of Parmenides who believed that motion was an illusion. This was upheld by Aristotle who continued to dominate Western thought until the end of the Middle Ages. The Catholic church agreed with Aristotle that the universe was static and the very concept of a void threatened the foundations of the Christian belief, there

9.   Seife, C. ibid

could be no such thing. Aristotle was certain when he claimed, 'Nature abhors a vacuum' based on his belief that there is no such thing as an empty volume of nothingness, and therefore space can have no volume. Instead, he believed that 'space is just the surrounding surface of objects.'[10] This was in turn founded on the concept of the abundance of the universe that is ultimately full. Due to Aristotle's towering intellect, his belief that an empty void was impossible went unchallenged for fifteen hundred years. As Wertheim has pointed out: "Moreover, this Aristotelian abhorrence of the void translated neatly into the context of medieval Europe, for Christianity also had a theological tradition of an abundant Creation – a universe that God had created full."[11] Incidentally, this is most likely where our strange idea that the Earth's resources are infinite came from – the truth is only just beginning to dawn on us that wanting everything, we might end up with nothing.

Wertheim argues that this story of zero contradicts our modern view that rather than thinking that religion constantly followed sheepishly behind new scientific breakthroughs, the converse is often true, especially in the science of physics.[12] In 1277 the Bishop of Paris, Stephen Tempier, published a decree condemning 219 philosophical views. His forty-ninth decree stated that 'God was unable to move the universe on account that it implied the existence of a void.' Many bishops thought that must be wrong because they believed God's powers were limitless and God could, 'of course' move anything. Tempier's decree resulted in an uproar which challenged scholars to finally admit that

10.   Wertheim, M. (1999). The pearly gates of cyberspace: A history of space from Dante to the Internet. New York: W.W. Norton. p.100

11.   Wertheim, M. (1999). ibid. p.101

12.   Wertheim, M. (1999). ibid. p.103

perhaps the heavenly objects did move and so a void may be possible.[13]

The Fear of the Void

While the second meaning of zero is a number in its own right, it came long after the first concept of a placeholder, and it is associated with deeply held philosophical and religious beliefs that are shared by many religions around the world. Zero not only has an association with the void and the abyss, but is bound to the mysterious cycle of life and the infinite eternity of spiritual reality. In the middle ages zero, or the void, continued to be associated with Satan and hell, and was often referred to as the abyss. It is a fear that lingers today as we imagine our world ending with nothing left, as extreme weather sucks us beneath the waves of sea level rise and human life is forever snuffed out by darkness.

For the medieval Christians physical reality was an illusion, and conversely only the spiritual realm was real. Wertheim argues this is why the art of the middle ages did not bother to depict a three-dimensional perspective because the size of the characters in the paintings was determined relative to their celestial importance and was not intended to depict proximity to the viewer or optical correctness.[14] By the end of the middle ages there was an increasing emphasis on the physical reality of the senses and the empirical research into nature that followed. This was the dawning of the Enlightenment and the modern scientific methodology. Yet the ancient void still loomed large in the imagination of these early scientist. The myths

---

13.   Wertheim, M. (1999). ibid. p.103
14.   Wertheim, M. (1999). ibid.

and legends from Greek and Roman antiquity, were mixed with the legacy of the *Old Testament* and were a major part of their school and university education. You might think that all this superstition has been erased by modern science and technology, however, you only have to do a quick scan of the Marvel universe to realise that some of our most popular cultural artefacts still depends on our subconscious belief in them. Zero lurks in the shadows even while it is disguised in the modern garb of binary computer code. It was only 500 years ago that disorder and chaos were not only associated with evil, but they also threatened to overthrow the authority of the Church and the *'divine right of Kings'* to rule. It was the void that was seen as the precursor to chaos and the symbol of anarchy, self-rule, and the loss of the power of the monarchy and the Church. Even today, the fear of the void, the fear of a power vacuum, and the fear of political instability are regularly deployed by those wanting to hold on to power. According to Carl Jung it was around the early part of the eleventh century that Christian scholars began to associate Earth with the Devil and Heaven with God – the Devil had created our World.[15] In the middle ages the ancient belief in the procreative nature of the void was set to one side and chaos assumed the characteristics of darkness, emotions, Satan and un-man-ageable feminine fertility. The patriarchal Church supplanted the positive attributes of the pagan underworld, such as the fertility rites of Persephone and the secret cult of Dionysus, and replaced them with the stories of Hell and the Devil's torment of souls.[16]

According to Seife: "Most ancient peoples believed that only emptiness and chaos were present before the universe

---

15.   Jung, C. G., Jaffé, A. (1989). Memories, dreams, reflections. p.365

16.   See Joseph Campbell, (1991). The masks of God: Creative mythology

came to be. The Greeks claimed that at first Darkness was the mother of all things, and from Darkness sprang Chaos. Darkness and Chaos then spawned the rest of creation. The Hebrew creation myths say that the earth was chaotic and void before God showered it with light and formed its features...The older Hindu tradition tells of a creator who churns the butter of chaos into the earth, and the Norse myth tells a tale of an open void that gets covered with ice, and from the chaos caused by the mingling of fire and ice was born the primal Giant. Emptiness and disorder were the primaeval, natural state of the cosmos, and there was always a nagging fear that at the end of time, disorder and void would reign once more. Zero represented that void." [17]

Joseph Campbell in his book, *Creative Mythology*, identified the ancient Sumerian serpent-god as the ultimate mythical lord of the watery abyss the giver and taker of life. In "classical mythologies he was Hades-Pluto-Poseidon, and in Christian mythology he is, exactly, the Devil." [18] In Western culture, as early as 2,100 BCE, in *The Epic of Gilgamesh*, the abyss predated the concept of hell, the place of the dead. The word *abyss*, in the Greek ἄβυσσος, meant bottomless, unfathomable, boundless. Gilgamesh anticipated the *Old Testament* tale of Noah and the Deluge and the search for eternal life in the dark watery depths of the abyss that had drowned everyone but the immortal Uta-napishti and his wife. Gilgamesh is obsessed with gaining eternal life and descends into the abyss to attempt to retrieve a sea plant that will give him immortality. However, as in *The Book of Genesis* a cunning serpent beat him to it.

---

17.   Seife, C. ibid

18.   Campbell, J. (1991). The masks of God: Creative mythology [volume 4. Penguin/Arkana. p.17

The Epic of Gilgamesh opens with a verse explaining how the King Gilgamesh sets out on a hero's journey to retrieve the lost knowledge of the world before the Deluge.
"He who saw the Deep, the country's foundation, [who] knew … , was wise in all matters!
[Gilgamesh, who] saw the Deep, the country's foundation, [who] knew … , was wise in all matters!
[He] … everywhere …
and [learnt] of everything the sum of wisdom.
He saw what was secret, discovered what was hidden, he brought back a tale of before the Deluge."[19]

This wisdom brought the order that his authority imposed on the chaos of pre-civilisation; before agriculture and the city-state. It was the law; a patriarchal world of men and dominion over the land, the trees and the animals. But it was also the symbolic split between humans and nonhumans, male and female, darkness and light. This departure from the mysterious unity of polar opposites became the modern quest and challenge for all of us, to reunite with nonhuman nature. This is our hero's journey, the dangerous descent into the primordial abyss to regain the archetypal singularity when opposites are reunited, and paradox is regained. We are informed by the symbols and mandalas of our primaeval past that in order to psychologically heal ourselves and our world we must retrieve the wisdom of fertility and life from the goddess who reigns with the creator from the Deep. This may seem like psycho-babble to you but our split with nature seems to be where it all started to go wrong.

This strange and ancient wisdom could seem oddly out

---

19.   See George, A. R. (Ed.). (2003). The epic of Gilgamesh: The Babylonian epic poem and other texts in Akkadian and Sumerian. London; New York: Penguin Books.

of place in the 21st century, and yet it hints at the origins of our current troubles and what we might do to adapt to climate change; i.e. before the next Deluge, caused by global warming; and the rising abyss caused by melting polar ice caps. Metaphors and myths never went away and speak to us in our dreams, our intuitions, and our creative imagination – teaching us how to adapt and create new worlds.

Ouroboros – the immortal snake

In his book, *China's Cosmological Prehistory*, Laird Scranton detailed the remarkable similarity between many of the ancient cosmological descriptions of world creation. He begins by considering the ancient Dogon people of Mali, in West Africa. As in the Chinese Book of Changes or *I Ching*, the Dogon used symbolism and mandalas to express their belief, *'as it is above so it is below'*, to reflect the connection between the macrocosm of the heavens and the microcosm of earth. Throughout many cultures the circle represented the heavens, and the square, the physical reality of earth. In Chinese culture the male and female forces, ying yang, have a striking resemblance to Ouroboros; the immortal snake was the symbol of the eternal cycle of creation that was beyond time and space. Scranton wrote:

"As in Egypt, Africa, India, and Tibet, Chinese concepts of creation begin with the notion of undifferentiated watery chaos, from which all material things ultimately emerge. This watery mass is said to exist either outside of or prior to the beginning of time and is thought to be the ongoing source of all creation."[20]

20.   Scranton, Laird. (2014) China's Cosmological Prehistory: The Sophisticated

In Norse mythology the serpent Jörmungandr encircles the planet Earth, defining a boundary around the equator separating the known North to the unknown South, and the land above and the watery abyss below. The snake's circumference defines the known world inside the circle of its massive girth and the void that is space in the universe beyond. Like many cosmological myths, the mythology of Ouroborus describes the conjunction of life, death and rebirth embodied in an immortal snake that is both a metaphor and paradox – made from the Earthly element carbon, but also a transcendent being that is composed of the fifth element that does not appear on the periodic table but offers an immortal life. For many ancient cultures, the spiritual reality of the abyss was as tangible as the reality of living beings, such as snakes, and those worlds were conceived as portals. For them after life, death and the immortal were reunited by the chaos born of the void.

The separation of mind and spirit, human and inhuman nature, man and woman, happened later and became the law during the Enlightenment, becoming ingrained and hegemonic in the 21st century. The bifurcation of the void and humans, zero and carbon; order appearing out of chaos became baked into the philosophy and then the common sense of everyday people as those theories eventually were regarded as scientific laws.[21] Yet, the paradox of the Ouroborus reminds us of the contradiction of *zero carbon*. Just as many scientific theories started as intuitive hunches or ideas based on imaginative concepts, Ouroborus is a metaphor for our examination of our

---

Science Encoded in Civilization's Earliest Symbols . Inner Traditions/Bear & Company. Kindle Edition.

21.   See Prigogine, I., Stengers, I., & Prigogine, I. (1984). Order out of chaos: Man's new dialogue with nature. Toronto; New York, N.Y: Bantam Books.

illusory imprisonment in a singular world defined by economics. It is essential that we understand the ancient wisdom that shaped our laws of energy and thermodynamics. Zero carbon is a formula, an algorithm that must be balanced by both sides of the equation. We cannot just burn carbon and expect zero results. This paradox was understood by Epicurus (341-270 BCE) who understood the notion:

"...that nothing came from nothing and nothing could be reduced to nothing, that is, that all human production involved the transformation and conservation of matter."[22] This is also bound up in the myth of the immortal snake, and is the basis for Chaos Theory,[23] and how chaos that appears to lack order, can achieve stability and predictable order out of noise following the bifurcation of opposites.

Elemental carbon appears to be immortal. You may think this a strange turn of phrase as immortality conjures up gods, myths and legends, but as I have attempted to show in the story of zero, scientific reductionism has been built on a bedrock of ancient stories, narratives and dreams; a world comprised of the collective unconscious of humanity, and our genetic ancestry. What is more the physical world (the nonhuman majority) have shaped our primitive psyche based on our material dependencies and so we have not changed much up to the present. As Bruno Latour succinctly put it, 'We have never been modern'.[24]

Carbon appears immortal because in our universe individual carbon atoms keep-on-keeping-on. One of the major archetypes of creation, knowledge, and

---

22. Foster, John Bellamy. Marx's Ecology: Materialism and Nature. Monthly Review Press. Kindle Edition.

23. See Gleick (2011); Prigogine, I., & Stengers, I. (1984)

24. Latour, B. (1991) We Were Never Modern.

the circle of life, is the snake, the serpent, taniwha, or dragon. This creature can be seen as a paradoxical metaphor that provides us with a way to examine and analyse our own paradoxical reality and our connection to this planet. A snake is an animal that is found in many geographical regions and is part of many mythical and religious cults. It is both an actual creature and a mythological archetype. Like all living things on this planet, carbon is one of its major components, but this very physical part of its nature occludes our philosophical understanding of the snake and its immortal, mythological legacy. From a chemical perspective carbon is the sixth element of the periodic table, but prior to our scientific understanding, there were the alchemical and mystical descriptions of creation and the cycle of life, death and rebirth.

The mythical creature known as Ouroboros is a snake that forms the shape of an O or the figure zero. It is a snake that forms a circle by curling back on itself to bite its tail. Because the snake is capable of shedding its skin it symbolised immortality. The symbol of infinity can be formed by twisting the zero into two halves and leaning it on its side, thus ∞. The psychoanalyst Carl Jung believed that humanity suffered from a separation of the psyche and that individuation could only be achieved by reuniting the two halves of that psyche. Jung's long-time colleague M. von Franz described Jung's process of individuation as the unification of the ego and the unconscious. It was also the reunification of the anima, or the woman within the man's psyche, and the animus, the man within the woman's psyche. According to Jungian psychology, the archetypal symbols revealed in the hero's journey represents the process of individuation.[25] Jung wrote:

"The alchemists, who in their own way knew more about the nature of the individuation process than we moderns do, expressed this paradox through the symbol of the Ouroboros, the snake that eats its own tail. The Ouroboros has been said to have a meaning of infinity or wholeness. In the age-old image of the Ouroboros lies the thought of devouring oneself and turning oneself into a circulatory process, for it was clear to the more astute alchemists that the prima materia of the art was man himself. The Ouroboros is a dramatic symbol for the integration and assimilation of the opposite, i.e. of the shadow. This 'feedback' process is at the same time a symbol of immortality, since it is said of the Ouroboros that he slays himself and brings himself to life, fertilizes himself and gives birth to himself. He symbolizes the One, who proceeds from the clash of opposites, and he therefore constitutes the secret of the prima materia which ... unquestionably stems from man's unconscious."[26]

The history of Western philosophy and culture have been overwhelmingly written and interpreted by men. As the historian and philosopher, Richard Tarnas, has pointed out, how we think and act have been shaped by the masculine psyches of men like, Socrates, Plato, Aristotle, Aquinas, Copernicus, Bacon, Newton, Locke, Hume, Descartes, Nietzsche, Kant, Darwin, Marx, Jung and Freud

---

25. The hero's journey is often associated with Joseph Campbell who was influenced by Carl Jung's belief that a similar myth can be found in all the great religions and primitive legends, such as *The Epic of Gilgamesh* and the monomyths of the Buddha, Moses, and Christ. See my book, *Worldbending: a survivor's guide for those who want to think and act creatively about our future.*

26. Carl Jung, Collected Works, Vol. 14 para. 513

to name only some of the influential thinkers who established Western hegemony in the 20th and 21st centuries. Tarnas sees this as humanity's most pressing project, a crisis that is ultimately a masculine crisis, the ego death of those who see culture and technology in male terms. Since the dawn of agriculture and civilisation, around 12,000 years ago, the dominance of men has forced the male and female psyche apart. Tarnas wrote: "But this separation necessarily calls forth a longing for a reunion with that which has been lost—especially after the masculine heroic quest has been pressed to its utmost one-sided extreme in the consciousness of the late modern mind, which in its absolute isolation has appropriated to itself all conscious intelligence in the universe (man alone is a conscious intelligent being, the cosmos is blind and mechanistic, God is dead)."[27]

Tarnas sees this man as all alone, in an existential crisis, in a universe without meaning and in a time and space that is ultimately unknowable. This pessimistic worldview held by man has become his own world prison, it is this pessimism that both builds his world, and at the same time breaks this world. The guilt, both masculine and predominantly white, is also his debt to the planet that is forcing a transformational change.[28] "i.e., in a man-made environment that is increasingly mechanistic, atomized, soulless, and self-destructive. The crisis of modern man is an essentially masculine crisis, and I believe that its resolution is already now occurring in the tremendous emergence of the feminine in our culture...in virtually every

---

27.   Tarnas, R. (2010) The Passion Of The Western Mind: Understanding the Ideas That Have Shaped Our World View (pp. 442-443). Random House. Kindle Edition.

28.   See Eula Biss, (Dec.2, 2015) 'White Debt'. The New York Times

intellectual discipline, but also in the increasing sense of unity with the planet and all forms of nature on it, in the increasing awareness of the ecological and the growing reaction against political and corporate policies supporting the domination and exploitation of the environment, in the growing embrace of the human community, in the accelerating collapse of long-standing political and ideological barriers separating the world's peoples, in the deepening recognition of the value and necessity of partnership, pluralism, and the interplay of many perspectives."[29]

The hero's journey, is today, not just seen as a masculine quest by Jung and Campbell but a metaphorical map outlining the steps that the individual needs to take in order to achieve individuation. They saw the individual as an alchemist whose ultimate goal was not the transmutation of lead into gold, a chemical process, but the primaeval reunification of the psyche predicted in the myths of cosmological creation and the singularity at the beginning of time – the big bang. This archetypal cosmology, Jung believed, accounts for the commonality of human mythology and explains why an ancient African race, the Dogon, has much in common with ancient Chinese beliefs, the cosmology of Māori, and early Egyptian death cults. The Ouroboros is a paradoxical creature that is 'all in one', two snakes, mortal and immortal, in China it is represented by the Tao, the yin and the yang, the dark and the light, male and female. The male represented the heavens, and the circle, while the female represented the square and the earth. In ancient cultures, through to mystical alchemical cults, the squaring of the

---

29.   Tarnas, Richard. ibid. pp. 442-443

circle was a symbolic representation of the union of the psyche, the male and female gods. In Māori cosmology, Tanē, the god of the forest, and son of Rangi and Papa, carried knowledge in three baskets from the gods to people on earth. The half circles that forms a whole circle, are the handles of the basket that represented the infinite heavens, and the square of the basket, represents the earth mother, Papa.[30] Divine knowledge squares the circle, the ying and the yang. It was a common symbol in ancient Egyptian iconography and the earliest example was found in the tomb of Tutankhamen. In the ancient cultures of the Dogon, Māori, Chinese, and Egyptians it was believed that technological know-how and gifts of civilisation were imparted by ancestor-gods who taught humanity. According to Laird Scranton in his book on *China's Cosmological Prehistory*:

"Even in ancient Egypt there was a tradition in which the Egyptian hieroglyphic language was understood to have been a gift to Egypt from their mythical gods. In Buddhism, it was understood that Buddha had imparted knowledge to mankind in archaic times. The Dogon also associate eight mythical ancestors with specific civilizing skills and provide specific details about how they were taught."[31]

It would seem to be more than a coincidence that the **O** twisted into an **8** is the symbol for infinity ∞, and immortality flipped upright. The Ouroboros is a mystical snake that flickers between two states, the virtual and the actual much like matter and anti-matter, metaphorical carbon and anti-carbon particles. In a mythical sense the

30. Scranton, L. (2018) ibid. p.47

31. Scranton, Laird. (2014) China's Cosmological Prehistory: The Sophisticated Science Encoded in Civilization's Earliest Symbols. Inner Traditions/Bear & Company. Kindle Edition.

circle of the Ouroboros eating its tail, has an alchemical meaning, it is hiding its snake-twin that can only be found through the search for the Philosopher's Stone – the mysterious journey in search of the meaning of life and death. The Chinese also regarded 8 as a magical number symbolising wealth and fortune and the double 8, 囍 (shuāng xǐ), meaning *double happiness*. In Chinese mythology there were 8 original Emperors who were also dragons. In Western culture, the snake or Ouroboros evolved into the winged dragon of medieval mythology that protected mounds of treasure and the archetypal princess. These myths again relate to the hero's journey and their descent into the Netherworld of chaos where they come across the Goddess and the monster of the abyss, Ouroboros, guarding the gold and gems, who will kill the hero if they fail in their trial to reunite with their feminine psyche.

In 1975 a book was published by Robert K.G. Temple called, *The Sirius Mystery*. Temple had brought to the West the Dogon cult and uncovered how Dogon priests were somehow aware of astronomical facts long before modern telescopes had confirmed their beliefs were true. According to Temple:

"Sirius is the brightest star in the night sky and was a center of focus in the myths of many ancient cultures. The Dogon knew that Sirius is not one star but two—that the bright star of Sirius (referred to by astronomers as Sirius A) has a dark, small, dense companion dwarf star (Sirius B). They also knew the correct orbital period of fifty years for the two stars. Temple presented this knowledge as evidence of a possible alien contact in ancient times."[32]

---

32.   Scranton, L. (2014) ibid.

This cosmological myth of human creation is also deeply bound up with our modern archetypal notions of imagination and creativity and the circular 8 stages of the hero's journey. The Ouroboros is also the symbol of knowledge attainment and the warning that the misuse of that knowledge can end in misfortune and death. In Norse mythology, Oroboros was a serpent called Jörmungandr, one of three children of the trickster Loki, also associated with knowledge, technology, and like Satan, the serpent, a gatekeeper to the tree of knowledge.

Jörmungandr grew so large that he/she encircled the Earth, and like the South American myth, formed a world-disc around the ocean that defined the knowable world from the unknowable abyss, beyond and below. In the myth, once Jörmungandr, the World Serpent, releases its tail from its jaws, Ragnarök happens – a series of events that shares a common apocalyptic narrative; the end of the world when it is submerged by water. When the abysmal water subsides (as in Noah's flood) the world will be born again fertile and abundant to be repopulated by two human survivors.

In *Worldbending*[33] I discuss the *'Price of Success'* the 6th stage in the hero's quest, a dangerous moment in the mythology of creation and only the hero who has followed their heart and instinct will survive the ordeal of the abyss. This all-important moment for the hero can also be seen as a metaphorical warning for us today. We can think of fossil fuel extraction in terms of these subterranean myths. We may be looking at the holy grail, the Philosopher's Stone, the unimaginable wealth buried beneath the sea of the abyss but to rescue the Goddess we must acknowledge the

---

33.   For more on the importance of the Hero's Journey see Worldbending, chapter 1, The Original Sin-thetic.

price of success and be wary of the mercurial snake who is guarding the black oily gold we are so greedy to extract.

"In the Roman version of the hero's journey it is Pluto who abducts Persephone to be his wife, and queen as the goddess of death. Pluto, or Plouton, was the name of the god of the underworld, and Pluton was conflated with the word, Ploutus, meaning wealth."[34] The question for humanity is whether that treasure is material or philosophical – is it going to be more oil (an energy resource in another form), or a new way of thinking about our worlds beyond carbon.

In the Greek translation of the Hebrew Bible, the abyss translates the Hebrew words *tehom* meaning deep, and *tsulah* meaning sea deep, deep flood; Tehomot was the Semitic dragon that dwelled in the cosmic abyss. This was one and the same ancient serpent or snake that ate its own tail, Ouroboros. This snake that formed the figure **O** was the mandala of alchemy, the pre-scientific belief in search of the Philosopher's Stone and the transmutation of lead into gold. This mandala symbolised the circle of life, death and rebirth. The fear of zero and the void related to the mathematical mysteries born from the beginning of the universe when it was a void, just before the chaos. In Jungian psychology, these ancient images are still prevalent although hidden in the collective unconscious of our modern society. Carl Jung delved deep into the mythological past in order to attempt a reunification of the modern individual's psyche. Alchemy and the myth of this snake eating itself provided him with a powerful metaphor in order to understand how we think? In his essay *"The Religious and Psychological Problem of Alchemy,"* Jung writes:

34.   Excerpt From: Pete Rive. (2019) "Worldbending." Apple Books.

"The self is a union of opposites par excellence."[35] If we deny zero, or the void, we are denying not only our creation myths but our creative solutions to our wicked problems.

The Powerful Myths of Zero

In Western culture zero was associated with the dark and deathly void that was constantly at war with the infinite creation of the *'One true God'*. Throughout the history of the planet and the history of religions and cultures there have been variants of these universal beliefs that formed the foundations of the great religions on Earth. Humans imagined worlds and universes that were sometimes very different from each other, and yet almost more striking are their similarities. The awe and wonderment of the life cycle was shared by ancient Egyptians, ancient Greeks, Romans, Christians, Hindus, Jewish, Dogon, Māori and ancient Chinese cosmologies.

The mythical snake, Ouroboros, also symbolised the 'hero's journey' with no clear beginning depending where the storyteller decided to pick up the plot. Even before the ordered world of the hero there was primaeval chaos and darkness.[36] The hero must descend into the dark abyss, the void, the Netherworld, or hell to try and retrieve the goddess of fertility, and life itself. This journey was never easy, even for the hero, who like Gilgamesh, was part God and part human. A mere mortal would never return from the underworld, it was the Land-of-no-return, the Waters of Death. As Dante described the Gates of Hell in his

---

35.   Cited in, Psyche and Singularity: Jungian Psychology and Holographic String Theory, by Timothy Desmond

36.   Read more about the hero's journey in *WorldBending* by Pete Rive, Chapter 2, The Original Synthetic.

Inferno they were adorned with the warning , *'Lasciate ogni speranza, voi ch'entrate.'* or *'Abandon All Hope Ye Who Enters Here'*. Gilgamesh who sets out to gain immortality by descending the watery depths of the abyss is told by the ferryman Uta-napishti:

'No one at all sees Death,
no one at all sees the face [of Death,]
no one at all [hears] the voice of Death, Death so savage,
who hacks men down.
Ever do we build our households, ever do we make our nests,
ever do brothers divide their inheritance, ever do feuds arise in the land.
'Ever the river has risen and brought us the flood, the mayfly floating on the water.
'On the face of the sun, its countenance gazes, then all of a sudden nothing is there!'[37]

This mythological story of the Deluge was the forerunner of Noah and the ark and spoke to the ancient people of the fertile crescent about the ever-present threat of flooding. More than that, it spoke of the regenerative and life-giving nutrients deposited on the banks of the rich alluvial soil that gave the fertile crescent its name and provided the breadbasket of early agriculture. The watery abyss not only brought death but also life. According to Timothy Desmond in his book, *Psyche and Singularity*:

"Jung explains that "the Sanskrit word mandala means 'circle.' It is the Indian term for the circles drawn in religious rituals." In the Tantric school of Hindu thought

---

37.   See George, A. R. (Ed.). (2003). The epic of Gilgamesh: The Babylonian epic poem and other texts in Akkadian and Sumerian. London; New York: Penguin Books.

the god Shiva as world-creator is usually depicted in the center of mandalas."[38]

Thus, zero is a paradoxical mandala that is at the same time a symbol of world ending, and world building. This mythical paradox was a reassuring reality for those who pondered the mystery of life, and death and the eternal cycle of the universe, but by the time we came along in the 21st century zero was no longer considered a paradox. It had been converted into a clearly defined and unambiguous binary state, a coded message telling a machine to switch off, awaiting its twin number, one, the code to turn the machine on again. This black and white world appears to be empty of mystery, mysticism, or even metaphor but the story of zero is not over. The ontology, or reality of its existence, remains unknowable and deeply philosophical even as we stare into the black screen of a sleeping computer, one made from zeros and carbon that seems to bend our digital world.

38. Timothy Desmond. (2018) Psyche and Singularity: Jungian Psychology and Holographic String Theory

2

The Story of Carbon

If zero is the black and deathly void, then its twin is carbon the *giver of light*, the definition of life, the only '*one*'. Life on Earth, as we know it, would be impossible without carbon. Zero is carbon's polar opposite, and yet '*zero carbon*' has come to be known as the slogan for our survival inverting the good and the evil. Zero was once feared, associated with the chaotic dark and bottomless abyss, and only carbon promised life, light and unity, but that has now all changed. Today, we fear what an abundance of carbon will bring – too much in our atmosphere and we will all cook, too much in our water and life will be burnt by acidification, too many of one carbon-based life form (let's say humans) and that species will consume too many resources and other species will die. However, the world is not black and white, at least not in a binary way, good or evil, off and on. Life is a hybrid system of both digital and analogue processes and yet for 60 years or more we have focused on the analogy of digital computation. According to Daniel et. al and their work in synthetic biology: "The reality is that cells use a hybrid approach to information processing. In some cases they use digital yes-or-no decisions, but in many cases cellular signals are analog, with levels of gradation. More exotic, little understood signal-processing techniques involving noise and other forms of signal probably also contribute. And on top of that, a complex chemistry exists that continuously reassembles the cell in real time."[1] Simply thinking that life and our brains are on/off – yes/no switches is misleading and a lot of the way life works is not run as a digital system of switches but rather as analogue potentiometers smoothly operating like light dimmers adjusting the flow of energy. Biochemistry flows like a stream, after all the chemicals are suspended in water, and so elements from the periodic table move from high concentrations to low concentrations acting not as individuals but as 'hyperobjects'[2] that have group characteristics, not singular personalities. There are no one to one relationships but rather many to many. We have to consider many narratives to fully comprehend life and the ecology that supports it. The reductive scientific methodology must be augmented with other ways of looking at the world. As the historian Lynn White wrote in a famous ecological paper in 1967, 'Our ideas are part of the ecosystems we inhabit.'[3]

---

1. Daniel, R., Rubens, J., Sarpeshkar, R. et al. Synthetic analog computation in living cells. Nature 497, 619–623 (2013). https://doi.org/10.1038/nature12148

2. See Morton, T. (2013). Hyperobjects: Philosophy and ecology after the end of the world. Minneapolis: University of Minnesota Press. Morton defines a hyperobject as an object beyond human time and scale that we have no visceral comprehension of.

3. L. White, 'The Historical Roots of our Ecologic Crisis', Science, p. 1203. Cited by Massey, Charles. Call of the Reed Warbler: A New Agriculture – A New Earth (p. 291). University of Queensland Press. Kindle Edition.

Simplistic reductionism is inherited from the scientific methodology that was passed down from the Enlightenment and the philosophy of science, that occludes our view of the truth. Carbon by itself does not exist so it does not make sense to reduce our understanding of the Earth by simply narrowing our view to just a chemical or physical understanding of carbon. Reductionism and simplified models have been extremely useful in explaining our world in gross terms but models are never the whole picture. *Zero Carbon* makes as much sense as a flip slogan such as *Zero One*. Of course, Zero Carbon is shorthand for another slogan '*carbon neutrality*' meaning net-zero carbon emissions and has been extended to the other so-called Greenhouse Gases or GHGs and they are evaluated on their carbon dioxide equivalence. '*Net zero*' means that any emissions are balanced by absorbing an equivalent amount from the atmosphere."[4] However, because of agricultural emissions it is thought that net zero emissions will be impossible or politically unpalatable in that sector and so it will mean negative emissions in other areas such as energy and transport.

Language is a dynamic technology that morphs and adapts according to the power-politics of the day. In 2006 the *New Oxford American Dictionary* made the phrase, '*Greenhouse Gases*', the Word of the Year. Acronyms, phrases, and even slogans are useful in compacting complex issues and yet it can also be responsible for obfuscation, confusion, lost meaning, and even deception. *Zero Carbon* as a slogan is a useful call to action but it is also obscuring our exploration of new worlds – it negates the essential role that carbon plays in the story of life. Of course, carbon is

---

4.   See https://eciu.net/analysis/briefings/net-zero/net-zero-why

not the only element of importance but without it, our life form and the plants and animals we recognise, would not exist.

The Birth of Carbon

First appearing as a gas, carbon was converted into a solid and then returned to a gaseous form to continue its never-ending life cycle. However, there was a time when carbon did not exist and it was born. True, we still don't really know what happened before the Big Bang or even before the earliest event horizon at 10^{-43} seconds because just as we can't look into a black hole, the secret of the very first moment of cosmological creation (and even before) may be beyond our ability to ever know. There is, however, the consensus amongst scientists about the creation of the oldest and lightest members of the Periodic Table that were first born into our universe after that first massive explosion. The first inhabitants of that first world were hydrogen, helium, lithium and beryllium. These can only be made with the extraordinary heat and pressure of the Big Bang. The other 88 heavier elements were all born by first fusing lighter elements together in the foundry of the stars, within these nuclear reactors. It is via this incredible stellar alchemy that our ancient ancestor carbon was born and it bequeathed to us, and every living thing on our planet, LIFE. Just as in the ancient creation stories from Africa, Australia, New Zealand, Egypt, and the biblical tale of Genesis, carbon assumes the mythological status of a god that lives forever in our universe. Carbon is the second most abundant element in the human body, and the fourth most abundant element in the universe after hydrogen, helium, and oxygen.

Carbon was born within the heart of a star and catapulted throughout the universe in a fiery explosion when the star burned through all its fuel and died in the blinding flash of a supernova. This instantaneous nuclear detonation is around the total equivalent energy produced by our Sun over a 10 billion year period. It took an astronomical amount of energy to create carbon and that stellar energy is stored in the carbon atom that reacts with other elements to pump life into all living things on our planet. Quite simply: zero carbon = zero life (as we know it). According to the latest and most authoritative scientific knowledge, the first stage in the process of life requires a nuclear fire of such intensity that it can first burn hydrogen and fuse it into helium known as nucleosynthesis. It is only from massive stars, at least ten times bigger than our Sun, that those other elements heavier than helium can be created and was necessary before Earth had the requisite elements for life. Carbon was born in such a massive star, created from three helium atoms fusing together at one billion degrees Celsius – it is this fire that fuses the foundation blocks of life.[5] The etymology of the word carbon come from the Latin word 'carbo' meaning charcoal or coal and hints at the incendiary nature of both its genesis; the very origins of life; and our destructive desire to burn it to create new life.

The Carbon Cycle of Life

Carbon does not exist as a solitary thing, we cannot know its precise position at any one point in time and space. Even as an atom it is not atomistic, in other words, it never

---

5.   https://science.jrank.org/pages/2413/Elements-Formation-Manufacturing-Heavy-elements.html

exists in the wild as a single atom by itself, it always keeps company. Their company can include not only other carbon atoms but molecules and chemical compounds such as CO_2 and CH_4 i.e. carbon dioxide and methane. To overstate the point, even these compounds do not exist by themselves in the wild. Carbon, in its many guises, has played many roles in the evolution of the universe and has been both an active player and a bit-part actor in the evolution of life on Earth. According to the evolutionary biologist, David Sloan Wilson, we should reconsider evolution as the primary lens when we regard not just urgent ecological issues but social, economic and political problems. Biological, physical and cultural evolution must be considered in terms of the network of actors, or as the French philosopher, Bruno Latour calls them *actants* so that we consider not just humans but nonhuman actors.[6] Scientific reductionism has become the dominant worldview when it comes to thinking about our interaction with carbon and yet this obscures our evolutionary ecological vision and negates the way we see the world and possible solutions to the anthropogenic problems, or the wicked problems caused by humanity. Wilson writes:

"An evolutionary worldview provides a refreshing alternative to these reductionistic traditions. Multilevel selection theory tells us that analysis should be centered on the unit of selection."[7]

I really do not know if there is a standard unit of selection for carbon but I do know that it is not one carbon atom and that most of our interactions with the carbon

---

6. Actant is a term coined by the French philosopher Bruno Latour used in his Actor Network Theory to describe things that act upon others in an interconnected network.

7. Wilson, David Sloan. This View of Life . Knopf Doubleday Publishing Group. Kindle Edition.

actant is on a massive scale even if it is invisible. Take for example a seemingly insignificant bottle of Coke. It is estimated to contain something like 10^{24} molecules of CO_2. That is an astronomical number and is more than the number of cells in your body which is 10^{13}, or even more than the number of seconds since the Big Bang 13.772 billion years ago or 10^{18} seconds. Carbon is what the ecological philosopher, Timothy Morton, refers to as a *hyperobject*, an object or thing beyond human scale, of such enormity in time and space that it is viscerally beyond our comprehension. There are carbon atoms around today that are spread to the very edges of the universe (if such a thing exists) and are at least 13 billion years old. What is more 'hyperobjects' may even exist at close to the smallest scale theoretically known to science at the Planck length of 1.6 x 10^{-35} metres, far smaller than a carbon atom and unobservable to us. Yet, we talk of zero carbon as if we know what carbon is and can eliminate it. There are many conceptual worlds but the one that currently rules the world is inherited from the Enlightenment 300-400 years ago and still views our universe as a mechanical thing, albeit it is no longer thought of as a clock but now as some gargantuan computer – a digital machine that is the repository for all information and intelligence in this universe.[8]

What I am talking about here may seem philosophical, and surely it is, but the point I am trying to make is that given scientific reductionism and our predominantly mechanistic view of reality there is a tendency to think about our world as if it was only made up of discrete atoms.

---

8. See Greene, B. (2011). The Hidden Reality: Parallel Universes and the Deep Laws of the Cosmos. Chp. 10 Universes, Computers, and Mathematical Reality.

According to the Australian agricultural ecologist, Charles Massey, the way to understand and repair our world is to take a more organic and holistic view of our planet and that includes the carbon soil cycles. "The key to it all has been to listen to the land, respond to its needs, be prepared to continually change your approach, and to constantly try new things."[9] We have neglected other aspects of carbon's biography because we have only thought of it as part of the scientific world and failed to creatively experiment with other worldviews. Jessica Hutchings, a Māori food and soil activist, in her book, *Te mahi oneone hua parakore : a Māori soil sovereignty and wellbeing handbook,* begins with a stated political position: "The indivisible relationship between tangata [people] and whenua [land] as a starting point.' and by extension *oneone* 'soil'...'This volume turns the tables on understanding soil as an economic resource to understanding soil as part of the interconnected universe that makes up indigenous realities in Aotearoa."[10]

This approach can be understood philosophically, and spiritually as seeing soil as an agent, an active *'thing'*, an *'actant'*, and in Māori culture, the goddess, Papatuanuku. The chemistry, physics, or even the biochemistry of carbon is not the end of the story and, just as philosophy was once the origin of theories about carbon that evolved into scientific theories, we can enlist philosophy again to open up new worlds in which we can co-exist with the 6th element on the periodic table. When we consider carbon we can only ever really appreciate its life cycle when we know how it acts and reacts within systems and, because

9.   Charles Massy, (2017) Call of the Reed Warbler: A New Agriculture - A New Earth.

10.   Edited by Jessica Hutchings and Jo Smith. (2020). Te mahi oneone hua parakore : a Māori soil sovereignty and wellbeing handbook. p.7

it participates in both organic and inorganic, or living and nonliving systems, it makes no sense to simply reduce it to one carbon atom or molecule, it is not helpful to only think of it in the abstract. Hutchings reflects: "on how kaupapa Maori organics can provide a guide for living beyond the fractured industrialized food systems that surrounds us."[11] The life cycle of carbon in all its guises will help us to creatively imagine new worlds, or rediscover old worlds in a new context.

The concept of an atom was conceived by the ancient Greeks and assumed that these invisible components made up everything in our world. This was Plato's and Aristotle's philosophy based on Democritus and the Greek word *atomon*, that means uncuttable or indivisible. Reality was comprised of two essential principles the atom and the void, so for our purposes we can see these as carbon and zero.

"Greek atomism had posited a universe made up of invisibly small, indivisible particles moving freely in an infinite neutral void, and creating by their collisions and combinations all phenomena. In this void there was no absolute up or down or universal center, every position in space being neutral and equal to every other. Since the entire universe was composed of the same material particles on the same principles, the Earth itself was merely another chance aggregation of particles and was neither at rest nor at the universe's center. There was therefore no fundamental celestial-terrestrial division."[12]

Beginning in the Enlightenment philosophers and

11.   Hutchings & Smith (eds.) (2020). ibid.

12.   Tarnas, Richard. The Passion Of The Western Mind: Understanding the Ideas That Have Shaped Our World View (p. 265). Random House. Kindle Edition.

scientists reacquainted themselves with this ancient Greek theory of atomism and despite no observable evidence or even indirect validation of the atomic theory, it became the bedrock of our physics, chemistry, biology and cosmology. Chemists in the 18th century, such as Antoine Lavoisier, Joseph Priestly and Humphry Davey were the first to model the carbon cycle and others discovered the process of photosynthesis. Since the beginning of the 20th-century scientists have gained empirical evidence of the evolutionary life cycle of carbon and its essential role in all life cycles known to humanity. While we can only speculate about the origins of carbon (before the Big Bang), within the modelled confines of the Earth system we now have a carbon life cycle that has almost unanimous scientific consensus.

If for now, we ignore the carbon elements that enter our simplified closed world model via the cosmos, carbon could be said to begin its life cycle as gas, and for our simplified purpose, that gas could be carbon dioxide, CO_2. It is then consumed by plants and using the energy from our star, the Sun, enters a biochemical reaction with another actant, chlorophyll found in plants, and phytoplankton such as cyanobacteria thought to be responsible for turning Earth into an oxygenated planet between 2.5 and 2.3 billion years ago – otherwise known as the Great Oxidation Event that depleted methane levels and probably caused the first ice age. As the Earth scientist, Matthew Warke noted it is important we understand the evolution of life so we can recognise other habitable worlds: "...and as we continue to change our atmosphere through rising anthropogenic greenhouse gas emissions, and consider schemes to mitigate climate change by directly removing greenhouse gases from the air, it is important that we understand the extremes of how Earth's climate has shifted in the distant past. "[13]

Collaborative Evolution

It was once thought that primitive life on Earth did not appear until around a billion years after the Earth was first created 4.543 billion years ago. However, this theory could be revised based on *putative fossilized microorganisms that are at least 3,770 million and possibly 4,290 million years old* that had probably found a habitat in the hydrothermal vents on the seabed found in Canada. While the evidence remains contentious, because it is hard to determine whether the microscopic shapes could have been caused by geology, there has also been evidence in Australia of stromatolites that provide scientists with the best *smoking gun* so far that life began at least 3.5 billion years ago.[14] The theory of a primordial soup is that life on Earth was enabled by the conditions making it possible to *cook up early microbial life* and is supported by traces of graphite, a form of carbon. These have been dated to have been present 4.1 billion years ago, only a few million years after the Earth was created approximately 4.5 billion years ago. Another possible ingredient of this life-giving soup was likely to have been cyanide that could have hitched a ride on a meteorite at around the same time.[15] The earlier dates for first life on

---

13. Matthew Warke (June 3, 2020) The Conversation. "Billions of years ago, the rise of oxygen in Earth's atmosphere caused a worldwide deep freeze. https://theconversation.com/billions-of-years-ago-the-rise-of-oxygen-in-earths-atmosphere-caused-a-worldwide-deep-freeze-139722"

14. 'Smoking Gun' Evidence Dates Some of Earth's Earliest Life to 3.5 Billion Years Ago By Mindy Weisberger - Senior Writer September 30, 2019. https://www.livescience.com/earliest-signs-of-life-on-earth.html

15. Cyanide-Laced Meteorites May Have Seeded Earth's First Life By Stephanie

Earth are supported by the evidence that primitive life can exist in harsher environments and possibly planets elsewhere in the Universe.

Whether or not life began 4.1 billion years ago or not, there is general consensus amongst the scientists that the first life on this planet was probably a simple celled microbe known as a prokaryote that evolved through symbiosis to become a more complex microbe, a eukaryote. This is known by the technical name as endosymbiosis or symbiogenesis and is the generally agreed process of the early evolution of life on Earth. This once contentious theory has contributed to a groundswell amongst biologists who have elevated symbiosis to one of the most influential theories in biology today. In their review, *A Symbiotic View of Life – we have never been individuals*, Gilbert, Sapp & Tauber (2012) wrote:

"Symbiosis is becoming a core principle of contemporary biology, and it is replacing an essentialist conception of "individuality" with a conception congruent with the larger systems approach now pushing the life sciences in diverse directions."[16]

In her 1970 publication Lynn Margulis: "convincingly argued for a theory of microbiological symbiosis, called endosymbiosis, that has profound implications for epistemology and knowledge sharing and is based on RNA and DNA analysis that shows that micro-organelles do not out-compete each other, but rather consume and retain the original code (or shared knowledge) within a symbiotic ecology. The link was then made between this biological

Pappas June 28, 2019. https://www.livescience.com/65821-cyanide-laced-meteorites-fueled-life-maybe.html

16. Scott F. Gilbert, author, Jan Sapp, author, & Alfred I. Tauber, author. (2012). A Symbiotic View of Life: We Have Never Been Individuals. The Quarterly Review of Biology, (4), 325. https://doi.org/10.1086/668166

epistemology and human knowledge by Maturana and Varela (1992) and further supported by Montague's (2007) theory of human decision-making."[17]

Margulis and her son, Dorion Sagan made the case for the evolutionary status of microbial symbiosis and alert us to the ideological and mechanical shortcomings of concepts like *survival of the fittest* and *red in tooth and claw*. They wrote:

"Without microbes, life's essential processes would quickly grind to a halt, and Earth would be as barren as Venus and Mars. Far from leaving microorganisms behind on an evolutionary ladder, we are both surrounded by them and composed of them. The new knowledge of biology, moreover, alters our view of evolution and chronic, bloody competition among individuals and species. Life did not take over the globe by combat, but by networking. Life forms multiplied and grew more complex by co-opting others, not just by killing them."[18]

Margulis and Sagan imagined humanity as a massive global network, what we might understand as one of Morton's *hyperobjects* that is linked genetically and biologically by symbiotic microorganisms that communicate biochemically. This radical reconceptualisation of evolution, supported by research that has shown we are mostly composed of alien microbial DNA, should cause us to stop and think about our unbreakable connection with the nonhuman networks of other life forms and inorganic matter.

---

17.   Rive, P. B. (2012). Design in a Virtual Innovation Ecology: A Cybernetic Systems Approach to Knowledge Creation and Design Collaboration in Second Life. Retrieved from http://researcharchive.vuw.ac.nz/handle/10063/2747

18.   Margulis & Sagan, 1997, p. 78

Imagining Carbon Networks

The Enlightenment is praised for its liberation of the individual spirit and a flourishing of the intellectual imagination. Philosophers, scientists and artists cast off two thousand years of Greek, Roman, and Judeo-Christian dogma to look at the world a fresh, and discovered by turning their heads on the side that right before their eyes were worlds that they had never imagined. Of course, the legacy of previous beliefs, legends, and myths did not go away, but they became submerged in their subconscious and that legacy has been inherited by us today. Many writers have talked about the dawning of a new age, but ironically while some imagine a technological utopia on other planets, others have been guided by new biological and philosophical revelations that encourage us to reconceptualise our world, not as a uni-world, but as a multiverse made up of multi-worlds here on Earth.

The ancient philosophical quandaries: *'who am I?'* and *'what is the meaning of life?'* are still alive and kicking today in science and philosophy. It is by engaging our imagination and creativity that we are discovering that uncertainty, paradox and contradiction are not to be feared but followed. This book could have been just about carbon, however, I wanted to give you a sense of how carbon is just one element in the story and to use it as an example of how a deep and transdisciplinary dive into some *thing* like carbon can reveal our entanglement in a weird and aesthetically queer multiverse. Carbon can not be reduced to a single molecule in order to understand it, it is enmeshed in a carbon network, among an infinite number of networks of other *things*. I am going to give a brief sketch of one small part of the carbon network to try and illustrate how a simple shift of perspective can enlighten the imagination and unleash creative energies that have been moribund by outmoded thoughts, theories, and behaviours.

I would like to direct your attention to one of the smallest, and also the most exceptionally large living carbon kingdoms that is; fungi. This is a separate and recently discovered kingdom, distinct from plants, animals and archaea, and are closer to the kingdom of animals than plants. The kingdom of plants is wholly dependent on the carbon network of fungi to survive, without fungi plants would not exist. With a name like a wizard, Merlin Sheldrake, is a mycologist who has studied the strange worlds of fungi since he was a boy. In his words he has studied fungi for such a period and so intensely he has been 'imprinted' by the fungi, and they have infected his imagination. As Sheldrake describes it: "Fungi inhabit enmeshed worlds; countless threads lead through these labyrinths. I have followed as many as I can, but there are crevices I haven't been able to squeeze through no matter how hard I've tried. Despite their nearness, fungi are so mystifying, their possibilities so *other*. Should this scare us off? Is it possible for humans, with our animal brains and bodies and language, to learn to understand such different organisms? How might we find ourselves changed in the process?"[19]

Sheldrake tells us that plants only managed to shuffle out of the water about 500 million years ago, all thanks to their friend fungi, and today 90 per cent of plants depend on mycorrhizal fungi. They can be so small that we can't see them with the human eye, and yet *Armillaria* is a fungi

---

19.   Sheldrake, Merlin. (2020). Entangled Life (Kindle Locations 601-604). Random House. Kindle Edition.

which is among the largest organism on this planet. The current record holder amongst this species was found in Oregon and weighs hundreds of tonnes and covers a massive 10 square kilometers and we can only estimate its age at somewhere between 2,000 and 8,000 years old.[20]

The fantastic world of fungi should fascinate us all because it may even contain the evolutionary answer to our survival in a world we have made toxic, hot and increasingly uninhabitable for many species, including our own. Fungi can breakdown toxic substances and can be used to clean up hazardous waste. It can be used to break down pollutants in oil spills in a process known as 'mycoremediation' and even be used to remove heavy metals from soil. Fungi produce around fifty megatons of spores each year which means they are the biggest living particles in the air and are an essential component of rain, sleet and hail. In this world that is being threatened with drought and starvation we need to understand that even the air we breathe is a living thing that is enmeshed in a network of 'actants'.[21]

Fungi act as a shared network for trees, known as the 'Wood Wide Web' and is remarkably similar to the Internet, except better. Underground a fungal mat acts as a super highway that allows the sharing of nutrients, moisture, chemical and electrical communications between plants in their ecosystem. Sheldrake mused that this interconnected symbiotic network has had a profound impact on the way he sees the world – the individual (including himself) has become decentred and more like a

---

20.   Sheldrake, M. (2020). ibid.

21.   See Sheldrake, Merlin. (2020). Entangled Life p.11. Random House. Kindle Edition.

node in a symbiotic network of *things*, both living carbon enabled and inorganic *things*.

Biochar, Fungi & Super Soil

The longevity of carbon in our atmosphere is of grave concern because even if we stopped pumping carbon gases and other long lived GHGs right now scientists have run a simulation that shows that the oceans could stop absorbing carbon and so our planet could continue to heat for hundreds, if not a thousand years.[22] Therefore, it is not sufficient for governments to aim for net zero carbon emissions by 2050 because if we don't remove these gases from the atmosphere millions of future generations of humans and nonhumans will die and be threatened with extinction.

According to Hawken (ed.) in the book, *Drawdown*, there is an ancient process of waste disposal that could sequestrate almost a gigaton of carbon, removing it from the atmosphere and burying it in the ground. This would not only slow global heating, but it would also benefit the lively carbon network of fungi, soil microbes, plants and animals. Ironically, in the country that is home to the so-called *'lungs of the planet'*, the Amazon, farmers are releasing millions of tons of carbon into the air as they burn forests turning biomass into carbon dioxide, and methane so they can plant soybean crops, and graze cattle. Yet, going back in time before Portuguese colonisation ancient societies once got rid of their organic and inorganic leftovers such as fish bones, livestock manure, and broken pottery by baking

22.   Even if emissions stop, carbon dioxide could warm Earth for centuries
        https://www.princeton.edu/news/2013/11/24/even-if-emissions-stop-
        carbon-dioxide-could-warm-earth-centuries

them underground without exposure to oxygen.[23] The process, pyrolysis, created a biochar they called *terra preta* or black earth. This helps to create super fertile soil in which fungi and microbes thrive, and at the same time holds carbon under the ground for hundreds, and sometimes thousands of years. Wiedner & Glaser have outlined three physical and chemical ecological benefits of biochar.

1. Biochar has a porous structure that is resistant to microbial degradation, and is therefore responsible for long-term carbon sequestration.

2. Chemical groups on the edges of biochar can contribute to enhanced soil quality by providing electric charges that allows soil nutrients to be attracted to the site.

3. The physical structure of the porous biochar enables a massive water holding capacity by increasing the surface areas and creating a home for the microorganisms such as fungi.[24]

Therefore, biochar, which is also created naturally: it contributes to slowing global heating; it improves soil fertility; and it retains water and nutrients. In the Amazon these nutrients would otherwise be leached away as the trees and the biomass are burnt off and the tropical rains flush them into the rivers. Just 1 gram of biochar can have a surface area of approximately one thousand to two

---

23. This was common throughout the ancient world and Māori also practiced this beneficial waste disposal in Aotearoa.

24. Katja Wiedner and Bruno Glaser (Martin-Luther-University Halle-Wittenberg, (2013) The chapter 'Biochar-Fungi Interactions in Soils' included in Soil Biogeochemistry, von-Seckendorff-Platz 3, Halle, Germany.

thousand seven hundred square metres. The tiny pores act like a negatively charged magnet that draws in the positively charged elements such as calcium and potassium.[25]

Mental Symbiosis

Sheldrake's research in the Amazon, and his contemplation of his own scientific practice, drew him to a research project that was examining the creative process of science and how the imagination might unlock new ideas using LSD. He realised that he was entering a recursive feedback loop in which his research question relating to a jungle flower, *Voyria*, that lives without photosynthesis, and might be surviving with the help of his beloved fungi. His inspiration came from an acid trip, and that LSD was a synthetic drug similar to the fungus *Claviceps purpurea*)."[26] For Sheldrake LSD represented a *'fungal solution to his fungal problems.'* LSD and the drug psilocybin, $C_{12}H_{17}N_2O_4P$, popularly known as *'magic mushrooms'* share a similar chemical structure and are known to not only have an hallucinogenic effect but to also induce wellbeing and connection with a universal network. Sheldrake points out that throughout human history: "psilocybin are fungal molecules that have found themselves entangled within human life in complicated ways exactly because they confound our concepts and structures, including the most

25. Hawken, P. (Ed.). (2017). Drawdown: The most comprehensive plan ever proposed to reverse global warming. New York, New York: Penguin Books.

26. See Britannica. John Philip Jenkins. "LSD was a synthetic drug that can be derived from the ergot alkaloids (as ergotamine and ergonovine, principal constituents of ergot, the grain deformity and toxic infectant of flour caused by the fungus *Claviceps purpurea*)." https://www.britannica.com/science/LSD

fundamental concept of all: that of our selves. It is their ability to pull our minds into unexpected places that have caused psilocybin-producing 'magic' mushrooms to be enveloped within the ritual and spiritual doctrines of human societies since antiquity."[27]

Years of training in the scientific method, painstaking observation and precise data collection obscures the creative impulse and imaginative leaps that are a fundamental part of science, the arts, and the humanities. Allowing the mind to see new *things* can liberate us to imagine new theories and experiment with new practices and concepts. Sheldrake documents his own creative process but admits: "There was something embarrassing about admitting that the tangle of our unfounded conjectures, fantasies and metaphors might have helped shape our research. Regardless, imagination forms part of the everyday business of enquiring. Science isn't an exercise in cold-blooded rationality. Scientists are – and have always been – emotional, creative, intuitive, whole human beings, asking questions about a world that was never made to be catalogued and systematised."[28]

It is towards these worlds of imagination and myth that we need to look to rekindle novel and creative solutions to our world that continues to try hackneyed theories and methods of change. LSD may be a modern synthetic drug but it is derived from ancient rituals even when humans were only indirectly aware of fungi and attributed it to the wrath of the gods. The Romans had a designated deity responsible for the effects of fungi, *Robigus*, the god of rust, who was honoured with the festival, *Robigalia*. It is posited that ergotism caused by a fungal rye disease baked into

27.   Sheldrake, M. (2020). ibid.

28.   Sheldrake, M. (2020). ibid.

bread may have been responsible for hallucinations, and convulsive behaviour attributed to communication with the gods.[29]

It is as if we have sanitised science deliberately cleansing it of all those mucky, messy ideas that were once associated with myths and legends, and yet, at the same time we have thrown out the baby with the bath water sluicing the wisdom of those cosmological archetypes down the drain.

The Science of Serpents

Our true wealth and treasures are immaterial and hidden beneath the sea of our collective subconscious. They are the source of our imagination and creativity, our ability to build, bend and break worlds. Ironically, the oily fossilised treasure that we continue to steal from under the nose of the sleeping dragon originated from a dream by a German chemist in 1865.

Benzene is a colourless and explosive constituent of crude oil and an essential component in the manufacture of ethylbenzene and isopropyl benzene for the production of high octane petrochemicals. Benzene is what gives petrol its sweet smell and what gives the family of carbon and hydrogen compounds the name aromatic hydrocarbons. The sweet smell of benzene is originally derived from benzoin resin that was the basis of frankincense, that has been made for 5000 years, and was used in perfumes beginning in the 16th century. Carbon in many of its forms has provided the foundations of life as we know it – it is the

29.   Miedaner, T., & Geiger, H. H. (2015). Biology, genetics, and management of ergot (Claviceps spp.) in rye, sorghum, and pearl millet. Toxins, 7(3), 659–678. https://doi.org/10.3390/toxins7030659

basis of our cultures and civilisations. It should come as no surprise that there are hints about our origin as a species and our fate as humans remains buried in those ancient cultures.

The mythology of Ouroboros illustrates the modern conjunction of science and the arts that is often lost and buried under a stack of scientific papers and hegemonic arrogance. The Endless Knot of Ouroboros was formed out of the twisted circular zero forming the figure of **8**, and the symbol **∞** or an eight on its side, meaning infinity. It could also be seen as the double **O** made from the DNA double helix – the twin snake unified by twisting the self back from the deformed shape that separates the male from the female psyche. The mortal snake is largely comprised of carbon and yet is paradoxically immortal because of that carbon. It is also ironic that carbon-based fossil fuels are known as non-renewable energy sources because in fact most of the carbon that was created at the birth of the universe will exist forever. The physics of this also connects to the second law of thermodynamics that states that energy can neither be created nor destroyed.

Benzene's extraction from coal tar began in 1845 when Charles Mansfield invented the first industrial production of benzene that was important for the dye industry and is a byproduct of coke and coal gas.

However, in 1865, Friedrich August Kekulé, had been puzzling over the chemical structure of the carbon compound known as benzene. It was eventually mythology that unlocked his imagination illuminating how 6 carbon atoms could form a ring. Kekulé's recalled his dream:

"I was sitting, writing at my text-book; but the work did not progress; my thoughts were elsewhere. I turned my chair to the fire and dozed. Again the atoms were gambolling before my eyes. This time the smaller groups kept modestly in the background. My mental eye, rendered more acute by the repeated visions of the kind, could now distinguish larger structures of manifold conformation: long rows, sometimes more closely fitted together; all twining and twisting in snake-like motion. But look! What was that? One of the snakes had seized hold of its own tail, and the form whirled mockingly before my eyes. As if by a flash of lightning I awoke; and this time also I spent the rest of the night in working out the consequences of the hypothesis."[30]

The cyclic formation of benzene was confirmed in 1929 through crystallography by Kathleen Lonsdale. It was the mysterious process of creativity and imagination by which the serpent Ouroboros had rewarded the hero, Kekulé, with the knowledge of organic chemistry, aromatic hydrocarbons and the black gold we know as fossil fuels. While benzene was used to reduce 'knocking' and wear and tear in engines, today the amount that is added to petrol has been considerably reduced, and is now only 1% of petrol due to its proven carcinogenic nature.

In its natural form, before the synthetic form of benzene, the benzoin resin comes from the storax species of tree, *Styrax officinalis*, and the word benzoin probably comes from the Arabic word meaning *'Javan Frankincense'*. Perhaps the origin of the legend of the snake and the Tree of Knowledge in the *Book of Genesis* was the Arabian belief that small winged snakes guarded the source of benzoin resin and frankincense. The snakes that were commonly found circled around the tree could only be driven away by burning the storax tree that gave off highly toxic benzene

---

30.   Read, John (1957). From Alchemy to Chemistry. pp. 179–180. ISBN 9780486286907

and formaldehyde or methanal – another organic compound of carbon. The mythology of alchemical carbon; fossil fuels; the hero's journey; and the genesis of humanity are inseparable from our understanding of life and death on Earth even though we might have forgotten the origins of creativity, imagination and the birth of the cosmos.

Non-renewable energy sources are actually renewable, just like Ouroboros, carbon-based fuels are immortal, it is just that their life cycle is beyond the lifespan of humanity who may not even exist as a species by the time the serpent reawakens from guarding its black gold and kills the hero who has failed their quest to achieve immortality. It will take millions of years for carbon to cycle through all its forms to return to its concentrated essence in oil, gas and coal but unless we try to understand the nature of its being, its ontological nature, and our philosophical connection with carbon we will be ignorant of not only this world but worlds beyond carbon. Carbon is part of our story and it is an ultimately unknowable hyperobject that is beyond our time and space.

3

Zero Humans

Whatungarongaro te tangata, toitū te whenua

the land is permanent, people disappear

As I write this chapter, my neighbourhood, in a peaceful
valley of trees, is unusually silent. We live on the edge of a
leafy park in Parnell in Auckland, Aotearoa New Zealand.
Despite only being a twenty-minute walk to the CBD we
are lucky to have native birds, such as the melodious Tui
and cheeky Pīwakawaka, so it is normally very tranquil.
However, during this entire week, it has been much quieter
than usual – no traffic noise, almost no one out on the
street, and even the busy Port of Auckland is silent as I now
look out my open window at the dockside cranes and the
freighter that no longer belches black bunker smoke into
the blue sky. It's like a scene from Vincent Price's movie,
The Last Man on Earth (1964) *'the end has come'* but in colour.
For some the silence is eerie but I find it an idyllic break
from the usual rip, shit, and bust that creates the

cacophony of capitalist consumerism. Yet, for many, the end of fossil fuels could spell the end of capitalism and would result in the end of civilisation, and by association maybe even the end of humanity. Today is Saturday, and for more than a week, since Thursday 26th March 2020 the whole of our country has been in level 4 lockdown as we try and shake the global pandemic coronavirus, COVID 19. The irony of this global pandemic is that it has enforced a local realisation on many of us; the very local has been brought back into sharp focus as our ability to travel internationally, or even nationally have been severely restricted. As someone who used to contribute a massive amount of GHG emissions, flying around the world, I feel little loss and much gain from realising the importance of the land beneath my feet and the quietness of fewer planes and helicopters. This book is a project that documents my deep reflection on my own country and our context in a globalised consciousness. The gentle peace that has descended on this world is a whimsical reminder that we do have the ability to radically change our behaviour overnight and that the Anthropocene still threatens to kill many more humans and nonhumans than the virus could ever achieve. It is also a reminder that humans have made the world much noisier, and more noise signifies more heat, more waste. Think about how quiet the planet once was and would become again with zero humans? Could the pandemic reduce the global population to a more sustainable number and so ensure the survival of both the humans and the nonhuman majority?[1] There is a lot that we can learn from the past. The number zero recalls the dark past when there was no thing, no humans, and only

---

1. See my book 'Worldbending', Chp. 10 and the short story, 'The Silent Worlds of Earth'.

blackness. For most of human culture it was believed that at the dawn of the cosmos there was only chaos and disorder. To speak of the void was to speak of no senses, there was no thing to see, hear, or feel – it was a prehistory that predated reason. It was only relatively recently that we discovered that that prehistory is much longer and richer than we ever imagined. Humanity was an emergent afterthought.

The Egyptian Deep Freeze

Palaeontology has revealed the amazing prehistory of the times before humans and before our recorded history, a time of zero humans. At the end of the Ordovician, around 445 million years ago there occurred dramatic climatic swings from tropical temperatures at the poles to a sudden ice age causing the second worst mass extinction in the history of the planet. As Peter Brannen pointed out in his book, *The Ends of the World:*

"There was the draining of the seas, the cooling of the tropics, the distance between the continents, the oxygenation of the deep, and the collapse of the food chain. But we're not done killing this world yet. There was still one final coup de grâce. Directly on top of the glacial rocks of North Africa and Saudi Arabia are black radioactive shales."[2]

As Brannen explains it was the end of one world that created the radioactive oil and gas shales of North Africa and the Middle East. During the Ordovician there were pulses of destruction as the planet violently changed from a superheated hot house with sea levels 60 metres higher

2. Brannen, P. (2018). The Ends of the World: Volcanic Apocalypses, Lethal Oceans and Our Quest to Understand Earth's Past Mass Extinctions

than the ice age that followed as the sea contracted. The second mass extinction on the Earth killed millions of tons of marine biomass that floated to the bottom of the oceans eventually turning black from lack of oxygen and turning into the opaque blackened oil and gas that drives today's overproduction and consumption. This is regarded as one of the world's most important petroleum sources. Those species that managed to survive the first catastrophic pulse were then killed off by the next disaster. Just like our own ice age, there were waves of heat and cold lasting around a million years, with glacial advances and retreats and then at the end of the Ordovician ice age a global heating that flooded the world's continents with melted ice. The end of that world was the end of a period of massive evolutionary diversification as life was rapidly swallowed by the deathly abyss and the surviving species approached a watery zero population. Brannen wrote:

"Sometimes black shales are about as close to an SOS as one finds in the fossil record—a grim notice that oxygen is running dangerously low. They're black from being suffused with carbon from the dead sea life that sinks and settles to the bottom of the ocean, where it can't oxidize or decay. There it's left undisturbed to accumulate on a lifeless, anoxic seabed. And there it stays until it's discovered half a billion years later by a curious species of primate determined to dig it up and burn it." [3]

What is apparent from the deep freeze that would have hit the previously hot zone around Egypt and the middle east at the end of the Ordovician was that GHGs from massive volcanic, and other seismic activity, deposited dead paleontological treasures deep beneath the sea, and

3.   Peter Brannen, The Ends of the World: Volcanic Apocalypses, Lethal Oceans and Our Quest to Understand Earth's Past Mass Extinctions"

embedded them in rock. Our fossil fuel obsession is now simply releasing the carbon that was sequestrated for millions of years beyond a human time frame that amounts to a geological explosion of gas into the atmosphere. The temperature and the planet's climate changed radically when that happened, and today as the oceans run out of oxygen we are likely to see a climatic rerun of what happened last time with catastrophic consequences. Zero oxygen will move life closer to the void and only those organisms that thrive on high carbon gas concentrations will survive. [4]

The fertile crescent which included the area we know as Egypt, has also been called the 'cradle of civilisation' as the invention of agriculture enabled nomadic people to create cities and hierarchical power structures. Throughout the fertile crescent mathematics became a tool to rule and administer the areas under the control of the various rulers. Mathematics has been the technology of our world building, bending, and now the tool of world breaking.

The Egyptians were some of the earliest to use mathematics to do just that. More than 5,000 years ago even before the pyramids, Egyptians invented a decimal counting system where pictures were used instead of numbers to regulate the ownership and administration of the agricultural lands surrounding the bountiful Nile river.

According to Seife:

"The Egyptians' innovation of the solar calendar was a breakthrough, but they made an even more important mark on history: the invention of the art of geometry. Even without a zero, the Egyptians had quickly become masters

---

4.   See Climate change: Oceans running out of oxygen as temperatures rise
https://www.rnz.co.nz/news/world/405058/climate-change-oceans-
running-out-of-oxygen-as-temperatures-rise

of mathematics. They had to, thanks to an angry river. Every year the Nile would overflow its banks and flood the delta. The good news was that the flooding deposited rich, alluvial silt all over the fields, making the Nile delta the richest farmland in the ancient world. The bad news was that the river destroyed many of the boundary markers, erasing all of the landmarks that told farmers which land was theirs to cultivate. (The Egyptians took property rights very seriously. In the Egyptian Book of the Dead, a newly deceased person must swear to the gods that he hasn't cheated his neighbor by stealing his land. It was a sin punishable by having his heart fed to a horrible beast called the devourer. In Egypt, filching your neighbor's land was considered as grave an offense as breaking an oath, murdering somebody, or masturbating in a temple.)"[5]

From the earliest myths of cosmological creation human culture has shared archetypes relating to depths of the watery abyss and the paradox of its life giving essence. Mathematics eventually encoded the narrative of events that told the story of both world building and world ending. Flooding, tsunamis and tempests were once believed to be plagues visited on humanity by angry gods. It was not until the 14th century that the word plague was used in the translation of the Bible chapter, Exodus. According to the Catholic historian and scholar, Gregory Elder, there were 70 references to plagues and pestilence in the Hebrew version of the bible. The most commonly used hebrew term was *maggephah*. "It can mean slaughter, a disease, a strong blow or a death blow. Another Hebrew term is debher, which more explicitly means a disease, and in the Hebrew Scriptures this is normally associated with divine anger.

---

5.   Seife, C. ibid.

Sometimes this is translated as pestilence. There are several Greek words in the New Testament that refer to plagues, the most common being plage, which means a blow, a wound or a disease. Another term is panoukla, which means a disease or illness. These words are often tied to the wrath of God, which inflicts the plagues."[6] The most well known plagues are the 10 recorded by Moses on Mount Sinai. An angry God inflicted them on the Egyptians because the pharaoh would not release the Israelites from 400 years of slavery. Interestingly, scientists have shown that there is good evidence to suggest that these plagues coincided with natural disasters caused by warmer climates. Ehrenkranz & Sampson "propose the root cause to have been an aberrant El Niño-Southern Oscillation teleconnection that brought unseasonable and progressive climate warming along the ancient Mediterranean littoral, including the coast of biblical Egypt, which, in turn, initiated the serial catastrophes of biblical sequence — in particular arthropod-borne and arthropod-caused diseases."[7]

There Be Monsters

While the flooding of the Nile brought disease and death to the Egyptians, water was also worshipped as the giver of life in many myths and religions embodied in monsters, spirits, and deities. Swimming was an integral part of ancient Egyptian and later, Greek lifestyles, as reported by

6. Elder, G. (2018) Professing faith: What's the meaning of the plagues in the Bible? https://www.redlandsdailyfacts.com/2018/05/18/professing-faith-whats-the-meaning-of-the-plagues-in-the-bible/

7. Ehrenkranz, N.J. MD, & D.A. Sampson. (2008) Origin of the Old Testament Plagues: Explications and Implications. YALE JOURNAL OF BIOLOGY AND MEDICINE 81 (2008), pp.31-42.

Plato, Socrates and Aristotle. However, with the advent of Christianity, the pagan sea nymphs and watery deities were demonised to usurp those ancient beliefs. The myths of Orpheus, Hades, Persephone, and Dionysus morphed into the legends of Christ and the resurrection. By the middle ages many Europeans had simply forgotten how to swim, a period described by the French historian, Jules Michlet, as 'one thousand years without a bath', the Europeans went 'fifteen-hundred years without a swim'. There was another good reason to fear the watery abyss as drowning was a common occurrence.[8] The myths, legends, rituals and biblical commandments became the laws and regulations to command and control the common people. The hero's journey was not for them, only the demigods with the divine right of kings could be expected to survive the descent into the abyss in order to claim the treasures of eternity and return alive. The Greeks believed everyone must know their place, and ordinary humans were definitely not gods.

The sea and the forests were part of the collective imagination, a thin line between civil society and a chaotic wilderness without laws and without people. From the Sumerian tales of Gilgamesh, down through the legends of the Greeks, 'nature' was seen as a dangerous realm in which anarchy and undisciplined passions would reduce humanity to uncontrollable beasts that were no longer human. The laws of man did not apply to nature, and for those who wandered or were reckless enough to deliberately cross the line into the wilderness, whether it was a forest, or across the ocean, they were all warned that beyond civilisation, 'There be monsters!' and zero humans.

---

8.   Chaline, E. (2018) How Europe Learnt to Swim.
     https://www.historytoday.com/miscellanies/how-europe-learnt-swim

According to Professor Diane Purkiss the ancient Greeks didn't like the wilderness very much, and were profoundly unenthusiastic about the place. The Greek god, Pan, was the god of the wild, and the origin of the English word 'panic'.[9] The physical realm was infused with the spiritual reality, and so for an individual to wander outside the gates was to be suspected of being an outlaw, or a heretic who had turned their backs on the holy orders. The wilderness was the void, the dark place where monsters roamed. In the forests beyond the cities was also a place of exile and where people who were deprived of arable land were forced to live. Therefore, the wilderness was also associated with economic failure, poverty, and deprivation. Zero, was the nothingness that came before debt and the negative rents owed to rapacious landlords. The authoritarian role of the Church and the King was reinforced by numerous myths, legends and harsh punishments, if not death.

From these ancient morality tales the collective unconscious of the 21st century would inherit the fear of the economic void, or zero population growth, propagating the economic moral panic of governments, wealthy elites, and corporate GHG emitters. For those who have the temerity to question the authority of the powerful who have taken us to the edge of this current existential crisis, their anti-authoritarian behaviour is labelled mentally ill. According to the psychologist, Brian Levine, who writes for the blog, *Mad America*, the so-called epidemics of ADHD and ODD, or Oppositional Defiance Disorder, relates directly to an American authoritarian culture that wants to suppress dissent. He writes:

"In an earlier dark age, authoritarian monarchies

---

9.   From the documentary series, Myths and Monsters, (2017) ep. 3 The Wild Unknown.

partnered with authoritarian religious institutions. When the world exited from this dark age and entered the Enlightenment, there was a burst of energy. Much of this revitalization had to do with risking skepticism about authoritarian and corrupt institutions and regaining confidence in one's own mind. We are now in another dark age, only the institutions have changed. Americans desperately need anti-authoritarians to question, challenge, and resist new illegitimate authorities and regain confidence in their own common sense." [10] For Dean & Altemeyer, America has fallen into an *Authoritarian Nightmare* under Donald Trump who is using citizen outrage over police brutality to incite race riots and white supremacist militia to take to the street. All of this was straight from the authoritarian playbook and was designed to help get him re-elected in the 2020 election by those who fear the chaos and void from the loss of law and order – command and control. [11]

In the age of the pyramids the Egyptian pharaohs used mathematics to ensure they continued to own and rule over the fertile agricultural land. This was despite the common occurrence of a deluge that repeatedly burst the banks of the Nile and destroyed the land markers, creating chaos, and threatening their rule over their subjects with the fear of an anarchistic void. As Seife explains:

"The ancient pharaohs assigned surveyors to assess the damage and reset the boundary markers, and thus geometry was born." [12] While Egyptian mathematics and

---

10.   Levine, B. (2012) Why Anti-Authoritarians are Diagnosed as Mentally Ill. https://www.madinamerica.com/2012/02/why-anti-authoritarians-are-diagnosed-as-mentally-ill/

11.   See Dean, J.W. & Altemeyer, B. (2020) Authoritarian Nightmare: Trump and his followers. Melville House.

12.   Seife, C. ibid.

geometry was famed throughout the ancient world they still did not have a use for zero because they were a very practical people and were not interested in the abstract or philosophical application of mathematics. The Greeks, however, were the opposite and they raised mathematics to a higher art form, but still they did not invent zero. "Zero came from the East, not the West."[13]

Creation Myths of the Void

In the Christian story of world creation, before the world begins, and before humans 'ruled the Earth', the story starts with the establishment of the world in which order is established. *The Book of Genesis* starts as follows:
[1.1] In the beginning when God created the heavens and the earth,
[1:2] the earth was a formless void and darkness covered the face of the deep, while a wind from God swept over the face of the waters.
[1:3] Then God said, "Let there be light"; and there was light.

The Roman poet Ovid wrote the poem, *Metamorphoses*, (8 CE) telling the story of World creation, beginning with chaos.

"Before there was earth or sea or the sky that covers everything, Nature appeared the same throughout the whole world: what we call chaos: a raw confused mass, nothing but inert matter, badly combined discordant atoms of things, confused in the one place."

In the earliest extant story (2100BCE) of a civilisation the king Gilgamesh gained wisdom from his hero's journey, descending into the abyss, the watery grave and mysterious

---

13.   Seife, C. ibid.

birth place of eternal life. He restored the laws and ancient rituals and offerings to the gods. After the Deluge had destroyed all the temples and washed away all of the agricultural gifts of the Gods, it was Gilgamesh who restored order out of the chaos following the abyss. The disjointed translation reads:

"He who saw the Deep, the country's foundation, [who] knew … , was wise in all matters!
[Gilgamesh, who] saw the Deep, the country's foundation, [who] knew … , was wise in all matters!
[He] … everywhere …
and [learnt] of everything the sum of wisdom.
He saw what was secret, discovered what was hidden, he brought back a tale of before the Deluge.
He came a far road, was weary, found peace, and set all his labours on a tablet of stone.
He built the rampart of Uruk-the-Sheepfold, of holy Eanna, the sacred storehouse."[14]

Agriculture, and therefore human population growth, only really became possible when the Earth moved into a new epoch, the Holocene, around 12,000 years ago. From out of the freezing death grip of the Ice Age, humanity emerged from the chaos of volcanic eruptions and the seismic turbulence that spewed millions of tons of carbon dioxide into the atmosphere resulting in a warmer planet with plenty of water. Where there had once been cold, and sterile wastelands, that could only support a very small number of humans and nonhumans, there was now a riot

---

14. George, A. R. (Ed.). (2003). The Epic of Gilgamesh: The Babylonian epic poem and other texts in Akkadian and Sumerian. London; New York: Penguin Books.

of life exploding through the thaw and delivering an abundance of food. To those species lucky enough to have lived through this procreative transition the difference must have seemed as stark as the contrast between the white snow and the black soil. Where once there was only frigid death now vast areas of land opened up before them. Those primitive people must have counted their blessings and given thanks to the Gods. However, it was from the creation myths of India and China that the West eventually came to embrace zero, the void, and infinity.

The Harmony of Opposites

As in the Western legend of Ouroboros, the East gave us the Buddhist mandala of Yin and Yang. The two are similar in their cosmological significance which has a deep psychological meaning reflected in the polarity of zero and one, the void and creation, and the birth and death of the universe. The harmony of these opposites signified harmony for the macrocosm reflected in the singularity of the microcosm. This view of the cosmos sees that no one state can remain alone for long – zero and one, zero and carbon, zero + humans must all come together to attain peace and unity.

The creation mythology of Māori is complex and intriguing, once again illustrating the inchoate origins of life beginning with a dark chaotic void. The genesis begins with the Earth Mother, Papatūānuku, and the Sky Father, Ranginui, they were bound in such a loving embrace that there was no daylight between them and the universe was completely dark with no light. The gods Papatūānuku and Ranginui had many children including, Tane, the god of nature. Tane their son twisted and turned in the darkness until he could see small specks of light. These were the children of his brother Tangotango, and so Tane asked his brother if he could have these luminescent children to ornament their mother. Tangotango agreed to gift their mother the Shining Ones which Tane arranged. However, Tane was dissatisfied with the cold pale light that the stars gave off and so Tangotango said, "Take my last child, Te Ra", the brightest and warmest star, the Sun.

"Then Tane threw Rangi far above, and with him Te Ra, so that the sun's rays would shine with kindly light on the Earth Mother, and on the gods and their children for ever."[15] Following the myth of the creation of light, the stars and the separation of the Earth Mother and the Sky Father by their son, Tane, Papatūānuku was turned over onto "her face so that she could no longer see her husband lest she continue to grieve for him. Tane then adorned his mother with plants and trees, and finally created from her body the first woman."[16] Like other creation myths there was a supreme being who existed before the stars, the sky and the earth gods. The god, IO, existed in the void and in 'darkness brooded over the all-pervading water.'[17] IO issued the command to bring light. The fertile waters of the abyss separated the Sky Father and Earth Mother. According to Reed the legend of IO was not commonly known because just as the Greeks and Romans feared to speak the name of the powerful creator so too was it tapu, or sacred on pain of death to speak the name of IO. According to the writings of George Grey (1853): "Ranginui and Papatūānuku continue

15.   Reed, A.W. (1964). Māori Fables and Legendary Tales. A.H. Reed & A. W. Reed. Wellington. Auckland. Sydney.

16.   A.W. Reed, (1963). An Illustrated Encyclopedia of Māori Life. A.H. Reed & A. W. Reed. Wellington. Auckland. Sydney.

17.   A.W. Reed, (1963). ibid.

to grieve for each other to this day. Ranginui's tears fall towards Papatūanuku to show how much he loves her. Sometimes Papatūanuku heaves and strains and almost breaks herself apart to reach her beloved partner again but it is to no avail. When mist rises from the forests, these are Papatūanuku's sighs as the warmth of her body yearns for Ranginui and continues to nurture mankind."[18] The extraordinary similarity of these archetypes is more astounding when you consider that New Zealand was so isolated from the ancient worlds of the Greeks, Romans, Egyptians and Chinese and that Māori mythology was not written down but was an oral tradition supported by carvings, weaving and other arts.

It was Francis Yates, the historian who wrote about the Greek tradition of the Memory Palace.[19] While Plato warned that the written word would weaken our memory, the Greeks used these Palaces to maintain the oral tradition of oratory. The orator would stand in various places in the Palace and learn long verses of poetry. To recall these verses they would imagine they were standing again in those locations. Māori oratory was also used in rituals and poetic rhyme was a mneumonic to recall the ancient myths and legends that instructed the tangata whenua (people of the land) on how to look after the land as the kaitiakitanga of the Earth mother, Papatūanuku. The enduring cosmological archetypes express the unity of opposites and the heart wrenching separation of the Sky Father and Sky

---

18. Rangi and Papa. Wikipedia. Retrieved 7 Nov. 2020.
     https://en.wikipedia.org/wiki/Rangi_and_Papa based largely on the
     writings of a Te Arawa chief, Wiremu Maihi Te Rangikāheke, who is the
     author of much of the material in George Grey's Nga Mahi a nga Tupuna
     (Grey 1971), originally published in 1853 and later translated into English as
     Polynesian Mythology (Grey 1956). This is only one of many versions of the
     creation myth.

19. Yates, F. (1966). The Art of Memory.

Mother. The universality of creation myths and cosmologies share many similarities.

"These include a principle of duality and the pairing of opposites. These principles are expressed symbolically in a number of different ways, such as opposing colors, like black and white, or opposite sexes, male and female. We see these same principles reflected in certain classic oppositions, such as the notion of the separation of earth and sky."[20]

Unlike the Western tradition in which chaos and disorder came to be associated with evil and death, the Eastern tradition viewed these cosmological states to be an essential part of creation and therefore zero was not feared even when it was associated with the void. Rather, "[In the] Chinese notion of chaotic disorder (luan), early Daoists posit a type of chaos that is to be cultivated rather than feared."[21]

Creative Deconstruction

The alchemical mandala of Ouroboros may seem strange, unscientific, and even downright silly to the modern mind, however, it may also reveal truths about our imaginative capabilities or constraints when we try and imagine worlds beyond a *Zero Carbon* construct. If the lesson of the eternal snake is its power to be able to consume itself, to cannibalize itself in order to reproduce with itself and be reborn as itself – then is the modern mind capable of the

20. Scranton, L. (2014). China's Cosmological Prehistory: The Sophisticated Science Encoded in Civilization's Earliest Symbols (1 edition). Inner Traditions.

21. Scranton, Laird. China's Cosmological Prehistory: The Sophisticated Science Encoded in Civilization's Earliest Symbols . Inner Traditions/Bear & Company. Kindle Edition.

same trick – and if so, how? Jung understood the incredible difficulty faced by mere mortals who can not ever see themselves beginning the hero's journey. If this is a dangerous and treacherous adventure for an individual, what hope is there that a plucky band of heroes would agree to set out together? The story of zero is also the hero's circular journey, from order descending into the caverns of chaos and mysteriously reborn in a new world. The abyss that threatens humanity with extinction is also the mysterious realm of our salvation if we have the courage to face it and to adapt and evolve with it.

Some degree of creativity is within the grasp of everyone and in fact any living thing can be said to be creative by a broad definition that states: *'creative'* is *'relating to or involving the use of the imagination or original ideas to create something'*. The ability of single cell microbes to *'imagine'* or foresee the future by propelling themselves towards sources of energy or food, evolved into the circular metabolic loops of imagination and creativity. This does not imply that the single celled bugs had consciousness but their primaeval and homeostatic processes sensed their worlds, and it was their feelings that sustained life even in organisms that did not have neural networks or minds.[22] Damasio and others have shown how these primitive organisms are capable of proto-design, a chemical *'imagining'* or sensory blueprint of a future world that sustains life, or as Maturana and Varela called it autopoiesis.[23] This is clearly not a design function exclusive to humanity and as we expand our ecological literacy we

22. Damasio, Antonio. The Strange Order of Things (p. 12). Knopf Doubleday Publishing Group. Kindle Edition.

23. Maturana, H. R., & Varela, F. J. (1980). Autopoiesis and cognition: The realization of the living. New York: Springer.

will perceive the biological basis of human understanding is connected to the same way as all living things 'know' and and build worlds. "Every act of knowing brings forth a world. ... All doing is knowing, and all knowing is doing. ...We have only the world we bring forth with others..."[24] From an evolutionary perspective, this is where our mind first started to emerge, beginning with feelings and evolving into emotions, reason and our values. These biochemical feedback loops have informed nonhumans and humans about survival that eventually became consciousness. According to Damasio: "feelings tell the mind, without any word being spoken, of the good or bad direction of the life process."[25] The belief that feelings are exclusive to humans has misled us about the nature of reality and our connection with nonhumans. Zero humans does not mean zero feelings, or indeed zero consciousness. As our minds evolved so did our cultures and our ability to learn, and remember the good, the bad, and the complex emotions that are embedded in our collective unconscious. It was no longer easy to decide whether a feeling or emotion was related to survival as an individual, as a small group, an organisation, or even a planet. Decisions were no longer straight forward, and so-called rational thinking simply muddied the waters as emotions preempted most actions. Ideologies and the 'manufacture of consent'[26] have tapped into our primal desires and fears, fooling us that economics was a physical reality rather than a mental construct. Individuals that have succumbed to groupthink

---

24.   Maturana & Varela. (1992). The Tree of Knowledge – The Biological Roots of Human Understanding. Shambhala Boston & London. p.25 & p.249.

25.   Damasio, A. ibid.

26.   See Marx, Freud, Bernays, and Chomsky who have all written about the process and role of propaganda in the manufacture of consent.

have become paralysed by fear. If we were to begin to bend the world into novel shapes would we face annihilation before we even built a new world order? The thought of even starting to imagine new worlds is decried by many as a betrayal of human happiness – the future uncertainty is too great. It is as if the end of our current economic construct will also herald the end of human existence. World breaking involves the deconstruction of many symbols that many regard as the ultimate expression of human creativity and technological progress. Without explicitly saying it, many consider that it is much safer to risk the probability of societal collapse than to face the uncertainty of new world models that may fail. Meanwhile, as the icy evidence melts and falls into the oceans many simply ignore the fact that the monoculture of global capitalism will result in massive seismic shifts of rock and human culture. It is becoming highly unlikely that we will either be able to hold on to the world as it has been, or prevent the catastrophic changes predicted by scientists. Yet, there is still hope that we might be able to mitigate anthropogenic global heating and perhaps even learn to adapt to a radically different planet. However, before we begin the tricky process of adaptation we have to acknowledge that business-as-usual is not a solution.

What is more, there is the question, is it even possible for humanity to reimagine new worlds beyond the ONE that we now assume is the only possible one? The common sense one – the economic one? Margaret Thatcher said of capitalism, *'there is no alternative'*. Yet, neoliberal ideology is fooling us into thinking this world is a natural state of affairs.

"Dunne & Raby in their book, *Speculative Everything: design fictions and social dreaming* (2013) champion Fredric

Jameson's claim that "it is now easier to imagine the end of the world than an alternative to capitalism."[27]

Whereas, new ideas in biology, complexity theory, the evolutionary theories of mind and Speculative Realism are suggesting that symbiosis is grounded in an urge for existential survival rather liberalism's politico-economic faith in individual property rights. A new wave of transdisciplinary research is suggesting that the scale of human cooperation and creative collaboration will determine the survival of this species.[28] Beyond the individuals, and small teams, corporations also face their own existential dilemmas. The delusion of economic growth has them fighting for their imagined lives – these are virtual organisms, mental constructs that do not directly depend on biological homeostasis for 'life'. Corporations are nonhuman, inhuman, and can almost exist with zero humans as long as the algorithms, and business rules of profit are automating their existence.

The Business of Biting Your Own Tail

In this exponential world that is accelerating beyond our common comprehension, we are increasingly faced with companies that we thought would be here forever but are being 'disrupted' by new and surprising technologies. Even Sergey and Brin, the founders of Google, have stepped down from their executive positions. Many people will have no memory of life before Google and yet its survival is not guaranteed.

27. Dunne, A., & Raby, F. (2013). Speculative Everything: Design, Fiction, and Social Dreaming. The MIT Press.

28. See Wahl,D.C. (2007) Scale-Linking Design For Systemic Health: Sustainable Communities And Cities In Context. International Journal of Ecodynamics. Vol. 2, No. 1 (2007) 1–16

Back in the early 90s when I started my digital editing company there was no talk of '*disruption*'. I was even asked to sell my company to a wealthy ad man within the first year of business, but I actually liked my work and thought it would provide for me and my family financially and creatively for many years, if not the rest of my life. I know how laughable that sounds now but many still don't realise their careers are about to undergo massive change very soon. Back then the Internet was just making waves and I had made a number of successful predictions about technology – but not about the one that really counted – computerization, otherwise known as digitisation. In the parlance of today's management consultancies, after ten years my world had been severely '*disrupted*'. For those of you who have faced the failure of a business you will be able to relate to the feeling that your sense of survival closely identifies with the survival of that business. Corporations, and even capitalism is no different. There is a fear that the end of the business is the beginning of the end – it may be irrational but the end of capitalism is feared by some to be the death knell of humanity. Today we are warned that the cunning corporation does not wait to be eaten by the nimble young snake but must cannibalize itself to survive. However, if a corporation tries to start a new company within the skin of the old snake, according to Ismail, who co-wrote *Exponential Organizations:* 'The "immune system" of the parent company will come and attack it.' With respect to survival, the old snake must more than ever be looking to slip out of its wrinkly skin and do something very new – to be reborn before it becomes someone else's dog food. Ismail goes on to say:

"Only go after new markets (to avoid the immune system response). If you want to transform an existing cash cow or leapfrog a current business unit, you need a stand-alone unit with a small team that is isolated and fully autonomous." He warns that it is better to spin out rather than spin into the existing parent company wedging yourself into the mothership. He wrote: "A new enterprise won't fit neatly anywhere and internal politics will ensue, especially if you are cannibalizing an existing revenue stream."[29]

Can a corporation remake itself, to eat itself from the outside in and the inside out? Like a goldfish, in a fishbowl, today's corporations only understand the water they swim in. Sure the goldfish can look out at the universe surrounding it, but it's experimentation and creativity beyond the plexiglass can be very limited. The corporation that tries to reimagine itself must extract part of its mind, give that new mind just the right amount of sustenance to survive, then ignore it, and genuinely hope it will develop into a healthy organism capable of thinking and acting for itself. However, even if the senior executives of the old snake organization can sell the snake oil that will enable a new company to spin out from the old – why would they? Who will benefit if the young snake succeeds? Will the deal be structured so that the profits return to the old snake? Will the young entrepreneurs of the spin-out succeed beyond all expectations and be rewarded with massive salaries, share options, even an IPO, and thus, dimming the spotlight that once shone on the old snakes? Or might the young snakes fail – no skin off the old snakes' nose – '*I told you so*', they will mutter. Hidden agendas, takedowns, internal politicking and under-resourced startups that look

---

29.   Ismail, Salim. Exponential Organizations: Why new organizations are ten times better, faster, and cheaper than yours (and what to do about it) . Diversion Books. Kindle Edition.

to the old serpent who has no intention of shedding its old skin, all of this favours a split personality. The C suite culture has become a series of political campaigns that encourages executives to offer 'election' bribes and promises they have no intention of keeping. Many design their campaigns to run for three years at the most. Six months to gather business intelligence, a year to execute a plan, and another 18 months to look for a new job. There may have even been enlightened board members and senior executives who are aware that the old snake's days are numbered, but rather than calculating the company's ground-zero, they prefer to fantasize about its immortal future. The hideous portrait of Dorian Gray is hidden away in an abandoned warehouse – the truth about their past crimes are secretly hidden in commercially sensitive documents and out of court settlements, while the old snake's portrait becomes uglier and more distorted by a cynical smile.

Just as the individual faces the frightening prospect of taking the hero's journey, teams can suffer from groupthink, corporations can resist the call to innovate, and humanity can be swayed by climate change deniers and the 'merchants of doubt'.[30] In summary, nothing really changes, but we should not forget that humans do not need corporations to stay alive, corporations still need humans to stay alive – zero humans means zero corporations.

Planetary Delusion

Beyond the corporate scale, we can detect a planetary

---

30.   See Oreskes, N., & Conway, E. M. (2010). Merchants of Doubt: How a handful of scientists obscured the truth on issues from tobacco smoke to global warming (1st U.S. ed). Bloomsbury Press.

delusion. The inability of humanity to act collectively and decisively towards global heating even when its existence is challenged. This has just been confirmed by the UN 2019 Climate Change Conference or COP25 in Madrid. According to Kera Sherwood O'Regan, on behalf of Indigenous Peoples Organisations, this was how they saw the outcome of the summit:

"We can't help but feel the irony of your refusal to include human rights and indigenous people's rights in Article Six [of the Paris agreement] when we know that market approaches have already directly harmed our communities. Our knowledge cannot be upheld if our rights are not upheld. You treat negotiations like a zero-sum game where you make deals behind closed doors, trading off our rights for the profits of the very corporations who caused this problem in the first place. But you forget that we cannot negotiate with nature."[31]

There has been repeated failure by the UN and IPCC signatory countries to reduce GHG emissions by anything close to what is necessary to keep the planet below 2°C, let alone 1.5°C. Just like some lumbering corporate dinosaur the entire planet is addicted to legacy economics and energy consumption that stops governments from making meaningful changes to how we live and work. Political horse-trading and backroom deals are dressed up to sound as if the politicians have made the best and most scientifically, and so-called *realistic* decisions to mitigate global heating. Sir John Houghton, co-chair of the Working Group I for the Third Assessment Report (2001), once put it: "Any move to reduce political involvement in the IPCC would weaken the panel and deprive it of its political clout.

---

31.   Read more here https://www.bbc.com/news/science-environment-50801493

. . . If governments were not involved, then the documents would be treated like any old scientific report. They would end up on the shelf or in the waste bin."[32] The conclusion, cannibalization of legacy business-as-usual is extremely hard on a personal, corporate, and planetary scale even when the survival of their own species is concerned.

In, Aotearoa, New Zealand, we have an ongoing love affair with agriculture and so successive governments have shied away from any policies to significantly decrease our biggest contributor to our countries GHGs. In total farming contributes 48% of the total GHG emissions. As the Interim Climate Change Committee reported:

"One thing is clear – New Zealand must take action to reduce agricultural methane and nitrous oxide because these gases form such a large proportion of our national greenhouse gas profile. There is often less focus put on nitrous oxide – but this is a potent and long- lived gas and must be a part of efforts to achieve a net zero target."[33] This is not just a problem for New Zealand, as pointed out by the UN's Food and Agriculture report, as long ago as 2006, most of the *overdeveloped* world protects and subsidises industrial agriculture. They stated that livestock generate more GHG emissions than the entire transportation sector. Livestock accounted for 9 percent of anthropogenic CO_2, 65 percent of anthropogenic nitrous oxide, and 37 percent of anthropogenic methane. The co-author of the report wrote:

32.   Union of Concerned Scientists. Published Jul 16, 2008 Updated Oct 11, 2018. The IPCC: Who Are They and Why Do Their Climate Reports Matter? https://www.ucsusa.org/resources/ipcc-who-are-they

33.   ICCC, 30th April 2019. Action on Agricultural Emissions: evidence, analysis, and recommendations.

"Livestock are one of the most significant contributors to today's most serious environmental problems."[34]

Ironically, farmers who reject the collective ownership of natural resources that are *'enclosed'* by their land titles, directly benefit from the commonly owned resources such as water, air, rock, flora and fauna. It is because these resources have been enclosed by private property rights that farmers in the past have argued that they are entitled to do what they want to the biosphere, hydrosphere, lithosphere, or atmosphere on their own land. The farming lobbies have used economic growth to argue for their private property rights and the neoliberal defence against government interference and regulation. This has been changing but only after years of wilful ignorance of the environmental cost to the common resources that extend well beyond the boundaries of even the biggest farm. Successive governments that trace their origins back to the anti-socialist, William Massey, and the conservative, pro-farming parties that preceded National, convinced farmers that they do not have to worry about common resources, and that collective responsibility is a concept closely aligned with communist dictatorships. There are a number of farmers and their lobbyists who are either genuinely bewildered, or cry wolf, claiming undue haste to force better farming practices. For those who have not followed the news from the *'townies'* it may well have come as a shock that farmers who have ignored global heating, water pollution, agricultural GHGs, and soil degradation have gone from *'hero to zero'*. Farmers who have been told in the past that they are responsible for New Zealand's positive foreign exchange earnings, and the city's wealth have been

34.   United Nations Food and Agriculture Organization. (November 29, 2006) "Livestock a major threat to environment".

encouraged to borrow big to pay for more fertiliser, and industrial agriculture, so it is little wonder they are depressed when they can no longer afford their debt repayments and they are accused of environmental vandalism.[35] Our agricultural land use in New Zealand is threatened by those farmers, and industrial livestock farms that are destroying soil quality, damaging water ways, depleting resources, and polluting the atmosphere that we all hold in common because of their self interests, and neoliberal ideology. Of course, as this continues it may ultimately result in reduced land productivity and the inability to feed as many people as before. Historically, the farming lobbies in most of the so-called developed countries has been loud, and powerful and they have managed to dissuade their governments from taking environmental action to prevent damage to the ecosystems and wild habitats. In New Zealand the ICCC stated baldly: "Globally, we are not on track to achieve the goals of the Paris Agreement. Yet almost daily, right here at home, we are presented with reports that underscore the reality of a changing climate – whether it be coastal erosion and rising sea levels, more intensive floods, loss of New Zealand's glaciers or the warming Tasman Sea."[36] Ironically, it is the fear of the number zero that is pushing us towards a modern Deluge. The sea is rising due to ice melt and warmer temperatures are bringing more rain, and more extreme weather which could possibly cause between 25-40 metre sea level rise within decades and certainly within

35.   Avery, D. (2017). The Resilient Farmer. New Zealand. Penguin.

36.   See ICCC. (30 April 2019) Action on Agricultural Emissions: evidence, analysis and recommendations.

hundreds of years,[37] and the death and displacement of millions of people from New York to Auckland.

New Zealand farmers have been typically conservative and have historically voted to support the National Party, against the more progressive Labour and Green parties. Since the disastrous *new right* policies of the 80s Labour party known as Rogernomics *market forces* has been the active political ideology of both National and Labour, much the same way as both the Republicans and Democrats in the US have an *religious* devotion to their own version of the *free market* ideology established by Reagan and Milton Friedman. New Zealand elections are often more about political branding and campaign theatre than social and economic theory that are often devised by bureaucrats and academics in the pay of the universities or ministries, such as Treasury. However, at the last election Labour and the Greens won and have tried to bring in policies that will curb GHG emissions. New Zealand farmers have become more vocal and have even taken to protesting, complaining about central government regulation that is out of touch with farmers.[38] As one farmer's protest billboard read: 'Zero Carbon = Zero Jobs". And yet there is a growing number of informed farmers who are looking to regenerative and sustainable farming practices.[39]

Once again the fear of the uncertain void threatens the political confidence of those that associate zero with the chaos of the abyss. As the psychologist Daniel Kahneman has pointed out uncertainty is far more frightening to the

---

37.   See Brannen, P. The Ends of the World.

38.   There was some evidence that farmers voted strategically for Labour to try and keep the Greens out. https://www.stuff.co.nz/national/politics/123125923/election-2020-rural-vote-swings-to-labour--or-did-it

39.   Avery, D. (2017). ibid.

psyche of individuals than risk. Politicians and lobbyists understand this and will often make up statistical risk analysis that will make people more comfortable. What makes people fearful are absolute numbers such as 0 or 1; zero growth or stating with 100% certainty that there will be a human species extinction. When propagandists make these non-scientific claims they are revealing their tactics to many – i.e. the tactic of fear mongering. However, when these statements appear in slogans and tweets by cynical political leaders they are deliberately disguised as confident facts, rather than emotional propaganda designed to sway voters based on voter self-interest and self-delusion. After the COP25 climate change summit, Adam Currie, of the youth climate organisation, Generation Zero wrote:

"We are tired of governments siding with the polluters. We are tired of our lives being negotiated away for money. The people are tired of being ignored while a handful of wreckers and bullies negotiate in bad faith. We know that until we get them out of power they will continue to sabotage our future."[40]

The Rebirth of the Void

Our world is now a runaway system built from zeros and ones, an exponential binary world that is heavily dependent on computerisation and virtualisation – turning physical things into coded processes. Many of these coded abstractions are covered in layer upon layer of algorithms that make the human users, in all their diversity, transparent, almost as if they are not there, and yet our system now depends upon them. Should the Earth be hit

40. Read more here https://www.bbc.com/news/science-environment-50801493

by an EMP (Electromagnetic Pulse) from a massive solar storm or weapon, (which is not improbable), experts predict that all electronics would be fried and that our world would become impossible. We would be cast back into the dark ages, not just one hundred and fifty years ago, before electricity and fossil fuels, but to an earlier age when we were just inventing the printing press and still did not have reliable water supply. An EMP would be *'ground zero'* for our binary dependent world. It would be the end of the world as we know it, and not just the lights would go out; the human population would shrink dramatically. One of the Ten Plagues of Exodus, and Revelations is darkness, and even adults are afraid of the dark with good reason.

We were slow to catch on to the power of zero and our appreciation of the void but the Enlightenment led us towards the light. In 1887 Albert Michelson and Edward Morley performed an experiment that surprised most people and proved that empty space was not filled with an invisible substance known then as *'ether'*, and so this only left the abysmal void in its place. However, according to Arnold Harms we still abhor a vacuum and must fill our lives with happiness and joy – "All human activity exhibits the primordial drive to fill the void within one's life." "There is something within the human heart that hates a void."[41] Filling gaps and creating something out of nothing is the creative urge of living things to fill the void because the void still haunts us like death.

So we can see that zero, the void, and the abyss are paradoxical concepts that illuminate the mysteries of creation and the terrors of darkness, nothingness, and death. By the end of the 19th-century reality was being

---

41.   The Void Within: An Inner Quest for Wholeness By Arnold C. Harms Ph.D.

reimagined and discovered to be something altogether different from the view of Aristotle and the Western scholars prior to the Enlightenment. The void was once again seen as a precursor to life and the science of Einstein and others described a very different reality that embraced the infinite and the calculus of the numerical zero. Mathematically statistics and probability became the bedrock of modern thinking and a binary universe was born. Yet, the void has not left us and is still embedded in our digital world and our dreams and desires have been programmed into the electronics that make the world tick. Despite inconclusive evidence that there was zero spacetime at the singularity of the big bang, (because there is no way to measure nothing), there is a very reliable working model that kicks in moments later that has established that without humans the void and zero would still exist. This may be an engineering kludge of theory and experiments that may not be proven by the cyclic cosmological model but it works for the most part until physics and mathematics breakdown as you get infinitesimally close to the zero point in spacetime at the very beginning of the universe.[42]

For students learning electronics and programming, there is Arduino that allows them to cheaply and easily learn the foundations of building the binary worlds, of zeros and ones, that now surround us. 'Hello World' is usually the first code that novice programmers learn. It is also known as a 'sanity check' to reassure the programmer everything is working as it should. The first world-building program they learn to write to test the Arduino circuit is

---

42. See Greene, B. (2005). The Fabric of the Cosmos: Space, Time and the Texture of Reality (New Ed edition). Penguin.

much like the ancient creation myths and begins with the void:

```
void setup() {
Serial.begin(9600);
Serial.println("Hello World!");
}
  void loop() {
}
```

Algorithms and computer protocols hide the myths and legends of our primaeval past encoded in the zeros and ones, simulating mirror worlds of our collective unconscious.[43] While the 21st century would seem to have embraced the number that was once denied, zero still lurks in the darkness jumping out when we least expect it – scaring us half to death. Zero still negates humanity. In 1997 a mistaken formula in the code running a state of the art US submarine tried to divide by zero and resulted in the USS Yorktown being scuppered. Around the same time, there was the Y2K bug that had tried to save computer storage by only having two-digit dates in the code. This meant that when the new millennium clocked over from midnight, changing from 1999 into 2000, the computers running FORTRAN did not know what to do, 99 would become 00 and all logic would go out the window. Two zeros!! Welcome to the abyss – the end of worlds.

The Pandemic – a numbers game

Just like the Spanish flu that ravaged humanity in 1918,

---

43. See WorldBending: a survivor's guide (2019) by Pete Rive, What Algorithms Want (2017) by Ed Finn, and Protocols: How control exists after decentralization. (2004)by Alex Galloway

this coronavirus, named COVID 19, is a social epidemic. It is our most prominent existential threat today and for those who let it, COVID 19 will suck them into the void, playing on their fear of zero humans. According to Laura Spinney in her book, *Pale Rider: The Spanish Flu of 1918 and how it changed the World* (2017), despite the flu infecting 500 million worldwide, and killing between 50 to 100 million people, the Spanish flu, has been overshadowed by the less deadly First World War that killed significantly less people with only 17 million victims.

Surprisingly, viral epidemics are a relatively modern innovation that only became possible 12,000 years ago when humans first started to cluster in agrarian communities. Epidemics like its bigger brothers, the pandemics, rely on numbers and social transmission. To state the obvious without humans there would be no human pandemic. The virus is a parasite that needs a body to survive, and more than one to avoid extinction. Once it has infected its host and used their DNA to reproduce, it must spread to a new host before the host dies otherwise the virus will never make it to first base, so they need a population size of critical mass to survive and spread within that species. Fortunately for the coronavirus, the 21st century has resulted in the perfect conditions for a pandemic with a human population explosion and the invention of supercities that cluster their host in close proximity to each other allowing fast and easy social transmission.

COVID 19 is not a flu, however, like all the influenzas, it is what is known as a zoonoses virus, which means it was spread between species, probably from bats to humans. Birds, and especially ducks, and pigs are other popular super-spreaders we like to get close enough to eat so they have given us other famous epidemics such as SARS and Swine Flu. Bats, birds, camels, and pigs provide the viruses with population reservoirs of asymptomatic animals with the virus but that are unaffected by their viral infection. The microbiologist, Peter Piot, is in awe of the bat because it never gets cancer and happily lives infected with a multitude of viral parasites without batting an eyelid. Piot thinks we should be taking far more interest in our batty friends rather than treating the animal with disgust as they may suggest new vaccines and cures for diseases such as COVID 19.[44]

Spinney in her book, *Pale Rider:* noted that scientists once thought that disgust was an exclusively human condition. However, it is now believed that disgust is a basic survival mechanism amongst almost all animals. Long before COVID 19, first animals, then humans, practised social distancing to avoid socially transmitted diseases from honeybees to lobsters and badgers all avoid others who are infected. The word quarantine was invented by the Venetians in the fifteenth century when they feared disease might return from the Levant and so they forced ships to stay anchored for forty days – a *quarantena* – before the passengers could disembark and rejoin society.

So by the time, the Spanish Flu came along in 1918 many modern cities were already familiar with the concept of 'social distancing' which was publicised and practised along with face masks. According to Spinney:

"In these modern cities, anti-infection measures had to be imposed from the top down, by a central authority. To pull this off, the authority required three things: the ability to identify cases in a timely fashion, and so determine the

44.   Piot, P. (2013). No time to lose: A life in pursuit of deadly viruses. W.W. Norton. New York; London.

infection's direction of travel; an understanding of how the disease spread (by water? air? insect vector?), and hence the measures that were likely to block it; and some means of ensuring compliance with those measures."[45]

Yet, despite repeated warnings from the likes of Ebola, HIV, Zika, and the scientists that warned us, the microbiologists, epidemiologists and the WHO, (World Health Organisation), COVID 19 happened and already over 7.5 million have been infected with a rising death toll.[46] On the 8th June, 2020 the New Zealand government announced that the lockdown was lifted but there were still on-going restrictions and another level 3 lockdown for Auckland on the 12th Aug. 2020. The main reason that New Zealand went 'hard and early' was that successive governments had run down the health system and our hospitals would have been overwhelmed, hence the strategy of 'flattening the curve' by restricting the number of cases and the number of hospital admissions by imposing a 2-month quarantine.

Recently, the Global Risk Report (2019), by the World Economic Forum warned that all around the World there is a repeated pattern of panic and neglect that has undermined pandemic preparedness. The report noted: "During and after every major outbreak, leaders are quick to call for increased investment in preparedness. Real progress often follows these calls— but as the effects of the outbreak fade, neglect sets in again until a new

---

45. Spinney, Laura. Pale Rider (Kindle Locations 1208-1210). Random House. Kindle Edition.

46. This was written before the first lockdown was lifted in New Zealand on June 8 2021. A year on my city, Auckland is in level 4 lockdown again. There are a total of 221,648,869 confirmed cases and 4,582,338 deaths as of the 8th Sept. 2021, with 5,352,927,296 vaccines does given. As reported on the WHO website https://covid19.who.int

outbreak erupts; this prompts a new burst of panic, in which time and energy may be wasted on unnecessary and potentially costly measures."[47]

As recently as 2018 the United States undertook a pandemic preparedness exercise with a wargaming scenario in which a terrorist group released a virus 'modified to combine a high case fatality rate with ease of transmission'.[48] This scenario is not dissimilar to the actual COVID 19 infection with a fatality rate 15 times more than the regular flu, and 10 times more infectious. How prepared were they? In this hypothetical scenario, the vaccine failed, and tens of millions of deaths followed, overwhelming healthcare, crippling the government and a stock market crash of 90%.[49] While this was hypothetical the Trump administration, rather than heeding the warning, delayed their response to COVID 19 despite hearing about intelligence reports about an outbreak in Wuhan, China in November 2019. As America's confirmed cases continued to grow, becoming the largest in the World, and deaths began to overwhelm hospitals in New York and other large cities, Trump tried to deflect responsibility back on the underfunded WHO. In fact, in the middle of the pandemic, he announced America would be withdrawing funding. Trump and other authoritarian leaders such as Bolsonaro of Brazil and the Chinese dictator Xia Jinping have helped to undermine public trust in the media and have politicised scientific reporting of

47. Global Risk Project 2019, World Economic Forum. https://www.weforum.org/reports/the-global-risks-report-2019

48. Global Risk Project 2019, ibid.

49. Johns Hopkins Center for Health Security. 2018. "Clade X Pathogen Engineering Assumptions". http://www.centerfor-healthsecurity.org/our-work/events/2018_clade_x_exercise/pdfs/Clade-X-patho-gen-engineering-assumptions.pdf

the pandemic. According to David Nabarro, professor of global health at Imperial College London who worked at the highest levels of the WHO for many years:

"The challenge for the director general of WHO is always to maintain the core values of public health even when this goes against some of the political priorities of elected leaders. It is not unusual for there to be some form of conflict. The challenge is to try to create an environment where the opportunity for people to share is maintained and they are not having to look over their shoulder in fear that they are going to fall foul of the political priorities of leaders."[50]

The governance of supranational organisations designed to try and prevent pandemics have suffered from populist leaders who have tried to politicise the COVID global pandemic as they desperately try to hang on to power by controlling the narrative and silence criticism or withhold data that might cause voter panic and corporate loss of confidence. As governments lurched from crisis to crisis in their own jurisdiction it led the World Economic Forum to conclude in 2019:

"The world is badly under-prepared for even modest biological threats, leaving us vulnerable to potentially huge impacts on individual lives, societal well-being, economic activity and national security."[51]

Globally the frequency of disease outbreaks has been rising steadily. Peter Piot, the former leader of UNAIDS, and microbiologist, has warned in 2018, 'There is no time to lose', that we are unprepared for the 'big one', and that the

---

50.   Health experts condemn Trump's halting of funding to WHO. Guardian. (15th April, 2020). https://www.theguardian.com/world/2020/apr/15/health-experts-fears-over-trumps-suspension-of-funds-to-who

51.   Global Risk Project 2019, ibid.

WHO is not fit for purpose. Between 1980 – 2013 there were 12,012 epidemics comprising 44 million confirmed individual cases that have affected every country in the world.[52]

The lack of preparedness against global infection has been more than apparent but when the WHO finally announced that COVID 19 was a global pandemic governments and citizens rapidly changed their behaviour and began self-isolating and social distancing. The astounding thing about the COVID 19 global pandemic is that while scientists and journalists have been talking about the existential threat of climate change to our planet for the past sixty years, it has until relatively recently been all but ignored by politicians and the public alike, and yet COVID 19 got almost immediate attention. Despite the justified criticisms of a number of international leaders who failed to move fast enough, their delay of a couple of months makes the action on climate change appear positively glacial.

Heating the Pandemic

The threat of global warming from GHGs, greenhouse gases, has been discussed by scientists going back to the Johnson Presidency. In 1965 LBJ's scientific advisors presented a report that warned that industry and consumers were burning ancient fossil fuels at such a rate that they were contributing to vast quantities of waste, pollution, and GHGs that will eventually threaten food production, and increase temperatures that could cause environmental catastrophes. Yet, despite continual warnings by scientists, it was not until the Climate Change

52.   Piot, P. (2013). No time to lose: A life in pursuit of deadly viruses.

Paris Agreement in 2016 that governments have begun to put forward their own plans to limit their GHG emissions. The attempts of the Kyoto agreement and Emission Trading Schemes have been a dismal failure and the Paris Agreement continues to be hung up on Article 6 which could either help mitigate GHG emissions or accelerate them. According to the website Climate Change News there are three parts to Article 6.

"Article 6.2 allows countries to strike bilateral and voluntary agreements to trade carbon units.

Article 6.4 creates a centralised governance system for countries and the private sector to trade emissions reduction anywhere in the world. This system, known as the Sustainable Development Mechanism (SDM), is due to replace the Clean Development Mechanism (CDM), established under the Kyoto Protocol.

Finally, Article 6.8 develops a framework for cooperation between countries to reduce emissions outside market mechanisms, such as aid."[53]

Each country has its own lobby groups who look to kick the can down the road and make an exemption for their sector – in New Zealand our biggest and most successful lobby is agriculture, and that has been excluded from our emissions limits for the third time until 2025. Inadvertently, while humanity seems incapable of changing their lifestyle to avert a climate disaster, history shows that pandemics and disease can force social changes that impact the climate. In the sixth century in the late Roman era, there was a bubonic pandemic, known as the Plague of Justinian, that killed around 25 million people. As a result of this drastic population reduction, forests grew

53.   See 'What is Article 6' https://www.climatechangenews.com/2019/12/02/article-6-issue-climate-negotiators-cannot-agree/

back and farms were abandoned resulting in sequestration of carbon and cooling of the climate.[54] Pandemics often followed colonisation as the indigenous people had no immunity to the diseases brought by the invaders. In the Americas, in the sixteenth century, the massive decline of the native populations caused by waves of pandemics was the probable cause of the Little Ice Age that lasted until the nineteenth century. The other lesson is that populations usually rise after a pandemic.[55]

Prior to COVID 19, bankers had begun warning us that the global financial system faces collapse due to climate change. Mark Carney, the Bank of England's governor, and François Villeroy de Galhau, the governor of the Banque de France wrote in an article published in The Guardian: "As financial policymakers and prudential supervisors we cannot ignore the obvious physical risks before our eyes. Climate change is a global problem, which requires global solutions, in which the whole financial sector has a central role to play."[56]

Then suddenly overnight the noise was gone, and the blue skies returned as one-third of the planet pulled over to the side of the road, parked their cars and returned to the sanctuary of their homes. According to the Asia Development Bank the economic impact to the global economy could go as high as US$8.8 trillion.[57] Those who have suffered the poor health of fossil fuel pollution have

---

54.   Spinney, L. (2017) ibid. Kindle location 302 of 5276.

55.   Spinney, Laura. Pale Rider (Kindle Locations 306-310). Random House. Kindle Edition.

56.   https://www.theguardian.com/environment/2019/apr/17/mark-carney-tells-global-banks-they-cannot-ignore-climate-change-dangers

57.   COVID-19 economic impact could reach 8.8 trillion U.S. dollars globally: ADB report. (15th May, 2020) http://www.xinhuanet.com/english/2020-05/15/c_139058712.htm

suddenly become aware of a better quality of air and life that they could have barely imagined before the virus. And yet, the developed countries of the world are desperate to get back to work and let the busyness return. Busyness-as-usual will bring back the noise, the smog, and global heating might not miss a beat.[58] During the Spanish flu carbon emissions around the world fell by 14% as industrial production and consumption slowed down. Medical experts think that the pandemic could continue until 2022, and even become endemic, therefore, we may see a fall in GHGs and the remote possibility that we might keep below 2°C before 2050. However, our ability to reflect on these gains and permanently adopt a degrowth lifestyle will require conscious decisions about our economy and how we extract and spend the planet's resources. In 1920, just one year after the pandemic production and pollution rose again by 16%.[59]

Each country that has signed the Paris Agreement on Climate Change has submitted its plan to reduce GHGs: "Nationally determined contributions (NDCs) are at the heart of the Paris Agreement and the achievement of these long-term goals. NDCs embody efforts by each country to reduce national emissions and adapt to the impacts of climate change."[60] The UN initiated a virtual online dialogue to maintain momentum during the COVID 19 pandemic. Since the beginning of the pandemic New Zealand has had a nationwide lockdown and an Auckland

58. 'It's positively alpine!': Disbelief in big cities as air pollution falls. https://www.theguardian.com/environment/2020/apr/11/positively-alpine-disbelief-air-pollution-falls-lockdown-coronavirus?CMP=share_btn_link

59. "Coronavirus is a tragedy – but it could be the wake-up call we need" https://gu.com/p/dg9z7/sbl (31st Mar, 2020)

60. See https://unfccc.int/process-and-meetings/the-paris-agreement/the-paris-agreement/nationally-determined-contributions

lockdown to 'flatten the curve' and prevent further infections and deaths. According to the Climate Action Tracker (CAT) website: "The CAT's New Zealand emissions projections for 2030 are 8% to 17% lower compared to CAT's previous projections in December 2019, and the 2020 projections are 14% to 23% lower. This difference in projections account for the impact of the pandemic on emissions and changes to government policy projections. The projected drop in emissions brings New Zealand closer to meeting its 2020 target, and within range of meeting the 2030 target. However, the COVID-19 impact is not a sustained reduction in GHG emissions and as the economy recovers emissions are expected to increase in 2021/22 if there is no strong push for a green recovery shifting investments towards lower carbon projects."[61] The government used the UN phrase to say they would 'build back better' and that this would be defined as their 'nuclear free' moment, (recalling the 80s when Labour grabbed global headlines with their 'no nukes' policy). CAT has reviewed New Zealand up until the 30th of July, 2020 and have ranked our NDC as insufficient to keep temperatures below an increase of 1.5°C and is most likely to go above 2°C. New Zealand would not be contributing their 'fair share' : "If all government NDCs were in this range, warming would reach over 2°C and up to 3°C."[62]

What does this tell us about our ability to affect our environment? The Anthropocene is a geological epoch of our own device; so, is this planet now our prison? Humanity seems to be like the proverbial Tyrannosaurus Rex caught in the headlights of our own fossil fuel transportation. We appear to be incapable of reasoning

61.   See https://climateactiontracker.org/countries/new-zealand/
62.   ibid

that it was LIFE that first decomposed and created that fuel, that now threatens life – like some primaeval clock – we are those dinosaurs now facing extinction. But surely if we know how to ignite all that solar energy that is concentrated in those fossil fuels we must also know how to break the cycle and defuse the bomb before it blows up in our face? Part of the problem is that most of us possess what is described as a 'socialized mind', one that is immune to change, a 'mindmeld' that is only capable of groupthink.[63] For many of us, there is no alternative to the status quo, there is no alternative to the capitalist consumer construct. We fear to unplug from the matrix because we have seen the movie and we know it is our only source of energy, no matter how toxic it is for ourselves and the nonhuman majority. The other problem is that we don't realise it is a mental construct because all the signs of climate change appear so physical. We are imprisoned by this strange paradox that elevates physicality over immateriality yet we are also enslaved by our socialized minds.

Empirical Reality

Since the Enlightenment, we started down a rocky road that was well signposted as 'REALITY', we trusted everything that we saw as true and leading us out of the darkness of superstition and the ideologies of those in power. Actuality was not just the scientific methodology that would deliver us heaven on Earth but it was to become the only logical and rational thing to trust. The Enlightenment put empirical 'reality' on a pedestal. What

---

63. Kegan, R., Lahey, L. L., & Kegan. (2009). Immunity to Change: How to Overcome It and Unlock the Potential in Yourself and Your Organization.

Western culture achieved technologically and economically gave its followers a massive sense of confidence and self-worth. We no longer thought we needed a map, in fact, we no longer noticed that the map has lain on the ground beneath our feet and that actuality is only one component of reality that is most likely beyond any hope of our ultimate comprehension. Our world(s) is/are our models – we made them up and we continue to renovate them every morning. Every time we interact with our physical environment we are using mental models, concepts, and philosophies we cannot explain and don't even recognise as constructed worlds – it is after all the 'common' sense of the socialized mind. The linguist George Lakoff has pointed out: "We are neural beings. Our brains take their input from the rest of our bodies. What our bodies are like and how they function in the world thus structures the very concepts we can use to think. We cannot think just anything — only what our embodied brains permit."[64] While our concepts are deeply contextual, shaped by our sensory response to the environment, it is also a deeply social environmental context. We don't want to upset the people we love, the people we work with, those we work for, and those who drip feed us with the lollies of success and status – to do so risks being called mad and you will probably end up medicated on antidepressants or antipsychotics.

It is therefore hardly surprising that after this social, political and economic collapse the powers-that-be will expend vast amounts of physical and immaterial resources on the reconstruction of the conceptual world that had crumbled away. The walls of the citadel will have to be

---

64.   "EDGE 3rd Culture: A Talk with George Lakoff". Edge.org.

shored up so the inmates could not escape. The good news for them is that throughout history there have been many examples of the inmates volunteering to grab a shovel, bricks and mortar and ensure that it was impossible for themselves and any other 'fool' who thought they could live outside the construct of that civilisation – the wilderness, the zero human zone. This is the civilisation that we have built and is the result of what David Graeber has called bullshit jobs and the *'guard economy'*.[65] We now know that after the Spanish Flu and World War I, consumption, and energy ignition didn't just return to *'normal'* but those who imagined the reconstructions after many more crises, including another World War, succeeded in accelerating down the same rocky road they then converted into an *'information superhighway'*. The reality was that Western culture was always designed to transport ideas, ideologies, methodologies and models. Most of the populations that helped build that superhighway were confused by the physical resources they extracted to lay the foundations that transported them. Eventually, they called it the Internet and, for many, the material world disappeared in a cloud of disembodied data. We need to recognise that the physical and immaterial worlds are inseparable and that there is a constant dialogue between not just the human worlds but the inhuman. To be a human being is not to float lonely as a cloud but to accept our physical and virtual place in the universe. We are not privileged colonists but co-exist with things much greater and beyond our comprehension. The Anthropocene is not just material – it is made of concepts, physics, and metaphysics, therefore, ethics and aesthetics can no longer be swept aside or ignored by

---

65.　See Graeber, D. (2018). Bullshit jobs: A theory. Allen Lane.

history, the present or the future. This pandemic is both actual and virtual, both a physical reality and a conceptual reality, something that has been caused by a thing and something that we have caused by the way we think and act. Understanding how it came about and the agents or actants involved will help us to be better co-hosts, and co-designers of our future worlds.

Why was it called the Spanish Flu?

Going back to the Spanish Flu of 1918 we can examine some of the resources, events and agents that built, bent and broke that world; to help us understand the worlds that followed and how we ended up here. You might assume that because it was called the Spanish Flu that Spain was the origin of the virus. However, there is strong, but inconclusive, evidence that this zoonotic disease came from migratory birds to Kansas, USA. Only one thing is agreed, that is the flu did not come from Spain.[66] So why was it called the Spanish Flu instead of its scientific name, H1N1?

According to Spinney the naming of diseases have a long and ignoble history based on xenophobia and conflict. In Spain, they gave the flu the pejorative name the 'Naples Soldier'. Trump attempted to attribute blame by calling the current pandemic 'China virus' and 'Kung Flu'. In 1918 there was insufficient scientific knowledge to name it H1N1, which stands for a genera of virus that contains the glycoproteins haemagglutinin and neuraminidase. Back in 1918, the virus was too small to be seen under a normal microscope and so the disease was thought to be bacterial. It was not until 1931 with the invention of the electron

66. Spinney, L. ibid.

microscope by German engineers, Ernst Ruska and Max Knoll, that the first images of viruses were published in 1940. During the First World War most of Northern Europe was embroiled in fighting, however, Spain remained neutral and kept out of it. It was during the Great War that governments truly embraced the awesome power of their most powerful weapon, *'fake news'* or propaganda. A young Eddie Bernays, who is credited as the father of PR and propaganda, had begun working for the American Information Office when he became exposed to the virulent spread of false stories that convinced young American men to enlist in the military to fight the German *'Hun'*. Bad stories about the war and the plight of their troops were suppressed and so when more people were killed by the outbreak of the flu than the war itself, this was definitely bad news and was censored in the countries at war. In neutral Spain, however, there had been an outbreak of the flu that spread by large congregations of churchgoers, public bullfights, and general uncontrolled social transmission. The Spanish King Alfonso XIII became gravely ill from the flu and the news spread fast throughout the World giving the impression that it had originated in Spain.

Just like the virus knowledge transmission requires warm bodies. According to Niall Ferguson, during the Great War:

"the German government had begun experimenting with what would prove to be the decisive, war-winning weapon. The idea was to destabilize the other sides' empires by unleashing an ideological *'virus'*. With the help from their Ottoman allies, the Germans sought to spark a jihad throughout the British Empire, as well as the

French."[67] The German's attempt to seed a Muslim jihad ultimately failed but where they did succeed was in Russia. The Bolshevik *plague* was unleashed via the transmission of Vladimir Ilyich Lenin from exile in Switzerland to Russia via a train sponsored by the Germans in the wake of the Russian Revolution in February 1917. The German plot was helped largely by Tsarist incompetence and on October 25th 1917 the Bolshevik army conducted a successful coup d'etat in a propaganda war won more by banners and posters than guns and bullets. However, it was not just the Bolsheviks who caught the ideological virus but the German soldiers who fought them and the communist plague quickly spread throughout Europe.[68] Communism was not the only *'disease'* spread by propaganda it eventually led to fascism, American imperialism, French neocolonialism, and other dangerous forms of government misinformation around the world. More people, more transmission, more misinformation, more disease. The Spanish Flu was in part a virtual and in part an actual disease spread by censorship and a contagious virus, it reduced the global population by between 50 and 100 million people, and is the worst human pandemic in the history of the world so far. COVID 19 may still rival this record and has equalled the Spanish Flu with a global wave of misinformation spread by an infodemic.

The Population Bomb

In 1968 Paul and Anne Ehrlich published their best selling book, *The Population Bomb*. Following World War ll there

---

67. Ferguson, N. (2017) The Square and the Tower: networks, hierarchies, and the struggle for global power. Allen Lane

68. Ferguson, N. ibid chap. 36, The Plague.

had been a population explosion around the planet. The Ehrlichs warned that overpopulation would result in a worldwide famine and resulting environmental carnage. Unexpectedly, while there are numerous examples of less developed countries, LDCs, who have suffered from famine, the so-called developed countries spread out around the globe in a *green revolution* that used deforestation and super-phosphate to artificially boost agricultural production of food. Population growth in the developed countries has slowed down but it has continued at a pace in the LDCs.

The documentarian Michael Moore has produced a film, *Planet of the Humans*, directed by Jeff Gibbs. It offers a concrete solution to Global Warming: "We really have got to start dealing with the issue of population ... without seeing some sort of major die-off in population, there's no turning back." But George Monbiot of *The Guardian* critiques their view. He wrote:

"Population is where you go when you haven't thought your argument through. Population is where you go when you don't have the guts to face the structural, systemic causes of our predicament: inequality, oligarchic power, capitalism. Population is where you go when you want to kick down."[69]

Much of the environmental damage of the Anthropocene has been caused by the wealthiest countries and within those countries, the wealthiest people on the planet and that amounts to 74% of global GDP but only counts 18% of the global population.[70] The argument that deforestation

69.   How did Michael Moore become a hero to climate deniers and the far right? by George Monbiot. (7th May, 2020) The Guardian, https://www.theguardian.com/commentisfree/2020/may/07/michael-moore-far-right-climate-crisis-deniers-film-environment-falsehoods

70.   Cited by Angus in *Facing the Anthropocene* taken from the IGBP report 2010.

has been caused by the impoverished masses is a cruel accusation by those who have enriched themselves at their expense. Colonialism and then globalisation has been at the root cause of local ecological damage as wealthy countries have extracted resources, cut down forests, and established farms, thereby encroaching on ecosystems that had once been relatively isolated from humans. As human populations have also grown in LDCs, interspecies transmission of viral diseases have become geographically possible. According to the World Economic Forum, *Global Risk Report* (2019), there are five major trends driving the increased frequency of disease outbreaks globally: 1) the surge of international travel, trade and connectivity making the spread of disease faster and more global; 2) urbanisation and high-density living with 68% of the world expected to be living in urban environments by 2050; 3) "increasing *deforestation* is problematic: tree-cover loss has been rising steadily over the past two decades, and is linked to 31% of outbreaks such as Ebola, Zika and Nipah virus"; 4) the WHO have warned that climate change could potentially accelerate and alter the transmission of infectious diseases such as Zika, malaria, and dengue fever; 5) human displacement due to war, sea-level rise, and poverty increase vulnerability to biological threats.

The growing inequities of the global economy in almost every wealthy economy besides Scandinavia hides the profit motives of the very wealthy known as the 1 percenters. Inequality within and between countries is something that should interest anyone who wants to avoid social instability.[71] Fundamentally, it is not just iniquitous to ignore the plight of people in less developed countries,

---

71.   See Turchin, P. The Age of Discord.

it is a systemic component of capitalism which is aided and abetted by its economic handmaidens. Mainstream economists have argued up until recently that even the horrendous possibility of global warming by as much as 6° C will only impact GDP by a few percentage points. Their economic models entirely ignore the environmental and social cost to humans or less developed countries who are destined to suffer the worst from global heating. It is hard not to see these economists as the racist lackeys of entitled capitalists.[72] According to Foster and York: "Lawrence Summers, who was Obama's top economic advisor, wrote an internal memo while chief economist of the World Bank, in which he stated: "The economic logic behind dumping a load of toxic waste in the lowest-wage country is impeccable and we should face up to that." He justified this by arguing: "The measurement of the costs of health-impairing pollution depends on the foregone earnings from increased morbidity and mortality. From this point of view a given amount of health-impairing pollution should be done in the country with the lowest cost, which will be the country of the lowest wages."[73]

The wealthy minority of developed countries consume around 80% of the planet's resources, produce the most amount of toxic waste, and burn the greatest amounts of fossil fuel generating an explosion of GHGs. The population growth in the LDCs is a very small part of the cause of the geological destruction of the Anthropocene.[74] When some consider the gross population of humans on

---

72. See Foster, J. B., Clark, B., & York, R. (2011). The Ecological Rift Capitalisms War on the Earth. Monthly Review Press.

73. Foster & York. (2011). ibid (Kindle Locations 1364-1369). Monthly Review Press. Kindle Edition.

74. Angus, I. (2016). Facing the Anthropocene: Fossil Capitalism and the Crisis of the Earth System. NYU Press.

the planet they often think of human units, individual people, the problem is that not all humans consume the same amount of energy. According to World Population Balance: "The point is that the population problem isn't just something 'over there' in 'those poor countries', where they may be having more children. A family in India would have to have more than 10 children to match the energy consumption of an American family with just one child!"[75] Agriculture and fossil fuels continue to drive the consumption of the biggest economies of the World and have enabled urban population densities such as we have never seen before in the history of the human species. The 'overdeveloped' countries could make a very significant contribution to population size by voluntarily reducing their family size to just one child.

The enormous human population has put an undeniable strain on the nonhuman majority and the Earth system, however, population control including sterilisation is not only unethical but ignores the more sensible approach of fair distribution of resources which will improve the quality of lives and lead to better educational outcomes. The equitable distribution of wealth between and within countries has been shown to have better health outcomes and has been symptomatic of declining population growth in wealthy countries and wealthy communities.

The human population has been described as a virus that has spread like a pandemic infecting rivers, the sea, air and the land; it is why this geologic epoch has been called the Anthropocene, meaning 'anthro' (human or manmade) 'pocene' (epoch). There are those who advocate that we should reduce the human population but examples such

---

75. Population and Energy Consumption cited by World Population Balance
https://www.worldpopulationbalance.org/population_energy

as China's compulsory 'One Child' policy has proven to be cruel and ineffective.[76]

A zero population target could disguise covert racism and fascist state interventions and would be ultimately unsustainable for our species. This is not a popular goal, however, *reductio absurdum* is sometimes a handy way of exposing ridiculous assumptions that many may hold unconsciously calling it '*common sense*'.

This past Friday my friends and I celebrated the end of 'lockdown' or Level 3 containment. We were excited to get together for a dinner party, but everyone felt the same, that they would miss the quiet of a city that stopped burning so much fossil fuel and consuming unneeded materials, piling up unwanted waste. Less noise signals less waste and a better environment for the nonhuman majority and the humans they coexist with. My friends and I acknowledge that our post-lockdown sadness is related to our comfortable lifestyle, we don't have a large family crowded into unhealthy houses, and have slightly less precarious income streams than some, but we are also not part of the superwealthy and don't aspire to it. We feel a change is coming, which is scary for some, but the world has to do it or we will keep accelerating towards the end of the world and possibly a world with zero humans.

Unfortunately, so-called '*affirmative design*'[77] has encouraged us to all believe there is a design solution to these wicked problems. As the ecologist, Dr Mike Joy says rather than always thinking about what we can do about climate change, we should just think about what we don't

76.   Lily Kuo and Xueying Wang in Shenyang (Mar. 2019) Can China recover from its disastrous one-child policy? The Guardian. https://www.theguardian.com/world/2019/mar/02/china-population-control-two-child-policy

77.   See Dunne & Raby. Speculative Design

have to do, in other words concentrate on doing less, consuming less, using less energy. Joy expressed his disappointment with the Climate Change Commission report for being too shortsighted and not taking in a longer time scale that requires a change of lifestyle. He sketched an alternative future in which we are returning to the land and living locally by 2040. The interviewer may have felt uncomfortable with Joy's gloomy picture of our future – if we do not change but he was just stating his opinion based on good scientific evidence – we are heading for zero humans and the sixth mass extinction.[78]

As the AI construct, Agent Smith, told us in *The Matrix*: "I'd like to share a revelation that I've had during my time here. It came to me when I tried to classify your species and I realized that you're not actually mammals. Every mammal on this planet instinctively develops a natural equilibrium with the surrounding environment but you humans do not. You move to an area and you multiply and multiply until every natural resource is consumed and the only way you can survive is to spread to another area. There is another organism on this planet that follows the same pattern. Do you know what it is? A virus. Human beings are a disease, a cancer of this planet. You're a plague and we are the cure."

---

78. Listen to an interview with Mike Joy on RNZ: Afternoons with Jesse Mulligan. (10)une 2021) Disappointment over the final Climate Change Commission report. https://www.rnz.co.nz/audio/player?audio_id=2018799160

4

Zero Growth

In Aotearoa, New Zealand, the Climate Change Response Amendment Act, popularly known as the Zero Carbon Act came into force on 14 November 2019. At that time the WHO had been informed of the COVID 19 outbreak in China, but they had not yet called it a pandemic. By the time the WHO recognised that this coronavirus was spreading rapidly around the world and was officially labelled a pandemic, economists were beginning to recognise that it could have a catastrophic impact on the world's economies and economic growth would be severely curtailed.

Right from the beginning when scientists began to realise that fossil fuels could cause global warming due to the heating effects of greenhouse gases it became apparent that to reduce GHG emissions would mean that the economy would have to slow down. This was simply understood because the laws of thermodynamics predicts that while energy can be neither created or destroyed, increasing entropy, or disorder, means that heat would be given off when it is converted from one form to another. In other words more production + more consumption = more heat + more waste. The climatologist, Reid Bryson asked congress in 1973 'how on earth could you stop using fossil fuels?' In the 1960s the US government knew they faced a stark choice – fossil fuels and affluence or zero carbon emissions, clean air, clean water, and less food and consumer products. By their economic measure the US government knew that without fossil fuel combustion there would be zero economic growth, the GDP would drop like a stone and the American Dream would be over. In terms of capitalism, this would be a world-ending moment. According to Tainter and Patzek if we were to eliminate fossil fuels our current lifestyle, otherwise known as the *deathstyle*, would totally disappear and the global population of humans would shrink by 67%.[1] Of course, what we have subsequently learned is that fossil fuels can no longer ensure affluence as we face peak oil and global warming has begun to have a destructive effect on the quality of life even in the most energy-intensive countries. Global heating is not the only anthropogenic geological disaster we are facing, ocean acidification, dangerously high nitrogen and phosphorus levels from fertilisers, sea level rises, and the sixth mass extinction of nonhuman and possibly the human species. The human population and economic theories of growth have been closely connected to our ability to store and consume solar energy stored in food, agriculture, fossil fuels, and technology allowing the

---

1.   Tainter, J. A., & Patzek, T. W. (2012). Drilling down the Gulf Oil debacle and our energy dilemma. New York, NY: Copernicus Books. Retrieved from http://dx.doi.org/ 10.1007/978-1-4419-7677-2. Cited in Pete Rive. "Worldbending." Apple Books.

population to grow beyond the planet's sustainable capacity.[2]

Economic growth and agricultural growth have long been held as articles of faith expressed clearly in the *New Zealand Encyclopedia* from 1966:

"In a broad sense the history of New Zealand is essentially the history of agricultural development, as a country which has always been so dependent upon its primary industries must evolve policies which encourage their expansion. Moreover, the rate at which the economy as a whole can grow is tied to the capacity of agriculture to expand."[3] However, despite some farmers and agricultural lobbyists who continue to push this barrow, there is a rapidly dawning realisation that overcropping, overstocking, and overpopulating the land will start to have a negative impact on the quality and yields produced by agriculture that artificially forces growth through fertilisers and animal hormones.

Ecological Economics

It is enlightening to consider the etymology of both the words, *economics* and another word often relating to sustainability and a system's approach to agriculture, *ecology*. It was Ernst Haeckel (1834 – 1919) who coined the word *Ökologie* or *ecology*, in his *Generelle Morphologie der Organismen* in 1866. According to Foster, he drew on the Greek root, *oikos* for household which was the origin for the word *economy*. Haeckel wrote:

"By ecology we mean the body of knowledge concerning

---

2. Tainter & Patzek, (2012). ibid
3. (Ed.) McLintock, A.H. (1966) An Encyclopedia of New Zealand. Retrieved from https://teara.govt.nz/en/1966/farming

the economy of nature—the investigation of the total relations of the animal both to its inorganic and its organic environment; including above all, its friendly and inimical relations with those animals and plants with which it comes directly and indirectly into contact—in a word, ecology is the study of all those complex interrelations referred to by Darwin as the conditions of the struggle for existence. This science of ecology, often inaccurately referred to as "biology" in a narrow sense, has thus far formed the principal component of what is commonly referred to as "Natural History."[4]

In the 19th century it was not uncommon for scientists to combine their interest in natural history and social history, or in the case of Thomas Malthus political economy and population studies. Haeckel is also infamous for his *scientific racism* and his theories of *social Darwinism* that were later picked up by Nazi propagandists based on Herbert Spencer's *survival of the fittest*. However, Haeckel did make a valuable contribution to early ecological thinking by examining ecosystems and biodiversity within zoology. Haeckel was part of a group of 19th-century scientists, such as Darwin, Huxley, and Lyell who applied evolutionary natural history and materialism to social science. Marx and Engels, who both read widely and deeply the scientific literature of the day, attempted to apply dialectical materialism to their political-economic theory, which they called *scientific materialism*. Some of those ideas were useful in considering a scientific approach to natural history and the application of some aspects of Darwin's evolutionary theory, but they also showed the potential

---

4. Foster, John Bellamy. Marx's Ecology: Materialism and Nature. Monthly Review Press. Kindle Edition.

dangers of applying theories of biology and zoology to human population control.

The human population is currently over 7 billion people, however, by some estimates in order for humans to be able to live a sustainable and harmonious co-existence with the nonhuman majority that number should be reduced to between 2-3 billion. According to World Population Balance this would enable all of those 2-3 billion to enjoy the current living standard of the average European.[5] However, even the idea of a sustainable capitalist lifestyle is an oxymoron as physical and emotional growth limits imply a hard material boundary and emotional response to the consequences that most economists and political theorists ignore. It is as if they are still trying to figure out Zeno's Paradox and how they can get infinitely close to zero without ever arriving – surely with the help of innovative technology capitalism is limitless? According to Foster et al. the ignorance of the historical material and political forces of capitalism means that past lessons are never learned and the fantasy can continue to run. They wrote:

"The proposition that unlimited economic growth under capitalism can and should be managed so as to generate a system of sustainable capitalist development (a view we call "Capitalism in Wonderland") rejects at one and the same time an understanding of capitalism as a historical system and the notion that nature itself is historically complex and contingent in ways that we are only beginning to understand. The great geological eras in the history of the planet are separated by massive die-downs in species."[6]

All over the world, governments, and even dictators claim to have the best interests of their public in mind. The

---

5.   https://www.worldpopulationbalance.org/3_times_sustainable

6.   Foster et al. (2010).

simple argument is that a growing economy will provide the greatest number of people with the best possible employment and the best possible living standards. Putting aside the distribution of wealth it has become a religious catechism, even in the former communist countries, that a growing economy is a necessity if people are to be happy. Therefore, when the penny finally dropped that fossil fuels were threatening life on our planet governments argued that economic health, i.e. *capitalism*, must be maintained at all cost to keep the populace happy. What they often don't discuss is who exactly will enjoy that happiness? Trickle-down economics has been proven not to work as inequalities have grown more profound[7] and still the economists fiddle at the edges while trying to apply their theories to environmental problems with trickle-down ecology. In short, they were faced with a paradox, if you simply must have your cake and eat it, even if it kills you and everything around you, then humanity must bring out the big guns – a cognitive weapon in the 'war on happiness', otherwise known as cognitive dissonance. In other words, if someone presents you with uncomfortable facts, such as inequality or environmental vandalism, you must keep smiling, appear to listen and reason, then simply deny the facts and present your own *'alternative truth'* – it is after all what you believe. This is not simply a case of the crude and idiotic propaganda of Donald Trump and his cronies but it is now part of the education curricula – they prefer to call it *market forces* because the word capitalism implies historical conflict.[8] It was the political economist Schumpeter who

---

7. See Piketty, T., & Goldhammer, A. (2018). Capital in the Twenty-First Century. https://doi.org/10.4159/9780674982918

8. See Foster et al. (2010); Rive, P. (2019). Worldbending - chapter 7, The Weaponisation of Education; Mayer, J. (2016). Dark money: The hidden

stated: "as recognized by all of the leading economists—Adam Smith, David Ricardo, Karl Marx, Thorstein Veblen, Alfred Marshall, John Maynard Keynes—down to the present, "Capitalism is a process [of accumulation and growth], stationary capitalism would be a *contradictio in adjecto.*"[9] Therefore, many politicians argue in order to maintain a happy polity all forms of capitalism, (including China's State Capitalism), must grow.

Fuelling Growth

Growth does not come without energy cost and by far the biggest consumer of energy today is China. Currently, the cheapest sources of energy are fossil fuels which means that China's is by far the biggest source of carbon dioxide emissions contributing 28% to the total global output. In its rush to *catch up* to Western economies China now spews 9.43 billion tons of carbon dioxide gas into the atmosphere, almost double what the next emitter, the US, that emits 5.15 billion tons. This represents a massive 54% increase since the Kyoto Protocols of 2005 while the US has reduced its output by -12.1%, India has surged with a colossal 105.8% with a total output of 2.48 billion tons. China is keen to show the World that it is going to make an effort to reduce its carbon emissions to zero before 2060, and peaking by 2030. However, according to the magazine, *Science*, "achieving carbon neutrality before 2060 will require drastically reducing the use of fossil fuels in transportation and electricity generation and offsetting any remaining

history of the billionaires behind the rise of the radical right. New York. Doubleday.

9.   Foster, York & Clark. (2010). The Ecological Rift: Capitalism's War on the Earth.

emissions through carbon capture and storage or planting forests."[10]

China has not revealed details of how it plans to get to net zero while determined to continue to aggressively grow the economy. What is well known is that the coal and nuclear power lobbies are powerful and includes high ranking party members who benefit from the large capital expenditure and power that these energy resources give them. Coal currently accounts for 58% of China's total energy consumption and 66% of its electricity generation. According to a study it was found that China's coal-fired generating capacity grew by about 40 gigawatts in 2019 to 1050 GW and another 100 GW is under construction.

While renewable energy sources are increasing globally the scale and demand for energy, while accelerating economic growth, means that even renewable sources can cause environmental harm when the whole supply chain is taken into account. The only way to consider any transition to alternative energy supplies is to consider the entire ecosystem and take a wider view of the resources, agents and events beyond just human interests.

Global supply chains are shrouded in secrecy with companies such as Amazon claiming that their logistics information is proprietary and contribute to their competitive advantage. The benefit of this secrecy is that it is possible to avoid the scrutiny of the public, hiding the environmental harm, neglect of human rights and child labour, and their carbon footprint that stretches around the planet. According to the journalist, Sarah Emerson, the secrecy that surrounds Amazon's supply chain has shielded

---

10.   Normile, D. (Sep.29, 2020) Can China, the world's biggest coal consumer, become carbon neutral by 2060? Science Mag.org
        https://www.sciencemag.org/news/2020/09/can-china-worlds-bigger-coal-consumer-become-carbon-neutral-2060

the company from awkward questions about who, how, when and what makes their top selling AmazonBasics battery.[11] The harmful waste from the production of these manganese dioxide AA alkaline batteries also has a ridiculous gross energy cost. Most of the energy to charge these batteries comes from fossil fuels and only a small amount from renewables. One study showed that: "it takes more than 100 times the energy to manufacture an alkaline battery than is available during its use phase." And when the entirety of a battery's emissions are added up — including sourcing, production, and shipping — its greenhouse gas emissions are 30 times that of the average coal-fired power plant, per watt-hour." [12]

Companies, and countries, like Amazon and China must be held to account by consumers who happily benefit from poor labour and environmental practices – these low cost consumer goods, hide harm. To be fair it also requires governments to enforce new open and transparent corporate laws that tracks and traces all resources, labour and transportation data. Many of the world's biggest emitters do not reveal the full extent of their contribution to GHG emissions because they only disclose those they are directly responsible for but their global supply chains are not revealed and are much much bigger by orders of magnitude. The oil and gas industry is just one obvious example as they report only what they emit, not the enormous extent of the entire global consumption and emissions that they fossil fuels enable. The just-in-time

---

11.   Emerson, S. (Oct 30, 2019) Unraveling the Secret Origins of an AmazonBasics Battery: One of Amazon's smallest and most popular products has a surprisingly large footprint. Medium OneZero. https://onezero.medium.com/unraveling-the-secret-supply-chain-behind-an-amazonbasics-battery-e7b9ead4d72e

12.   Emerson, S. (Oct 30, 2019) ibid.

supply chain has been interrupted by COVID 19, while also exposing exploitation and carbon pollution. In 2017 Amazon shipments produced 19 million metric tons of carbon which is the equivalent of five coal powered electricity plants.[13] The consumer demand for instant gratification is coming at a price to the environment as Amazon's Prime Day, with same day shipping in the US, is causing even greater carbon emissions. According to Rakuten Intelligence the time from 'click' to door delivery has dropped from 5.2 days to 4.3 days. UPS said in 2017 'e-commerce was leading it to make less-efficient deliveries, leading to "more miles, fuel, and emissions per delivery."[14] Meanwhile, 300,000, or 20% of all seafarers have been effectively enslaved on ships that are banned from entering 120 countries around the world because of the pandemic.[15]

In our busy lives it is easy to thoughtlessly expect fast delivery of consumer goods without giving it a second thought about the energy required to pay the very expensive environmental, and social bills. Companies and countries may hide the truth from us, but by ignoring it, out-of-sight-out-of-mind, we are complicit in their crimes. Our happiness has a cost and the more happiness we demand from a growing economy the greater the energy cost and the more brutal the social and environmental devastation.

---

13.   Emerson, S. (Oct 30, 2019) ibid.

14.   Irina Ivanova. (MAY 24, 2019) How free one-day shipping is heating up the planet. https://www.cbsnews.com/news/amazon-prime-day-one-day-shipping-has-a-huge-carbon-footprint/

15.   K Oanh Ha. (Sept. 21st, 2020) Worst Shipping Crisis Puts Lives and Trade at Risk. Bloomberg. https://www.bloomberg.com/news/newsletters/2020-09-21/supply-chains-latest-worst-shipping-crisis-puts-lives-and-trade-at-risk

The Politics of Happiness

It was not long after he was elected the 31st President of the United States that Herbert Hoover was faced by the shock of the 1929 stock market crash and the end of economic growth for some time. What happens when people are no longer happy, and their misery continues for days and then months? Happiness is followed by depression. Even as early as the 1930s politicians had begun to associate the material affluence of their citizens with happiness – it had become political. Hoover, with advice from the nephew of the psychologist, Sigmund Freud, Eddie Bernays, the father of propaganda and PR, had discovered that the sublimation of happiness could lead to votes. Hoover told a conference of advertising executives: "You have taken over the job of creating desire and have transformed people into constantly moving happiness machines, machines which have become the key to economic progress."[16]

Before Hoover, there was little understanding of what was really going on in the economy. In 1933 the next President, Franklin D. Roosevelt, hired Simon Kuznets to create national accounts – his ambition was to simplify all human activity that contributed to positive welfare calculated to produce one single figure. With only three assistants and five statistical clerks, Kuznets categorised and measured the American economy according to different sectors such as energy, manufacturing, mining and agriculture. He invented the economic measurement that has come to dominate almost all government policies around the world. Popularly known as GDP, or Gross Domestic Product, this simple percentage has become the

16. Curtis, A. (2002) Century of Self. Ep. 1 "Happiness Machines" Cited by Rive, P. (2019). Worldbending. Kindle Edition.

yardstick for measuring the success or failure of governments and their policies.

In the minds of many governments, it also equates to a measure of citizen happiness. Back then Kuznets was dubious about including expenditure on the military, the financial sector and the engine of all consumer 'happiness' – advertising. It turns out that what is measured, is by virtue of capitalist evaluation and quantification, regarded as valuable and what is not, is not. Nature in classical liberal and neoliberal political economics has long been regarded as a 'free gift' i.e. resources such as the water, the air, and the earth were once considered God's gift to humanity. Secular society retained this ethereal belief, even if it dared not mention 'his' name and so these gifts were not considered part of the economic equation but instead regarded as 'externalities'. However, the contents of the economist's hermetically sealed bubble has continued to expand to almost encompass the entire planet and suck up all the surrounding 'free' resources; humanity's God-given gifts. According to Foster et al. "One way to look at this is to see capitalism as a bubble economy, which uses up environmental resources and the absorptive capacity of the environment while displacing the costs back on Earth itself, thus incurring an enormous ecological debt."[17]

Kuznets, and economists like him, helped to build the worlds of modernity by categorising and defining data that is measured by governments and policymakers. Yet, while Kuznets wanted to measure activities that promoted wellbeing, his naivety failed to see that it was those sectors that are known to cause harm that the US government wanted to measure, precisely because they provided them

---

17.   Foster, York & Clark. (2010). The Ecological Rift: Capitalism's War on the Earth

with the tools of power, wealth, and re-election; so of course, Kuznets warnings were ignored. Therefore, today in the US and around the world those three activities: the military, that are responsible for death and injury; advertising, that promotes overproduction and overconsumption; and the finance sector, that has created abstract value, complex derivatives, and immaterial delusions of wealth decoupled from the environment, all count towards the government's 'happiness' index, GDP. But, we have to ask whose happiness is it anyway?

In 1939, Kuznets believed that he could condense all the relevant economic data points down into one number and that would indicate the growth or decline of the economy. Just one positive number would indicate success! The trillion-dollar question was what should be included and what should be left out? In 1936, Kuznets helped organise a conference on *Research in Income and Wealth*, the first time that the term Gross National Product, GNP, was first used. What Kuznets really wanted to reflect in the GNP figure was the welfare of the country and not just a gross tally of all its activity. He wanted to exclude illegal activities, socially harmful industries, the military, the financial sector, advertising, and government spending.[18] Unfortunately, Kuznets like most, if not all government departments, was pulled into the war effort. As a result of military obsessions his advice in 1937 that the national income statements should be designed according to an *'enlightened social philosophy'* and should discount activities that were detrimental or as he put it a *'disservice'* to society was ignored.[19]

---

18.   Pilling, David. The Growth Delusion. Bloomsbury Publishing. Kindle Edition.

19.   Pilling, David. Ibid.

In the US it was during the Second World War that industrial production accelerated under the explosive power of fossil fuels. As Barry Commoner put it "We know that something went wrong in the country after World War II, for most of our serious pollution problems either began in the postwar years or have greatly worsened since then."[20] It is no coincidence that military interests helped to accelerate global heating and environmental problems at the same time Kuznets' key indicators of social wellbeing and happiness were disregarded. The Cold War kept the aggression going and the fear of communist invasion bolstered the military budgets and all of that accelerated computational technology and fossil fuel consumption. Today, the US military is one of the biggest GHG polluters in the world, eclipsing many countries. According to the researchers from Lancaster and Durham universities: "An important way to cool off the furnace of the climate emergency is to turn off vast sections of the military machine," added Dr Neimark. "This will have not only the immediate effect of reducing emissions in the here-and-now, but create a disincentive in developing new hydrocarbon infrastructure integral to US military operations." They found these key facts relating to the US military's fossil fuel consumption:

- In 2017 alone, the US military purchased about 269,230 barrels of oil a day and emitted more than 25,000 kt- CO_2e by burning those fuels.

- In 2017 alone, the Air Force purchased $4.9 billion worth of fuel and the Navy $2.8 billion, followed by the Army at $947 million and Marines at $36 million.

---

20. See Barry Commoner. The Closing Circle: Nature, Man & Technology. Alfred E. Knopf, 1971. p. 140

- If the US military were a country, it would nestle between Peru and Portugal in the global league table of fuel purchasing, when comparing 2014 World Bank country liquid fuel consumption with 2015 US military liquid fuel consumption.

- For 2014, the scale of emissions is roughly equivalent to total — not just fuel — emissions from Romania. According to the DLA-E data obtained by the researchers, which includes GHG emissions from direct or stationary sources, indirect or mobile sources and electricity use, and other indirect, including upstream and downstream emissions.

- The Air Force is by far the largest emitter of GHG at more than 13,000 kt CO_2e, almost double that of the US Navy's 7,800 kt CO_2e.

- In addition to using the most polluting types of fuel, the Air Force and Navy are also the largest purchasers of fuel.[21]

Kuznets failed to get the military excluded from his GDP figures and apart from the research cited here there is little recognition of the planetary impact on global heating by the World's war machines that can only ever deliver the opposite of happiness. Ironically, many military advisors have recognised that global heating will contribute to increasing social and international conflict, so possibly they are aware of the positive feedback loop that supports militarisation. As war gets closer GDP goes up.

21.   Lancaster University. (2019, June 20). U.S. military consumes more hydrocarbons than most countries -- massive hidden impact on climate. ScienceDaily. Retrieved November 7, 2020 from www.sciencedaily.com/releases/2019/06/190620100005.htm

Measures of Wellbeing

In 2020 the New Zealand government announced their second Wellbeing Budget and the Prime Minister Jacinda Ardern informed the country that it was based on the Living Standards Framework using the Genuine Progress Indicator (GPI) instead of GDP. What this new measurement told us was that the country as a whole was only half as well off as we had previously thought we were. The Prime Minister, Jacinda Ardern claimed: "This Budget shows how we are positioning New Zealand for an economic recovery that will make New Zealand the best place it can be to live, study and work. All of this work has been done through a wellbeing lens that considers the needs of New Zealand's people and environment alongside our economy."

"It represented a critical first step for embedding wellbeing into the way we work, made meaningful progress towards breaking down agency silos and balanced the needs of present generations, at the same time as considering the long-term impacts for future generations."

"Achieving genuine and enduring change in the way Budgets and policies are developed takes time. We know that we cannot meaningfully address long-term problems like child poverty, inequality and climate change through a single Budget. This is why the Government committed to taking a wellbeing approach to Budget 2020 and beyond to build on the successes of our first Wellbeing Budget."[22] In support of this budget and policy shift Statistics NZ and the NZ Treasury are now publishing regular wellbeing and sustainability indicators that present dashboards, charts and tools to assist policymakers, bureaucrats, companies,

---

22.   https://budget.govt.nz/budget/2020/wellbeing/from-pm.htm

voters, and academic researchers to help improve things that matter in the lives of New Zealanders.[23] Wellbeing emerged as an explicit policy in the past decade arising from a dissatisfaction with purely fiscal measures and neoliberal assumptions about the *free market*. The economist Kate Raworth examines wellbeing as a means to move beyond fiscal metrics and treating economic growth and money as the only means of evaluating wellbeing. She writes: "Drawing on a wide array of psychological research, the New Economics Foundation has distilled the findings down to five simple acts that are proven to promote wellbeing: connecting to the people around us, being active in our bodies, taking notice of the world, learning new skills, and giving to others."[24]

In a recently published book, *Love You*, by Dr Girol Karacaoglu, the former Chief Economist for the New Zealand Treasury, Karacaoglu summarised the New Zealand government's policy shift: "

"There is the growing realisation that the more recent (say over the last thirty years) primary focus of public policy on material sources of wellbeing (economic growth and so on) is not delivering the other sources that people care about. In fact, these other sources of wellbeing (such as environmental quality, social connections, and the absence of poverty in the sense of deprivation) are deteriorating."[25]

While these indicators are much closer to the wellbeing ambitions of Kuznets they still retain economic growth embedded in their algorithms. The NZ Treasury undertook

---

23.   See Statistics New Zealand website,
      https://wellbeingindicators.stats.govt.nz

24.   Raworth, K. (2017). Doughnut Economics. Random House.

25.   Karacaoglu, G. (2021) Love You: Public policy for intergenerational
      wellbeing. The Tuwhiri Project. Apple Books. Wellington, Aotearoa, New
      Zealand.p.63

work to map the Living Standards Framework to the UN's SDGs, or Sustainable Development Goals. On many dimensions this is a laudable enterprise incorporating such objectives as climate action (goal 13), reduced inequalities (goal 10), zero hunger (goal 2), and no poverty (goal 1). However, goal 8 of the UN's SDG is 'decent work and economic growth' thus retaining the 'growth delusion' and this was then mapped by Treasury to the LSF. Interestingly, they do not mention the mapping of physical/financial capital to this goal of economic growth. This may leave the way open for New Zealand to diverge from the UN's SDGs and a conscious decision to rethink the implications of economic growth, or it may be wishful thinking. In their summary conclusion the Treasury wrote: "Most of the LSF relates to an SDG goal, but since a few domains do not – at least not closely – there is some sense in which the LSF takes a broader, and more New Zealand-specific, approach."[26] Karacaoglu in his book on wellbeing did recognise that economic growth as a concept was problematic. "Economic growth (income and other material sources of wellbeing) are necessary but not sufficient for wellbeing.

"In addition to their material comforts people care about many other aspects of life. Income can buy some of these other items (such as housing, education, …) but not all – for instance, social connections, civic engagement, good governance, clean environment. Furthermore, if we only focus on income growth, we may actually cause damage to other aspects of life that people care about, such as a clean environment (China), or work-life balance (South Korea)."[27]

---

26.   See Judd Ormsby (July, 2018) The Relationship between the Living Standards Framework and the Sustainable Development Goals. https://www.treasury.govt.nz/sites/default/files/2018-07/dp18-05.pdf

27.   Karacaoglu, G. (2021) ibid. p.108

What has been identified by Treasury is that adaptation and resilience to anthropogenic and geological change require more research. In their discussion paper, *Resilience and Future Wellbeing: The start of a conversation on improving the risk management and resilience of the Living Standards Capitals* the authors identified the importance of a National Risk Registry and the need to break down silos and increase collaboration. This also needs to be transdisciplinary so that together the collaborators develop new frameworks and disciplines. As Karacaoglu states: "

"economics, as a discipline, does not have a monopoly on framing and evaluating public policy. As I will argue, the wicked problems we are facing around the world require a multidisciplinary approach. There are all kinds of reasons in addition to economic ones, including moral and ethical, for which governments may choose to influence individual and community decisions and actions."[28] They may want to magnify public good and minimise public wrong.

The discussion paper made the point: "Moreover, there is often little backbone to the networks whereby risk managers across agencies collaborate with each other. Therefore, there is opportunity for improving the attention that is paid to the interconnectedness and cascading nature of risk factors. For example, decisions on land use impact not only on the resilience of our natural capital (eg, the absorption capacity of our biodiversity), but also on the resilience of our physical capital (eg, the capacity of our infrastructure to absorb natural disasters such as floods) and of our human capital (eg, through exposure to natural hazards). The generic resilience in society is thus significantly impacted by these decisions."[29] However, the

28.   Karacaoglu, G. (2021) ibid. p.81

29.   See Resilience and Future Wellbeing: The start of a conversation on

blind spot seems to be the recognition of the linkage of economic growth, financial capital markets, and anthropogenic climate change. In their first risk report, *National Climate Change Risk Assessment for New Zealand*, the Ministry for the Environment, failed to identify economic growth goals as one of the top 10 potential risks to New Zealand. However, they did identify the extreme consequences of the government's "Risk of maladaptation across all domains due to practices, processes and tools that do not account for uncertainty and change over long timeframes."[30] This broad warning is not given the highest risk ranking and yet surely our inability to adapt to the Anthropocene is what has accelerated all the other climate related risks they have identified. There appears to be a distinct inability to identify the interconnected risks of human behaviour with respect to growth economics and the environment. There has even been no causal relationship identified between the global supply chains, agriculture, urbanisation and disease. There was only one mention of COVID 19 in the 2020 risk report and there was no connection made between growth economics, the pandemic, and the threat of a heating planet.[31] Rather than seeing the current economic growth drivers as an existential risk the Ministry of the Environment has only identified the risks to business-as-usual – the threat to

---

improving the risk management and resilience of the Living Standards Capitals

30.   See (Aug. 2020) National Climate Change Risk Assessment for New Zealand pg.9. Maladaptation refers to actions that may lead to increased risk of adverse climate-related outcomes, including via increased greenhouse gas emissions, increased vulnerability to climate change, or diminished welfare, now or in the future. Maladaptation is usually an unintended consequence (IPCC, 2018).

31.   See Chp.3, Zero Humans, for a discussion of the five main trends driving pandemics and disease.

economic growth itself. They wrote: "The costs of climate change in New Zealand are already significant (Frame et al, 2018), and will only increase over time. Almost all risks detailed in this report impact on the economy and the Government's fiscal position, whether from loss of revenue or additional spending to adapt infrastructure, respond to health needs or recover from extreme events. The damages from, and costs of adapting to climate change will likely place a significant and growing financial burden on public authorities, who will be tasked with funding investments in adaptation, providing post-event relief and responding to health impacts."[32] The LSF and Wellbeing indicators face the challenge that the goal of economic growth has paradoxically contributed to a decline in wellbeing as the environmental and social limits have begun to appear, yet in the Wellbeing Budget 2020 it was reported by the government that: "Economic growth in New Zealand was robust prior to COVID-19." The consequences for the global economy of the pandemic has been a dramatic drop in growth and so there has been an emphasis on 'rebuilding' economies. However, there is little recognition that growth might have caused this pandemic disease. Still, there is some consciousness that our country could do things better as the budget states: "Resetting and Rebuilding – taking the opportunity to reset our economy, address longstanding challenges and chart a course to return to a more sustainable fiscal position." I am not quite sure what is meant by a 'return' to a sustainable fiscal position as our historic approach has been anything but sustainable. *Sustainability* is an ambiguous word and like *conservation* can imply ecological concern, or political and economic

---

32.   See (Aug. 2020) National Climate Change Risk Assessment for New
        Zealand pg.69

conservatism. The budget only mentions climate change 3 times and only with reference to the ETS and the Climate Change Response (Emissions Trading Reform) Bill.

A Circular Argument

I don't know of any government around the world who has dispensed with growth economics and yet at the same time most of those same governments have accepted that the Anthropocene is real and that humans are responsible for some of the worst environmental damages to the Earth, including global heating that has happened in the past two hundred years. Since the end of the Second World War the global GDP has increased year on year and almost every metric has shown an acceleration in growth, consumption, population, and of course, waste. It is an undeniable fact of the second law of thermodynamics that the more energy consumed to feed the growth, the more waste accumulates. Yet, most governments ignore this circular argument and believe that they can achieve the other SDGs of the UN while also pursuing economic growth. This is also promoted as the Circular Economy and its disciples preach a doctrine they call *'the good, or even great Anthropocene'*. In their *'Ecomodernist Manifesto'* a group of academics, activists, and scientists introduced their *'transitionary discourse'* as follows: "A good Anthropocene demands that humans use their growing social, economic, and technological powers to make life better for people, stabilize the climate, and protect the natural world."[33]

This remarkable conceit that believes that the human impact on the geological, biological, hydrological and atmospheric layers can be a good thing is seriously

33.   (April 2015). An Ecomodernist Manifesto. http://www.ecomodernism.org/

misguided. To put it bluntly the Circular Economy is a contradiction in terms that holds fast to the faith that humanity can have growth and sustainability (goal 8 of the UN's Sustainable Development Goals). In December 2015 the European Commission launched its plan known as, *'Closing the Loop – An EU action plan for the Circular Economy'*. This was a transition blueprint that boasted they would work towards a: "sustainable, low carbon, resource efficient and competitive economy . . . to transform our economy and generate new and sustainable competitive advantages for Europe." With the aid of organisations such as the Ellen MacArthur Foundation (EMF) and its Circular Economy network, CE100, some of the biggest companies in the world have joined with universities, cities and governments to *'lead'* the world in a *'radical transformation'* of the capitalist economy into a sustainable utopia, Capitalism 2.0. Companies such as Google, Cisco, Phillips, and in 2020, SAP, have all joined the happy throng to *'design out waste and pollution'* while using technology to grow an economy based on a range of *ecomodernist* methods promoted to the EU by the EMF (2015) i.e. ReSOLVE: Regenerate, Share, Optimize, Loop, Virtualise, Exchange. Intended to be a radical transformation of the European political economy. In the 21st century Schumpeter's concept of *'creative destruction'* has been renamed, *'disruption'*. As one New Zealand cheerleader for the Circular Economy, Louise Nash said, *'I disrupted myself.'*

In November 2018 Nash started a company called Circularity. She describes her business background as: "20+ years experience in global strategic brand development, human-centered design thinking facilitation and emerging disruptive technology." Her formal qualifications are: "Bachelor's degree in Economics, a Masters of Commercial

Law, Masters of Applied Technological Futures and an IDEO Certificate in Business Design."[34] At the heart of her offering is not so much radical transformation, but like many 'ecomodernists' she promises business-as-usual. In a nine week *sprint* with the assistance and support of Auckland's economic development agency, ATEED, Circularity and XLabs advertised their offering to New Zealand companies as: "Radically redesigning business for a resilient and regenerative future." To reassure those companies that they could continue to grow and *sustain* future profits they also advertised: "Growth with zero waste. Growth without environmental degradation."

Kate Raworth critiques both GDP and indiscriminate growth economics. As a disillusioned economist Raworth went looking for new answers but decided that economics was still the dominant paradigm and so she attempted to redefine it in her book, *Doughnut Economics.* She differentiates seven ways of thinking about economics – the 20th century way of thinking vs. the 21st century view of economics – the image of infinite GDP growth vs the doughnut economy – growth addicted vs. growth agnostic. She writes: "Today, we have economies that need to grow, whether or not they make us thrive what we need are economies that make us thrive whether or not they grow."[35] To many in business this agnostic approach is almost palatable and is why the circular economy is being entertained – it does not dispense with growth. Raworth returns to an ecological and biological metaphor to explain that nothing grows forever. She shows the graph image of the 'S' curve which is well known in population studies. Populations will start slowly then accelerate as they achieve

---

34.   https://www.circularity.co.nz/partners
35.   Raworth, K. (2017) Doughnut Economics.

an exponential growth rate as they consume more and more resources eventually hitting a limit and slowing down. This also happens with individuals as they reach maturity and stop growing; there is often a long phase prior to death. Eventually, any living thing will die and then return resources back to a common reservoir to be taken up as a component of either living or nonliving things. There will also be the inevitable byproducts of unused energy also known as waste.

Zero Waste

The first law of thermodynamics tells us that energy cannot be created or destroyed. This sounds like a perfect machine for the Circular Economy – a perpetual motion engine. Most of us know that such an engine is impossible. This closed loop equates to the concept of growth without waste but you may remember the next part of the first law which states – while it cannot be created or destroyed – it can only change form or be transferred from one object to another and that transformation is never 100% efficient but will result in a loss of heat – whereas, the perpetual motion engine will eventually run out of steam. Heat could also be defined as waste, as high proportions of it ends up as unused energy. So where does this unused energy, or waste go? It is explained by the Second Law of Thermodynamics that every bit of energy that is transferred will then increase the entropy of the universe and reduce the amount of usable energy available to do work. In ecological terms waste is also a natural part of all life cycles when organisms extract energy and nutrients from the environment and then excrete wastes that are recycled by other living organisms. Therefore, zero waste is nonsense as it does not occur in physical systems that transform energy into other forms and does not occur in ecological terms.

Definitions become important for understanding how we build worlds and the context for understanding how objects interconnect in complex networks. According to the Environmental Literacy Council we need to consider not just metabolic waste but the byproducts of material industrialisation: "humans produce an additional flow of material residues that would overload the capacity of natural recycling processes, so these wastes must be managed in order to reduce their effect on our aesthetics, health, or the environment. Solid and fluid, hazardous and non-toxic wastes are generated in our households, offices, schools, hospitals, and industries. No society is immune from day-to-day issues associated with waste disposal. How waste is handled often depends on its source and characteristics, as well as any local, state, and federal regulations that govern its management. Practices generally differ for residences and industries, in urban and rural areas, and for developed and developing countries.[36]

While there has been increasing sensitivity to toxic and hazardous waste in 'overdeveloped' countries (ODC), less developed countries (LDC) struggle to afford the safe disposal of their own waste and are forced to accept large volumes of waste imported from the 'first world'. A new and growing category of waste, known as e-waste, is regulated under the 1989 Basel Convention on the Control of Transboundary Movements of Hazardous Wastes and their Disposal, (the Basel Convention). Before this convention LDCs were often expected to accept: "the dumping of industrialized nations' toxic waste in exchange for

36.   Environmental Literacy Council. What is Waste? https://enviroliteracy.org/environment-society/waste-management/what-is-waste/

economic benefits the industralized nations provided".[37] In 1983 the OECD reported that a shipment of toxic waste crosses a national frontier once every five minutes, 365 days a year. Over 3 million tons of hazardous waste was shipped to LDCs between 1986-1988.[38]

Since 1990s the international law and regulation have not slowed the illegal shipment of toxic e-waste from our discarded consumer electronics, mobile phones, and obsolescent gizmos we throw away. In 2012 the International Labour Organisation reported that 80% of all e-waste produced is illegally shipped to LDCs.[39] The Global E-Waste Monitor 2020 reported that e-waste has surged by 21 percent in the past 5 years totalling a record 53.6 million tonnes or approximately the weight of 350 cruise ships the size of the Queen Mary 2.

This is in spite of technology companies signing up to the CE100 and their public commitment to 'zero waste'.

- In per capita terms, last year's [2019] discarded e-waste averaged 7.3 kg for every man, woman and child on Earth

- Europe ranked first worldwide in terms of e-waste generation per capita with 16.2 kg per capita. Oceania came second (16.1 kg) followed by the Americas (13.3 kg). Asia and Africa were much lower: 5.6 and 2.5 kg

37. Cusack, Marguerite M. (1990) "International Law And The Transboundary Shipment Of Hazardous Waste To The Third World: Will The Basel Convention Make A Difference?" American University International Law Review. Vol. 5. Issue 2. Article 8. Retrieved from https://digitalcommons.wcl.american.edu/cgi/viewcontent.cgi?article=1586&context=auilr

38. ibid.

39. Lundgren, K. (2012) The global impact of e-waste: Addressing the challenge. ILO. http://www.saicm.org/Portals/12/Documents/EPI/ewastesafework.pdf

respectively

- E-waste is a health and environmental hazard, containing toxic additives or hazardous substances such as mercury, which damages the human brain and / or coordination system. An estimated 50 tonnes of mercury — used in monitors, PCBs and fluorescent and energy-saving light sources — are contained in undocumented flows of e-waste annually[40]

Quite simply the Circular Economy doesn't work. There will always be waste with growth. Especially more heat and that is something that we are trying to avoid.

According to Circularity: In the Circular Economy growth comes from:

1. Designing products so that they can be repaired and reused over again – each product could be sold and resold several times in its life cycle

2. Creating services that enhance the customer relationship and enables a pay per use model

3. Using waste as a resource to produce new products and renewable energy sources[41]

These three growth engines may contribute to less harmful production and consumption. However, the Circular Economy and variations on this 'transitional discourse' such as 'ecomodernism' are problematic because they continue to advocate growth economics and technological design fixes for a broken capitalist system, whether it is 1.0 or 2.0. According to Escobar: "The forceful emergence

---

40. Global E-Waste Monitor 2020. Interntional Solid Waste Association. https://www.iswa.org/home/news/news-detail/article/-21c8325490/109

41. https://www.circularity.co.nz/projects

of transition discourses in multiple sites of academic and activist life over the past decade is one of the most anticipatory signs of our times. This emergence is a reflection of both the steady worsening of planetary ecological, social, and cultural conditions and of the inability of established policy and knowledge institutions to imagine ways out of such crises."[42]

In the *Ecomodernist Manifesto*, it is stated that: "we affirm one long-standing environmental ideal, that humanity must shrink its impacts on the environment to make more room for nature, while we reject another, that human societies must harmonize with nature to avoid economic and ecological collapse."[43] They maintain the long tradition that the bifurcation from nature is *natural* and humanity is now on its own technological path which is *'good or even great'*. In order to have less impact they advocate that human activities like farming, energy, forestry and settlement are intensified: "so that they use less land and interfere less with the natural world is the key to decoupling human development from environmental impacts. These socioeconomic and technological processes are central to economic modernization and environmental protection. Together they allow people to mitigate climate change, to spare nature, and to alleviate global poverty."[44] This is business-as-usual and is hardly changed from the view of the Enlightenment up until today. This was once a religious theme that believed humans were split from nature when God created the Earth and gave humans

42. Escobar, A. (2015). Degrowth, postdevelopment, and transitions: a preliminary conversation. Sustain Sci. 10:451-462. DOI 10.1007/s11625-015-0297-5

43. http://www.ecomodernism.org/

44. http://www.ecomodernism.org/

domain over the entire planet. Without irony or contradiction the Circular Economy has faith that: "through technological and policy innovation, we can "overcome environmental crisis without leaving the path of modernization" (Gibbs, 2006: 196; Mol & Spaargaren, 1993;): a form of modernization wherein market forces are the central agent in delivering change. We can thus become "both rich and green".[45] Intensification has been steadily building since the start of the Holocene 12,000 years ago, and has been the cause of soil degradation, pollution, ocean acidification, water shortages, pathology, to name just some of the wicked problems caused by the Anthropocene – it is not good, progress, or even a step in the right direction, it is responsible for a long list of seemingly intractable troubles for humans and the nonhuman majority alike. But it is nostalgic and popular to imagine the halcyon days when humans had a lighter touch on the planet. Degrowth may be unpopular for companies and governments but we are running out of choices and the more we kid ourselves that a tech fix is around the corner the greater the delay in averting disaster. Unfortunately, delay has replaced doubt as the number strategy to fend off climate regulation and cost to the biggest carbon emitters. Described as *magical thinking*' critics have pointed out that the worst culprits have long promised technological miracles will save us, such as sucking carbon from the atmosphere and burying it beneath the earth.[46]

---

45. Kersty Hobsona, Nicholas Lynch (2016). Diversifying and de-growing the circular economy: Radical social transformation in a resource-scarce world. Futures 82 (2016) 15–25

46. See Mann, (2021) *The New Climate Wars*, and Amy Westervelt (Sep. 2021) *Big oil's 'wokewashing' is the new climate science denialism.* The Guardian. https://www.theguardian.com/environment/2021/sep/09/big-oil-delay-tactics-new-climate-science-denial

Pop GDP

Under the previous government, led by Prime Minister John Key, of the National Party, GDP was sacrosanct and economic growth percentages were how they evaluated the country's success and their administration's policy success. It is well known that population growth will translate into GDP growth, however it is achieved, and the conservative *'party of business'* increased the population by encouraging more immigration. However, the negative consequences of this policy became apparent as more people put more pressure on scarce resources such as housing, water, energy, health, transport and education. Between 2000 and 2020 New Zealand added 1 million more people or 20% more consumers.

New Zealand has a relatively small population of 5 million and we have what might appear to be a slow population growth of 0.82% (2018). However, it is worth considering that in order to calculate how many years it will be before your country's population doubles you divide 70 by the growth rate which means that if the trends remain the same it will be 85 years or 2105 before New Zealand hits 10 million. To some, that might seem a long way off and yet we need to start considering longer time frames and to stop the insidious economic calculation that discounts the future for generations to come. This doubling of the population will be dramatic, if it does happen, as I reflect on the fact that it has doubled in 60 years since I was born in 1961 – the impact so far has been huge and the environmental cracks are beginning to show. Our capitalist growth economy has a built-in accelerator that means that not only is the population increasing but so is our speed of technological change and consumption.

The growth economists are advising governments and policymakers to forget about the future and to carry on business as usual because we will be wealthier and smarter when the next generation is turning 85. We are told that there will be technological fixes for all our problems then, which is much better than suffering from carbon withdrawal now which is how every government has avoided dealing with the problem in the present.[47]

While it is known that the global population growth has been declining from 2.2% 50 years ago to 1.05% per year today, that does not mean that the peak total that is expected around 2100 will not be massive and is predicted by the UN to hit 11.18 billion by the end of this century. However, research has shown that more people does not necessarily lead to famine, hunger, misery and premature death. This may come as a surprise to you so why do we think that that conclusion is just *common sense?*

The Dismal Science

In 1798 the Protestant cleric and early political economist, Thomas Malthus, published an anonymous work: *An Essay on the Principle of Population as it Effects the Future Improvement of Society; with Remarks on the Speculations of Mr. Godwin, M. Condorcet and Other Writers.* Malthus went on to publish six editions of his famous *Essay on the Principle of Population.* This book gave us his conclusion that a human population grows exponentially but that agriculture and our food supply will not keep up because it only grows

---

47. See Foster, J. B., Clark, B., & York, R. (2010). *The Ecological Rift Capitalisms War on the Earth* for a devastating critique of mainstream economists who deny or ignore the urgency of climate change and other anthropogenic disasters. Foster notes that when they do acknowledge climate change they prefer to delay solutions for future generations.

arithmetically. In other words, he argued that the human population would grow by an exponent such as 2 after each generation, e.g. 1,2,4,8,16,32,64 and so on; in contrast to agricultural crops that only grows 1,2,3,4,5,6,7,8. He was not the first to argue populations grow exponentially but his novel argument was that food supply would only grow arithmetically, yet this was presented with no evidence to support it.[48] What is more, Malthus, who was also known as a 'natural theologian',[49] believed that overpopulation would be constrained by misery and vice. Malthus held harsh beliefs that prosperity was not good for the Christian character and that sin and misery would surely follow from overpopulation based on his ethical utilitarian belief that God's given resources would provide the 'greatest happiness of the greatest number' – there was an optimum number that could not be exceeded without loss of virtue and happiness.[50] Malthus is credited with the dubious honour of promoting the 'dismal science', also known as economics. He convinced the English middle-class that it was not God's will to feed a poor *man* [sic], he argued further:

"and if the society do not want his labour, has no claim of right to the smallest portion of food, and, in fact, has no business to be where he is. At nature's mighty feast there is no vacant cover for him. She tells him to be gone, and will quickly execute her own orders, if he [does] not work upon the compassion of some of her guests. If these guests get

---

48. Foster, John Bellamy. Marx's Ecology: Materialism and Nature. Monthly Review Press. Kindle Edition.

49. He believed in God's providence but also that natural science revealed God's plan.

50. Foster, John Bellamy. Marx's Ecology. ibid,

up and make room for him, other intruders immediately appear demanding the same favour....."[51]

Malthus helped shape the English response to the famine in Ireland (1845-49) by arguing that there were two states of equilibrium for human populations – one example was in China where there was once a relatively equal sharing of land, yet that *'forced'* the population to starve, or the alternative state, as in England, where the aristocracy, gentry and middle-class were able to enjoy nature's *'mighty feast'* and *nature's* checks such as universal famine kept the population down.[52] The harsh treatment of Irish farmers by absentee landlords during the Irish potato famine was supported by Malthus who advocated in a letter to Ricardo in 1817 that rather than provide the poor with relief, Malthus, believed they should be evicted because:

"the land in Ireland is infinitely more populous than in England; and to give full effect to the natural resources of the country, a great part of the population should be swept from the soil into large manufacturing and commercial Towns."[53] The shocking result of those evictions of the Irish peasantry was a million deaths caused by greedy land agents and overcrowding due to unjust rentals.

So we should ask, should we be taking advice on population control and political economy from a religious ideologue who lacked not only compassion for the poor but presented no evidence to back up his so-called *common sense* belief? There is no evidence that increased population, on its own, will result in famine. In fact, over the past one hundred years per capita food supply has increased as

---

51.   Foster, John Bellamy. Marx's Ecology. ibid.

52.   Foster, John Bellamy. Marx's Ecology. ibid.

53.   Foster, John Bellamy. Marx's Ecology: Materialism and Nature. Monthly Review Press. Kindle Edition.

populations have grown, largely due to improvements in crop production and yields, and 'famine deaths have decreased, not increased, with population growth.'[54]

The uncaring inhumanity of Malthus and his laissez-faire economics helped to establish the liberal politics that infected ideologies, spreading around the world and convincing the aspirational and the wealthy that the poor suffered because of their wanton overpopulation and indolence. The same attitude exists today with respect to the rising numbers of homelessness people in the US, and sadly now in New Zealand. Yet, there were those that noticed that it was not just the interests of the upper classes that were causing famine but the depletion of soil lifted from Ireland with the export of potatoes and other agricultural crops. England and France both enjoyed a little bit of Ireland before the soil was washed from the crops into the muddy Thames and the Seine. Karl Marx was not just a political economist but an ecologist who noted the resulting air pollution from industrialisation and the environmental loss of Irish topsoil. He wrote: "Light, air, etc.—the simplest animal cleanliness—ceases to be a need for man.... The Irishman has only one need left—the need to eat, to eat potatoes, and, more precisely, to eat rotten potatoes, the worst kind of potatoes. But England and France already have a little Ireland in each of their industrial cities."[55]

Marx understood that capitalism and colonisation were causing a 'metabolic rift' in the global ecosystem as the metabolic cycle was disrupted by extractive industries and

54. Hasell, Joe. (April 03, 2018) Does population growth lead to hunger and famine? Retrieved from https://ourworldindata.org/population-growth-and-famines#licence 26th June, 2020

55. Foster, John Bellamy. Marx's Ecology: Materialism and Nature. Monthly Review Press. Kindle Edition.

agriculture. Humanity had been alienated from nature by capitalist agriculture, and industrialisation. In volume 3 of, *Capital*, Marx insisted that the: "excrement produced by man's natural metabolism," along with the waste of industrial production and consumption, needed to be returned to the soil, as part of a complete metabolic cycle."[56]

Marx had studied the works of Justus von Liebig the agricultural chemist, who noted as early as the 19th century that:

"Great Britain robs all countries of the conditions of their fertility" and had pointed to Ireland as an extreme example, Marx wrote, "England has indirectly exported the soil of Ireland, without even allowing its cultivators the means for replacing the constituents of the exhausted soil."[57]

By the 19th century, Europe was beginning to feel the pressure and stress from centuries of over cropping and too many farm animals. What made matters worse was the fact that not only was there less and less land available to exploit but that the arable land had been seriously depleted of nutrients, minerals, and the biomass from human and animal manure. The carrying capacity of the land had suffered proportionately to the increased extraction of the bio-chemical mass that produced soil fertility and allowed the land to be farmed. Capitalist farmers could literally not grow enough to supply the food, wood, and raw materials for the industrial explosion that was happening in their cities. Numerous enclosure laws forced peasants from the land in order to encourage them to move to the urban locations to provide the cities' factories with cheap labour.[58]

---

56.   Foster, John Bellamy. Marx's Ecology: Materialism and Nature. Monthly Review Press. Kindle Edition.

57.   ibid.

58.   Foster, J. B. (2000). Marx's ecology materialism and nature.

Colonialism was seen as the best solution to the lack of land, food, raw materials, and human labour, or slaves. By the early 1800s agronomy or soil science was beginning to inform landed gentry, farm managers, political economists, and entrepreneurs that the local depleted soil needed to be augmented with imports from offshore deposits of guano and other soil fertilizers. It was not just Britain that realised this but also Europe and North America. The ecological destruction that had led to soil exhaustion created an international trade war as countries competed for cheap sources of phosphorus, calcium, and nitrogen which are essential for growth and health in plants and animals.[59] It was known that ground-up bones could provide phosphorus and calcium to depleted soils. European farmers stole the dead remains from the battlefields after Waterloo and Austerlitz, and were reported to have also raided catacombs for bones to spread on their fields. As a result the price of: "... of bone imports to Britain skyrocketed from £14,400 in 1823 to £254,600 in 1837. The first boat carrying Peruvian guano (accumulated dung of sea birds) arrived in Liverpool in 1835; by 1841 1,700 tons were imported, and by 1847 220,000."[60] In the 19th century Britain in order to make up the shortfalls in food, and raw materials, including wood, cotton, and wool were sourced from the colonies such as Australia and New Zealand in order to fuel industrialisation supported by the principle of free trade. Colonisation required not just

---

59.   This includes the stimulation of root production, plant health, flowering, seed formation and fruiting. Cattle grazing on crops low in phosphorus are prone to diseases such as Rickets, reduced milk production, loss of appetite, and even death. See Veterinary Medicine (Eleventh Edition) 2017, Pages 716-844. https://doi.org/10.1016/B978-0-7020-5246-0.00011-5

60.   Cited by Foster, John Bellamy. Marx's Ecology: Materialism and Nature. Monthly Review Press. Kindle Edition.

shipping and logistics but advertising and promotion, something that was well suited to Edward Gibbon Wakefield (1796-1862).

The Land of Milk and Honey

Marx singled out the unscrupulous colonialist Wakefield for criticism because of his colonisation schemes, such as the New Zealand Company. According to Foster:
"Wakefield argued that the only way in which to maintain a cheap proletarian workforce for industry in the colonies was to find a way of artificially raising the price of the land. Otherwise workers would quickly leave industry for the land and set themselves up as small proprietors. For Marx, this pointed to the contradiction of the separation and estrangement of the population from the land that constituted the foundation on which the whole system of formally free labor rested."[61] Wakefield who had desperately wanted a political career of his own lacked the capital, so set about devising get-rich-quick schemes. He had come from a middle-class family of some wealth, and he had eloped with a 16-year-old heiress, and ward of chancery, Eliza Anne Frances Pattle. According to a biography by Miles Fairburn Wakefield used his "formidable powers of persuasion' to talk his way out of trouble and was granted the most generous settlement the chancery every made to a ward's husband of £1,500 to £2,000 yearly with a job promotion."[62] However, he found he did not have enough capital to achieve his ambition to

---

61.   Foster, John Bellamy. Marx's Ecology: Materialism and Nature . Monthly Review Press. Kindle Edition.

62.   This biography, written by Miles Fairburn, was first published in the Dictionary of New Zealand Biography in 1990. https://teara.govt.nz/en/biographies/1w4/wakefield-edward-gibbon

become a Whig in the House of Commons. On the death of Eliza during childbirth, Wakefield took up an appointment as the secretary to the Paris embassy. In 1826 he seized his opportunity and abducted another wealthy young heiress, a schoolgirl of 15 years old, Ellen Turner. Wakefield wanted to marry the young girl so that her wealthy father would be obliged to help him in politics. Wakefield had never met Ellen but lured her away from school using a false message saying her mother was dangerously unwell, he then tricked the underage, Ellen, into marrying him by telling her that her father had serious money troubles. Wakefield was subsequently arrested in Calais attempting to flee the country by agents of Ellen's parents. Wakefield served 3 years in the English prison, Newgate, and a special act of Parliament annulled the marriage causing a public scandal.

In gaol he studied political economy and utilitarianism. On his release, he contrived a new scheme to make money and convinced English politicians that the failure to attract labour, and sell land in Australia was due to the selling of vast tracts of land too cheaply to the wrong people. While still in prison he published, 'Cure and prevention of pauperism, by means of systematic colonisation', in which he argued, that to avoid repeating the problems of Australia, the land had to be sold to capitalists, and priced out of the reach of labourers. In Wakefield's mind the poor should stay poor in order to provide an eager agricultural workforce for the deserving capitalists, and only then could the poor be provided with enough incentives to labour and eventually save to buy their own land. According to Fairburn, Wakefield published his theory in *A view of the art of colonization* in 1849,that stated:

"the 'sufficient price' as its governing category. The 'sufficient price' was the price at which the Crown needed to sell its colonial 'waste land' so as to restrict the speed with which colonial wage-earners could become proprietors, as well as to build up a fund which would permit the greatest possible number of wage-earners to emigrate free of charge to the colonies."[63]

Wakefield followed up on a scheme to attract capital and labour to New Zealand, known as the New Zealand Association. After it went bust Wakefield formed the New Zealand Company in 1839, that used outlandish advertising and promises of wealth for both capitalists and labourers alike. The Church Missionary Society and the likes of the missionary Samuel Marsden, along with Britain's Colonial Office disliked the underhand tactics of Wakefield and the New Zealand Company that were misleading settlers and stealing land from Māori, claiming to have *legally* bought 20 million acres or one-third of all of New Zealand[64] with little more than cheap supplies of muskets, blankets, and as little cash as possible, at a cost of around a halfpenny an acre[65] so it could be sold at the highest profit.

Wakefield had high regard for capital accumulation and growth with little thought for ecology or indigenous cultures claiming that one of his aims in colonisation was to *'civilise a barbarous people'* who could *'scarcely cultivate the earth'*. Settlers were lured to New Zealand with false advertising and lies. Wakefield advertised windy Wellington, that was surrounded by steep hills, in the Company's prospectus as *'a place of undulating plains suitable for the cultivation of grapevines, olives and wheat'*. This may

---

63.   Fairburn, M. (1990), ibid.

64.   The Taranaki Report, Kaupapa Tuatahi, Chapter 5, Waitangi Tribunal, 1996, page 23

65.   Burns, Patricia (1989). Fatal Success: A History of the New Zealand Company. Heinemann Reed. ISBN 0-7900-0011-3

cause people who know Wellington to guffaw but for those who had utopian dreams of a South Sea paradise, and bought land sight unseen, the shock on arrival must have been very distressing after sailing halfway around the world. The advertising had worked and the numbers grew from 2,500 settlers in 1841 to 40,000 by 1843.[66] Wakefield attempted to buy the Māori land now known as Thordon at Lambton Quay in Wellington, the New Zealand capital, named after the Duke of Wellington due to his support of the New Zealand Company's colonisation scheme. Wakefield hastily signed up the purported Māori landowners with payment made in the form of iron pots, soap, guns, ammunition, axes, fish hooks, clothing—including red nightcaps—slates, pencils, umbrellas, sealing wax and jaw harps.[67] The land was later contested by land commissioner William Spain who attempted to settle the matter with compensation to local Māori. However, there was a major misunderstanding about communal ownership of Māori land and the inability for individuals to negotiate the land sales, and skirmishes resulted, and the matter was not settled until the Waitangi tribunal ruled in 2003 in favour of local Māori as the rightful owners.[68] Wakefield's ignorance of both cultural and agricultural ecology set the foundations for future mistakes to be made in New Zealand. Simple slogans without the backup of material evidence can have profound consequences. For example the words *progress* and *growth* that have been used by colonialists and salesmen to advertise a happy world gifted to the settler by *nature* but in fact was wealth stolen from bewildered indigenous people

---

66.   King, Michael (2003) The Penguin History of New Zealand. p172

67.   Burns 1989, ibid. pp. 113–117

68.   King, M. (2003) ibid. p182

and the commons such as land, water, air, fauna and flora, that were owned by nobody but held in common by every *thing* – human and nonhuman alike.

In the introduction to a university stage one New Zealand sociology text book, *Land of Milk and Honey?* the editors wrote:

"Since colonisation, New Zealand has been mythologised as a 'land of milk and honey' – a promised land of natural abundance and endless opportunity. In the twenty-first century, the country has become literally a land of milk and honey as agricultural exports from such commodities dominate the national economy. But does New Zealand live up to its promise?"[69] As the editor, Averil Bell pointed out the phrase 'land of milk and honey' first appeared in *Exodus*, in the *Old Testament*, describing a Jewish people living in exile depicting a utopian image of a homeland of agricultural abundance. In the 18th and 19th century for many of the desperate settlers who had left the *Old World* behind them, braving treacherous sea journeys halfway around the planet to come to the *New World* to seek their fortunes, there was no hope for a return.

Their descendants in the 21st century have discovered that it truly is a land of milk and honey with dairy becoming the country's export leader in the second half of the last century and now honey is rapidly becoming a valuable export. A New Zealand tourism campaign that used the slogan, *100% Pure*, has opened the country up to criticism as our agricultural practices, and tourism itself, have put increasing pressure on the ecology of the land as both have shown significant levels of growth in human visitors and

69.   Bell, A., Elizabeth, V., McIntosh, T. and Wynyard, M. (Eds.) (2017) A Land of Milk and Honey? Making Sense of Aotearoa New Zealand. University of Auckland Press. N.Z.

nonhuman exports. Considering that in the 1800s the North Island, in particular, was considered to exhibit poor soil fertility and low crop and animal yields. The South Island saw a gold rush in sheep wool but the North had to start clear-felling bush to grow grass for cattle and sheep. The British had embraced laissez-faire economics and had started championing free trade to welcome exports from their colonies and in turn, seek opportunities to export their industrial technologies back to them. Aristocratic absentee land owners profited from vast sheep holdings and the export of wool. Sheep numbers grew rapidly from 2,761,000 in 1861 to 20,233,000 in 1901 encouraged by wool exports and then refrigeration of meat exports to Britain. However, just as there appeared to be physical limits to growth in Europe due to soil exploitation, New Zealand farmers also faced their own limits. According to the *New Zealand Encylopedia* (1966):

"The growth of production in the 1920s, especially between the years 1925 and 1930, is of special importance in the history of New Zealand agriculture, as it was brought about in the main by the more intensive utilisation of the better-class ploughable country in the North Island. The principal agent in bringing about this change was superphosphate. Production expanded most rapidly in the Waikato and it was in this district that the increase in the usage of superphosphate was most marked. The importance of adequate supplies of super being available cannot be overemphasised, and the fact that its price was both absolutely and relatively lower than it was prior to 1914 gave an impetus to its use."[70]

There is no denying that fossil fuel fertlisers meant that

---

70.   (Ed.) McLintock, A.H. (1966) An Encyclopedia of New Zealand. Retrieved from https://teara.govt.nz/en/1966/farming

agricultural production in New Zealand boomed, yet its industrial production has been described as: "one of the modern world's biggest, dirtiest, and most time-honored industrial processes: something called Haber-Bosch."[71] As the science journalist, Robert Service, describes it: "Haber-Bosch led to the Green Revolution, but the process is anything but green. It requires a source of hydrogen gas (H_2), which is stripped away from natural gas or coal in a reaction using pressurized, super-heated steam. Carbon dioxide (CO_2) is left behind, accounting for about half the emissions from the overall process. The second feedstock, N_2, is easily separated from air, which is 78% nitrogen. But generating the pressure needed to meld hydrogen and nitrogen in the reactors consumes more fossil fuels, which means more CO_2. The emissions add up: Ammonia production consumes about 2% of the world's energy and generates 1% of its CO_2."[72] The massive amounts of nitrous oxide emissions are largely a function of the amount of nitrogen added to the land through fertiliser, urine and dung which is then washed into the rivers and oceans causing dead zones in which nothing lives. The leakage of nitrates into the water supply has been linked to high rates of colorectal cancer, with New Zealand having one of the highest rates of bowel cancer in the world. Recent research has estimated that the numbers of people in New Zealand potentially affected by high nitrate levels could be between 500,000 to 800,000. The Ministry of Health has followed the WHO guidance for safe levels of nitrates as 11mg/L of water whereas a major Danish study found a link

71.   Service, R.F. (2012). 'Ammonia as green alt energy - Australia.' Science. https://www.sciencemag.org/news/2018/07/ammonia-renewable-fuel-made-sun-air-and-water-could-power-globe-without-carbon
72.   ibid

with bowel cancer as low as 0.87mg/L. "The report's author, Jayne Richards, said nitrate in drinking water was "likely to be a significant contributing factor to colorectal [bowel] cancer rates in New Zealand and may be of a similar significance as the established risk factors of consumption of red meat ... processed meat, lack of physical activity and smoking".[73] There appears to be a consistent collusion between government agencies such as the Ministry for the Environment, MfE, Stats NZ and the agricultural lobby. The quiet backdown by Stats NZ with respect to what is reported as the *natural* level of nitrates at 3mg/l as opposed to 0.25-1mg/l seems to point to underreporting that might impact farming profitability. While Stats NZ and the MfE claimed the mistake related to lack of data, the water ecologist, Dr Mike Joy, responded: "a claim not to have enough data is the "best cop-out there is". "Funny how it is always a mistake in one direction that is downplaying the problem, never a mistake where it was claimed to be worse than what it is."[74] Joy says that there should be reliable, independent and scientific reporting, separate from the government regulator. "Given the failures of environmental protection and reporting through political and business lobbying, the need to keep independent scientific advice from political influence is clear".[75] Joy makes the analogy between the US tobacco lobby and the dairy promoters, backed by the government's pro-business department, pointing out that the New Zealand co-op

73.   RNZ. (22 February 2021) "Zealanders may have increased bowel cancer risk due to nitrates in water." https://www.rnz.co.nz/news/national/436879/up-to-800-000-new-zealanders-may-have-increased-bowel-cancer-risk-due-to-nitrates-in-water

74.   Williams, D. (12th July, 2021) Mike Joy wins battle over 'dodgy' water stats. Newsroom. https://www.newsroom.co.nz/mike-joy-wins-battle-over-dodgy-water-stats

75.   Williams, D. (12th July, 2021) ibid

Fonterra had attempted to downplay research into the carcinogenic properties of nitrates by funding a study that was not peer reviewed. Radio New Zealand reported: "The controversial study was funded 60 percent by Fonterra and 40 percent by the Ministry of Business, Innovation and Employment and explored whether the link being made between nitrates in drinking water and bowel cancer was valid.

It said those finding a link to cancer had not taken into account the role of food in nullifying the impact of nitrates. Lead author, ESR's Peter Cressey, said this called into question whether a link existed at all."[76]

What is more, there are other enormous social, political and environmental costs associated with the mining and supply of superphosphate from Western Sahara on disputed land surrounded by landmines.[77] The massive dumping of superphosphate on the land has resulted in poisoned streams and rivers and may even lead to heavy metal toxicity in the form of cadmium which can cause cancer and so the land could become unusable. There are very material limits to growth and as long as agricultural growth is for-profit then the cost to the environment may be ignored and become unacceptably high. In New Zealand, the fertiliser industry since 2011 has been self-managing cadmium levels, it is reasonable to be sceptical of the industry's ability to limit cadmium levels especially if it entails cost and limits their company growth.[78]

---

76. Radio NZ. (23 Sep. 2021) Experts question Fonterra-funded nitrates study. https://www.rnz.co.nz/news/environment/452100/experts-question-fonterra-funded-nitrates-study

77. Doherty, B. (Mar. 2020) 'West Saharan group takes New Zealand superannuation fund to court over 'blood phosphate'. The Guardian. https://www.theguardian.com/world/2020/mar/16/west-saharan-group-takes-new-zealand-superannuation-fund-to-court-over-blood-phosphate

78. Ministry for Primary Industries. Cadmium in New Zealand soils.

While successive New Zealand governments continue to see growth economics as a positive goal they will continue to turn a blind eye to the collusion between government agencies, departments, scientific researchers, universities and the agricultural, and horticultural industrial complexes. Meanwhile, New Zealand suffers from a strong anti-intellectual legacy that limits critical debate around agriculture that isolates and alienates anyone that should question land use and the 'holy cow' of this country's prosperity. The independence and objectivity of any coalition of those who are united in their faith in growth economics, and profit for the country from primary produce, when they are challenged by environmental concerns, is suspect. A group that included Beef + Lamb, DairyNZ, Federated Farmers, Irrigation New Zealand, and Horticulture NZ responded to the government's Zero Carbon Act with the following statement: "We are pleased that the government has recognised that it does not make sense to bring agriculture into the ETS and that we have a pathway to work with the government to develop a more appropriate framework."[79] Not only has the Zero Carbon Act delayed the introduction of carbon pricing for agriculture, when it was enacted it was seen to have no real application for GHG reduction. According to the international Climate Action Tracker website: "New Zealand lacks strong policies, despite its Zero Carbon Act.

https://www.mpi.govt.nz/protection-and-response/environment-and-natural-resources/land-and-soil/cadmium/

79. Skerrett, A. (24.10.2019) Government backs down on plans to include farmers in Emissions Trading Scheme https://www.newshub.co.nz/home/rural/2019/10/government-backs-down-on-plans-to-include-farmers-in-emissions-trading-scheme.html

The Act does not introduce any policies to actually cut emissions but rather sets a framework."[80]

An example of how political compromise and lobby pressure can become hidden within algorithms can be illustrated by software known as Overseer™. Overseer™ is a software application that was designed as an on-farm management tool. "The company responsible for developing and maintaining Overseer – Overseer Ltd – still sees it very much as an on-farm management tool. The company's Strategic Plan 2015-2017 envisions making Overseer "the trusted on-farm strategic management tool for achieving optimal nutrient use for increased profitability and managing within environmental limits."[81]

In 2016 a group known as the Biological Emissions Reference Group (BERG) was formed. It included the following companies who have little interest in reducing their profits from farming: Fonterra, Horticulture New Zealand, Fertiliser Association, Beef and Lamb, Ministry of Primary Industries, Ministry of the Environment, Deer Industry, Federated Farmers and Dairy. They pointed out in their executive summary from their 2018 report that: "Our Terms of Reference intentionally exclude developing policy advice or providing recommendations. However we did commission analysis to estimate the costs and barriers of hypothetical policy options to reduce emissions. The analysis did not consider how biogenic methane emissions from agriculture could be treated within a domestic emissions target." They went on to comment that: "OVERSEER is a suitable tool for estimating biological

---

80. See https://climateactiontracker.org/countries/new-zealand/

81. Parliamentary Commissioner for the Environment.(December 2018) Overseer and regulatory oversight: Models, uncertainty and cleaning up our waterways

emissions on farms. Its calculations are supported by current scientific understanding, are consistent with the National Greenhouse Gas Inventory, and can be used for different farming systems and management practices. This work did not assess the suitability of OVERSEER as a regulatory tool for biological emissions."[82]

However, the Commissioner for the Environment, Simon Upton, had this to say about Overseer[TM]: "I have come to the conclusion that in some important respects, Overseer does not meet the levels of documentation and transparency that are desirable in a regulatory setting. Important issues remain to be clarified concerning the uncertainty that attaches to its outputs. All of these things are resolvable if there is a desire to do so."[83] In 2020 Overseer Ltd. with backing from the fertiliser companies and others have begun to negotiate and work with regional councils in an attempt to get their application accepted as the primary tool to manage GHGs, and nutrient runoff. The likelihood of the governmental-industrial-agricultural complex developing an application that would report negatively on fertilisers, methane, nitrous oxide, and water is unlikely and the algorithms have a tendency to obscure and hide unpalatable political and economic assumptions as the vested interests seek to protect their profits.[84]

Real growth is not limitless, no plant, animal or physical thing continues to grow forever. They all eventually die or collapse or are transformed by the worlds around them –

---

82.   Report of the Biological Emissions Reference Group (BERG) (Dec. 2018)
      https://www.mpi.govt.nz/dmsdocument/32125/direct

83.   Parliamentary Commissioner for the Environment.(December 2018) ibid.

84.   As has been already noted the lobby group BERG has shown its hand in its
      promotion of OVERSEER using it to measure methane emissions on the
      farm without dramatically effecting profits. The CCC has heavily relied on
      BERG for evidence in their final report.

this is the organic life cycle that recycles nutrients. The focus on profit by unsustainable agriculture has meant that productivity, or overstocking, over cropping, and toxic forms of fertilisers have enriched industrial agricultural capitalists. According to David Korowicz: "Global food producers are already straining to meet rising demand against the stresses of soil degradation, water shortages, overfishing and the burgeoning effects of climate change. It is estimated that between seven and ten calories of fossil-fuel energy go into every one calorie of food energy we consume. It has been estimated that without nitrogen fertiliser, produced from natural gas, no more than 48% of today's population could be fed at the inadequate 1900 level. No country is self-sufficient in food production today."[85] There is the political dimension of soil sovereignty and the wellbeing of everyone who depends on food that has not been contaminated with toxins. Organic and biodynamic farming introduces a more mindful and intuitive component to food that is aware of both the spiritual and philosophical nature of being a respectful kaitiaki or guardian of the whenua (land). Hutchings has pointed out the dangers of both capitalist growth economics and unsustainable agriculture that relies on fossil fuels aided by computer algorithms that alienate humanity from their physical connection and intuition with the oneone (soil). There is a growing movement of First Nations who are beginning to recover the ancient indigenous knowledge about cultivating food.[86]

---

85.   See David Korowicz (2011) On the cusp of collapse: complexity, energy, and the globalised economy https://static1.squarespace.com/static/5ba7a97cb91449215ec18448/t/5ba906d5c830250c0ffia3fd/1537803990535/On+the+cusp+of+collapse.pdf

86.   Te mahi oneone hua parakore : a Māori soil sovereignty and wellbeing handbook / edited by Jessica Hutchings and Jo Smith

Silent Spring Growth

From the very beginning of early civilisations 12,000 years ago the energy output of communities was a function of the weather. Warmer weather stimulates growth and growth stimulates human procreation. As David Pilling wrote in his book, *The Growth Delusion:*

"The output of agricultural societies was pretty much a function of the weather. If rains were good, the harvest was good. If not, it wasn't. Nor, in this pre-industrial world, were there huge productivity gaps between one region and another. Most people were just scraping by. Thus the size of a region's economy was largely determined by the size of its population."[87]

After the inclement and unstable Pleistocene, otherwise known as the Ice Age, that lasted from 2,580,000 to 11,700 years ago, the planet Earth entered a new epoch, the Holocene, a relatively pleasant and stable period of climatic stability and warmth. According to Ian Angus, *Facing the Anthropocene*, this created the perfect conditions for growth, for plants, animals, and humans:

And "agriculture needs not just warm seasons, but a stable and predictable climate—and indeed, in just a few thousand years after the Holocene began, humans on five continents independently took up farming as their permanent way of life."[88]

Agriculture was one of the earliest indicators of an 'economy'. The origins of the word were Greek meaning something like household management. According to the Oxford dictionary, the definition of economy is:

87. Pilling, David. The Growth Delusion. Bloomsbury Publishing. Kindle Edition.

88. Angus, I. (2016). Facing the Anthropocene: Fossil Capitalism and the Crisis of the Earth System. NYU Press. Kindle Edition, location 936.

1. "the state of a country or region in terms of the production and consumption of goods and services and the supply of money."

2. "careful management of available resources."

While the history of agriculture extends back well before money the association of production, consumption, and the management of resources suggests a causal relationship and a historic chain of events that have culminated in the global crises also known as the Anthropocene. It was in the middle of the 13th century that the word 'cattle' meant money, land, income, taken from the Anglo-French word *catel* meaning 'property'; Old French meaning *chatel*, and Medieval Latin meaning *capital*. Agriculture has for a very long time had an association with the accumulation of energy in the form of land, and until it could be stored in a form that did not die, rot, or degrade its value was dependent on the fertility of the soil, on the land that was bought, stolen, and accumulated. *The Great Acceleration* of energy conversion and consumption began after the Second World War tying economics and agriculture to violence, weapons, fossil fuels and capitalism. The biologist Richard Levins put it succinctly:

"Agriculture is not about producing food but about profit. Food is a side effect."[89] Capitalism is inextricably tied to voracious consumption and growth that is inevitably dragging us towards the *'cliffs of despair'*. We are like lemmings who can see that we are all headed towards the edge of extinction and yet we have lost our own agency to resist or imagine an alternative. We have all been trained and indoctrinated to believe that *'market forces'* dictate our

89.   Angus, Ian. Facing the Anthropocene: Fossil Capitalism and the Crisis of the Earth System. Monthly Review Press. Kindle Edition.

addiction to the growth model, yet capitalism knows no bounds. Since the mid-80s New Zealand has remained under the sway of neoliberal economics, an ideology imported with little or no public debate and without theoretical rigour. Today, New Zealand continues to abide by the doctrine Milton Friedman's *free market* and be buffered by vested global forces it barely understands.

The response of governments around the world to the COVID 19 pandemic has varied from state to state, and yet there is a global recognition that there has been a dramatic impact on economies and most have responded with the injection of government spending known as stimulus packages to prop up companies and encourage consumer consumption. In other words they are doing everything they can to prevent zero, or negative growth. In the US the Republican backed government has spent almost 7 trillion dollars (as of today) to save airlines and other fossil fuel dependents from going out of business and this will most likely not stop there. As one commentator recognised *'bankruptcy is the same for capitalists as hell is for catholics'*. Yet all the previous talk of *'free markets'* and so-called Darwinian economics that proclaimed *'only the fit should survive'* has almost evaporated as the neo-liberal economists retreat from the fray mumbling something about *'of course the government should stimulate the economy and save capitalism'*. Capitalism is our 21st century religion and so our *'moral hazard'* today, according to those same capitalists. They argue that if we ignore Schumpeter's *'natural gale'* forces of *'creative destruction'* and prop up the fossil fuel beneficiaries then we will be faced with inefficient markets and technological innovation will be suppressed. This central neo-liberal paradox is ignored by governments and capitalists who continue to shrink away from the obvious consequences of free market dogma. If the fossil fuel industries, (including agriculture) lost their government subsidies tomorrow, according to neo-liberal theory, then those unprofitable businesses would no longer exist and global warming would rapidly decline. The paradox is that the system that created these businesses is the same system that feeds 'zombie' companies that should die rather than terrifying everyone by stomping around the planet as the 'walking dead'. Governments, ironically, continue to protect the wealthy elites by protecting their ill gotten capital – hardly a free market mechanism. As Foster et al. warns:

"What makes matters so serious is the inability of our social system to respond effectively to this planetary crisis. It is an inner characteristic of the capitalist economy that it is essentially limitless in its expansion. It is a grow-or-die system."[90] Capitalism is regarded much the same as the religions of the past as a natural state of affairs, as if economic growth is biological or according to the laws of physics. As the movie character, Chauncey Gardiner, played by the comedian Peter Sellers, explains that 'growth has its seasons' and rest assured 'there will be growth in the spring the economy is like a garden?'[91] However, if we are to really understand growth and decline in populations, economies, and a decline in waste, GHGs, and global heating, then we need to be much more proficient in understanding energy and knowledge flows in a global ecosystem. As Bruno Latour remarked, when he was asked about what would he change in a post-COVID world, he said:

"What we need is not only to modify the system of

90.   The Ecological Rift: Capitalism's War on the Earth by John Bellamy Foster, Richard York, Brett Clark

91.   From the movie, "Being There" (1979)

production but to get out of it altogether. We should remember that this idea of framing everything in terms of the economy is a new thing in human history. The pandemic has shown us the economy is a very narrow and limited way of organising life and deciding who is important and who is not important. If I could change one thing, it would be to get out of the system of production and instead build a political ecology."[92] Capitalism is not a process born of biology, physics, or chemistry, it is a culture invented by humans for humans, not nonhumans. We have to recognise that we have the agency to create a new system that avoids the dangerous paradox of capital accumulation that is out of kilter with life on this planet.

92. Jonathan Watts. (6/6/2020) The Guardian. Bruno Latour: 'This is a global catastrophe that has come from within'. https://www.theguardian.com/world/2020/jun/06/bruno-latour-coronavirus-gaia-hypothesis-climate-crisis

5

OOO Zero Worlds

It is highly unlikely that COVID 19 will end in the extinction of humanity. Our population explosion in modern times is almost exclusively made possible by sanitation, vaccines, and antimicrobial treatments. At the same time unecological land use, clear felling forests, synthetic fertilisers, refrigeration and global logistics have all, not only enabled food production to grow dramatically, and food distribution to become global, but they have created the stressed environments suitable to cause a pandemic in the first place. Add to these, another complexity is the entangled nature of these networks. Now we also know that those innovative treatments are also killing the very creatures that keep us alive and have helped us to evolve and adapt to a very dynamic world for approximately 4 billion years.

Ancient Remnants

Apocalyptic fears of death and extinction are literally part of our DNA; our cultures, myths and legends have all been pre-baked into our biochemical makeup. That is not to say that those hopes and fears have ossified or are static but it is more like these are the ingredients to be used in our improvised recipes of who and what we are? Scientists have begun to realise that epigenetics[1] provides us with hints about the origins of our phobias, (and I would also suggest our utopian dreams). There is evidence that anxieties and even depression could be genetically passed down from generation to generation. Could it be that the rise of anxiety and depression in a world threatened by existential annihilation is some kind of nasty feedback loop? This could be a partial explanation for what Freud and Jung called our *'collective unconscious'* or our *'ancient remnants'*; angst may have been encoded in our DNA as humans interacted with their environment, these fears could have been passed on by our ancestors. Epigenetics also suggests that this can happen in real-time while we are living and adapting to the world around us. Stress and anxiety suffered by ancestors can be passed to following generations, or from the mother to the foetus within the womb, or encoded within and during the lifetime of an individual.[2] This scientific description of adaptation and psychopathology (and surely also optimism) overturns the

---

1. "Epigenetics most often involves changes that affect gene activity and expression, but the term can also be used to describe any heritable phenotypic change. Such effects on cellular and physiological phenotypic traits may result from external or environmental factors, or be part of normal development." See https://en.wikipedia.org/wiki/Epigenetics

2. Lacal I., Ventura R. Epigenetic Inheritance: Concepts, Mechanisms and Perspectives. *Front Mol Neurosci.* 2018;11:292. Published 2018 Sep 28. doi:10.3389/fnmol.2018.00292

'central dogma' of molecular biology (proposed by Crick, in 1958) that our genetic program is predefined.[3]

The Biblical warnings of plagues, pestilence, floods, and locusts were real events that were periodically inflicted on primitive humans who attributed their misfortune to sin and disobedience to the gods. Today, these superstitions still linger in the modern psyche and it turns out that epigenetics may be the scientific explanation.[4]

We may well have replaced myths and legends with scientific explanations about reality without knowing about epigenetics, and yet we know from history that there have been human punishments for bad social, environmental, and economic behaviour which are today encoded in punitive law and so are the social norms of acceptable behaviour and our values. Sin has a new guise in the form of humanist values without religious ethics or morals. Yet, even anthropocentric humanism can shroud the real objects of our interest and it requires a new and speculative philosophy to see the strange qualia of 'thing related reality'.

OOO

This philosophy is called the OOO, the triple O, or Object Oriented Ontology. Each O could be thought of as a Zero, a world/void/world that beckons us but will always remain ultimately hidden. The object or thing will be illuminated and then shadowed by other qualities that come into focus just like an orbiting asteroid or black hole. We cannot fully know a thing, living or inanimate, because we cannot comprehend the complexity of its historic relationships

---

3.   ibid.

4.   See Lacal I. et. al, (2018) ibid, and Rive, P. WorldBending: a survivor's guide

with other things, whether they are another nonhuman; an environment; a human being; a fear; a hope; or an idea. The complex interconnectivity of everything is philosophically and scientifically essential to our understanding of the complexity of climate change that cannot be reduced into component parts but only understood as an emergent system. According to Wahl: "The holistic sciences study interactions and relationships in the behaviour of complex dynamic systems. The focus is on the underlying patterns and emergent properties. These sciences carry three important lessons beyond the bounds of traditional science. These lessons are:

i) we are living in a fundamentally interconnected world where local actions can have global repercussions,

ii) in such a complex and dynamic web of changing interactions and time delayed, multi- causal relationships it is of limited use and can have dangerous effects to isolate individual and simplistic, linear cause and effect relationships,

iii) the behaviour of complex systems is fundamentally unpredictable and uncontrollable beyond a very limited time period and scale."[5]

For now, I will try and show how paradox, complexity, and the interconnection of strange things, such as the COVID 19 virus, Zero Carbon, ecology, climate and humans connect in a network of agents, resources and events. Sheldrake noted that far from being a free agent in a world of our own making there is a multiplicity of *things* that we live in symbiosis with and the more we study their networks the more we see that humble organisms such as

5.   Wahl, D.C. (March, 2005) "Eco-literacy, Ethics, and Aesthetics in Natural Design: The Artificial as an Expression of Appropriate Participation in Natural Process" A paper presented at the European Academy of Design Conference. Bremen, Germany. Paper no.92

viruses and fungi are far more influential on our worlds than we might think. Sheldrake was asked if his years of studying fungi had meant that this organism had imprinted on him. He wrote: "I'm still not sure. But I continue to wonder how, in our total dependence on fungi – as regenerators, recyclers, and networkers that stitch worlds together – we might dance to their tune more often than we realise."[6] Simple models can often be mistaken for reality, but really all *things* have their own reality. To think that there is only one reality is a political illusion. Our ability to hold contradictions, uncertainties and inexactitudes in our head all at once will define our ability to adapt and pacify our determination to transform the worlds around us. We need more humility and comprehension of our ignorance to prevent simple questions leading to simple answers and potentially catastrophic technological fixes. My colleague, Jorn Bettin, wrote:

"*Civilised* humans are so self-absorbed that they conceptualise Earth as '*their*' planet without blinking an eye. It is impossible to paddle back from this extreme position without acknowledging the collective delusion induced by our '*civilised*' way of life.

How do we go about to construct ecological niches that contribute to the thriving of life on Earth rather than taking away from it? We have triggered the sixth mass extinction, and biodiversity is declining at unprecedented rates.

What ecological role do we want to play going forward? Note that we have successfully disqualified ourselves from the absurdly anthropocentric role of 'owner'. Are we still

---

6.   Sheldrake, M. (2020) Entangled Life: How Fungi Make Our Worlds, Change Our Minds and Shape Our Futures.

capable of relearning of how to engage with other species at eye level? We might be able to learn quite a bit from other less self-absorbed species."[7]

What is more this can also apply to non-living things – if only humanity spent more time contemplating the agency of all nonhumans and more time meditating on the context of that agency and our interaction with them then the harm would most likely be so much less.

A Brief History of Things

You might well be wondering what has philosophy got to do with our primary objective, the decarbonisation of human activity, or Zero Carbon. Simply put, philosophy has always preceded the way we think, the technology we have built, and the way we live.[8] It was a philosophy that invented the void, or zero, in the first place.[9] Speculative realism, or the OOO, gives us a philosophical tool to recraft our view of not just this world but to pick up our own cognitive hammer and build personal new worlds, or embark on worldbending the worlds we are already familiar with. According to this novel philosophy, in simple terms, every *thing*, whether it has a physical reality, or a conceptual reality, has some kind of agency. It is an agent that defines and is defined by a vast and complex network of other things and relationships that it affects and it is affected by. Speculative Realism or the OOO builds on the philosophy

7.   Bettin, J. (Oct.12. 2020) Nuturing Ecologies of Care. https://jornbettin.com/2020/10/12/nurturing-ecologies-of-care/

8.   Tarnas, R. (1991). The passion of the Western mind: Understanding the ideas that have shaped our world view (1st ed). New York: Harmony Books.

9.   My earlier book, *Worldbending*, is mostly about creativity and how Western culture has been shaped by the significant milestones in philosophy.

of Alfred North Whitehead who set about addressing the *'bifurcation of nature and humanity'*. Speculative realism challenges the 18th century philosophies of Descartes and Kant that put humanity at the centre of the physical universe. Their philosophies only made sense from the perspective of human consciousness.[10] However, this has not only resulted in worlds built exclusively for human consumption but has wrought untold environmental damage on Earth. Why is it that this human centric world view has caused so much harm? Being so self-obsessed has distorted our view of reality and installed a simplistic and narcissistic worldview that favours the minority, ourself, over the majority, the nonhumans. Simple questions lead to simple answers and ignores the wicked problems that need humility and pluralism beyond our vested interests.

From an anthropocentric perspective, it can be understood by pondering the truism that as individuals we can never truly know ourselves. If you consider yourself as an example you could not hope to understand the impact of every thought or experience you have had in your life, and you most certainly could not name and know every alien bit of DNA, RNA, and genetic code that makes up the nonhuman majority inside your body. We are a paradox in that we are more non human than human. The human body has more alien bacterial, and viral genetic code from other species than our own. Viruses make up the largest biomass on this planet and the total number is more than the stars in the universe i.e. more than 10^{28}, or cells in our body, 10^{18}. The total number of viruses estimated to be on our planet is 10^{33} and they are present in every single existing species. Mammals and birds are thought to host

10.   See Shaviro, S. (2014). The Universe of Things: On speculative realism. Minneapolis: University of Minnesota Press.

around 1.7 million undiscovered types of viruses but only just over 200 are known to cause human disease. We have not even begun to understand the important impact and complexity of bacteria and viruses with respect to the ecology of life on our planet, and life in the universe. At present we only know, or rather, can name around 3000 types of viruses on Earth, yet every day we are showered by 800 million viruses per square meter falling from the troposphere, and from as high up as the stratosphere 300 km high above us.[11]

With these astronomical numbers, we are faced with an incredibly difficult and complex problem in understanding the ontology, or reality of even one particular type of virus such as COVID 19. A simple question might be, 'How do we eradicate COVID 19?' A simple answer: "Nuke it with bleach." or "Medicate with Hydroxychloroquine". These absurd answers come from what might appear to be a sensible question but the reality of sustainable healthcare[12] and the survival of not just humans, but the nonhuman majority, will expose the dangers of simple questions. In the US the pandemic has joined a long line of things that have become weaponised. There was the Cold War, the War on Drugs, the War on Poverty, the War on Terror, and now the War on COVID 19.[13]

According to Eugene Thacker: "The threat is not simply

---

11. 'Billions of Viruses Are Falling to Earth Right Now (But That Isn't Why You Have the Flu)'(Feb, 2018). LiveScience Retrieved 2020 from https://www.livescience.com/61689-viruses-fall-from-sky.html; Moelling K and Broecker F (2019) Viruses and Evolution – Viruses First? A Personal Perspective. Front. Microbiol. 10:523. doi: 10.3389/fmicb.2019.00523

12. See this webinar series on Sustainable Healthcare and Climate Health. https://www.otago.ac.nz/news/events/otago738191.html

13. For a discussion about the inappropriate use of the war metaphor see this webinar https://healthcare-solutions.s23m.com/2020/08/22/trans-tasman-knowledge-exchange-part1/

an enemy nation or terrorist group, the threat is itself biological; biological life itself becomes the absolute enemy. Life is weaponized against Life, resulting in an ambient Angst towards the biological domain itself."[14] We are not at war with COVID 19, it is not the enemy, we have to learn what it means to co-exist with this virus – that is the theory of vaccination and is not simply eradication.

COVID 19 – A Simple Question of Survival?

As I am writing this today, according to John Hopkins University, there are 7,628,687[15] confirmed cases and 425,313 deaths worldwide – we still don't know if the world is approaching a pandemic as brutal as the Spanish Flu but substantial underreporting in LDCs suggests that we could be moving in that direction. The US COVID 19 pandemic has now surpassed the deaths caused by the Spanish Flu.[16] The US, that boasts it is the most technologically advanced and wealthy country in the world tops the number of cases and deaths due to COVID 19. Brazil, led by the 'Trump of the Tropics', Jair Bolsonaro, has just now overtaken the UK to become the second biggest number of cases and deaths.[17]

Human disease and death have always been political, socio-economic, and an environmental issue, as well as

---

14. Thacker, E. (2011) In the Dust of This Planet: Horror of Philosophy vol. 1

15. Every time I reviewed this chapter before publication I considered updating the latest number before publication, but the numbers are still going up and is better tracked in realtime; today the total number of confirmed cases is 16,189,581 with 647,846 dead. More than double in just over 1 month. Today, 8th Aug, 2020 it is 19 million.

16. Time. (20 Sept. 2021). COVID-19 Is Now the Deadliest Pandemic in American History. https://time.com/6099962/covid-19-spanish-flu/

17. Retrieved from https://coronavirus.jhu.edu/map.html 13th June, 2020 As this is being written during the pandemic, with no clear end in site, the number and order of countries will most likely keep changing.

medical, biochemical, and microbiological, to name just some of the disciplines required to understand the origins and possible treatments of diseases. This is especially true of the epidemiology and public health risks as globalisation has enabled viral infections to be transmitted easily via planes, trains and automobiles not to mention cruise liners and even spaceships. The outbreak of COVID 19, Ebola, SARS, and HIV to name just some of the epidemics that caused a crisis in public health have exposed the inadequacy of taking a narrow medical view of these outbreaks. According to the geographer Matthew Gandy:

"If we are to make sense of the current public health crisis, we need to explore interconnections between political, economic, and social developments that are ignored by the fragmentary emphasis of the biomedical sciences."[18]

The management of public health in the US, China, Brazil and the UK give us some disturbing evidence of the results of the lack of transdisciplinary approaches to the wicked problem of a pandemic, climate change, zero carbon and a myriad of other anthropogenic problems. We have a very limited understanding of the complexity and interconnected nature of urbanisation, decarbonisation, propaganda, and infectious diseases, their origin, transmission, containment, suppression, elimination and eradication. Very simple agents with very simple interactions can rapidly become a part of complex networks of multiple agents, resources, and events radiating out from nodes in the network, thus presenting exceedingly wicked problems.

---

18. Gandy M. Deadly alliances: death, disease, and the global politics of public health. PLoS Med. 2005;2(1):e4. doi:10.1371/journal.pmed.0020004

The Pasteurization of France

The OOO philosophers and a number of modern theorists such as the new materialists have embraced Bruno Latour's democratisation of the *'parliament of things'*. Latour who began his metaphysical research by examining the wider relationships of captive primates with the scientists, the lab and a decentered world of inanimate objects, gained notoriety from his book, *The Pasteurisation of France* (1984). Pasteur was no longer the centre of the narrative and had to contend with Erlenmeyer flasks, bacterial microbes, networks of General Practitioners, and politicians. Politics had emerged from the cell, and other researchers have followed Latour in exposing what Levi Bryant has called the *'democracy of things'.*[19] Enfranchisement is something that even inanimate objects confidently claim, they calmly occupy every space humanly imaginable, and even those recesses of the mind that no *thing* had previously inhabited. This philosophy critiques the world of the Enlightenment, a world that Immanuel Kant had surgically separated into two camps, human and nature, concept and reality. Speculative realism not only rejects the anthropocentric worldview but fights for the rights of all things, including the immaterial and conceptual to be taken seriously, and amongst these strange assemblages,[20] the 'nonliving' virus is just one of many things to demand a vote.

It is assumed by most evolutionary microbiologists that life began simple and then evolved into more and more

---

19. See Bryant, L. R. (2011). The democracy of objects (First edition). Ann Arbor: Open Humanities Press.

20. Assemblage theory is attributed to Deleuze and Guattari in which material systems self-organize. See Deleuze, G., & Guattari, F. (1987). A thousand plateaus: Capitalism and schizophrenia. Minneapolis: University of Minnesota Press.

complex life forms. It is argued by some virologist that challenging environmental conditions led to the adaptation and evolution of viruses that became increasingly complex and in turn became part of our cellular makeup and our DNA. Many viruses, such as COVID 19 are parasitic and rely on the internal ecology of their host to survive. Their ecosystem is effectively a network assemblage of relationships which descends to the biochemical interactions of RNA and DNA and at the same time ascends to the upper strata of the living host, its food and sustenance, the biospheric, atmospheric, hydrospheric, and lithospheric relationships that bind the assemblage of objects, including the immaterial, all together in a dynamic interplay. The litany of objects gathered together may seem absurd,[21] and yet the extraction of any of them or the addition of new objects can have unintended and unpredictable consequences. Bruno Latour researched this in his influential study into microbes and the assemblage of objects that were drawn together by Louis Pasteur's research. Latour wrote a book published in French, *Les Microbes: guerre et paix* , that translates into *Microbes: war and peace,* and was published in English as, *The Pasteurization of France.* Latour wrote:

"Try to make sense of these series: sunspots, thalwegs, antibodies, carbon spectra; fish, trimmed hedges, desert scenery; "le petit pan de mur jaune," mountain landscapes in India ink, a forest of transepts; lions that the night turns into men, mother goddesses in ivory, totems of ebony. See? We cannot reduce the number or heterogeneity of alliances in this way. Natures mingle with one another and with "us" so thoroughly that we cannot hope to separate them

---

21.   Absurd litanies and lists have become associated with Bruno Latour. See the Pasteurization of France, pp 205-6

and discover clear, unique origins to their powers." [22] This absurdist list is most likely intended by Latour to suggest no matter how scientifically informed we might be about the material causes of a disease or the life of a microbe there could well be immaterial aspects of which the scientific mind has no inkling, such as mythological *lions that the night turns into men* or *mountain landscapes in India ink.* Without a transdisciplinary curiosity and an appreciation of inhuman spacetime evolution, we might well miss clues about objects that weave a strange narrative around the origins of mysterious things such as the COVID 19 virus. The assembly of objects and their network of relations in a symbiotic community of things give rise to novel assemblages. What Latour realised is that this pandemic crisis can ignite the imagination and encourage adaptation even to economic models that long seem outmoded. Latour wrote in an essay about the pandemic:

"The first lesson the coronavirus has taught us," he wrote, "is also the most astounding: we have actually proven that it is possible, in a few weeks, to put an economic system on hold everywhere in the world..."[23] Perhaps viruses can help us to adapt as argued by Moelling and Broecker.

Coal, Canaries and Covid

According to Moelling and Broecker viruses may have even been the first living creatures on Earth and through their

---

22. Latour, B. *The Pasteurization of France.* Cited by Ian Bogost in *Alien Phenomenology, or What It's Like to Be a Thing* (Posthumanities Book 20)

23. Jonathan Watts. (6/6/2020) The Guardian. Bruno Latour: 'This is a global catastrophe that has come from within'. https://www.theguardian.com/world/2020/jun/06/bruno-latour-coronavirus-gaia-hypothesis-climate-crisis

symbiotic evolution eventually contributed to the adaptation and even the imagination of humans as well as the creativity of all other nonhuman living organisms.[24]

We often consider that there is a clear delineation between living and nonliving things and the definition of viruses was once considered to be nonliving because they cannot replicate by themselves. The virologist, Karin Moelling, greatly expands this definition, as he believes viruses occupy the evolutionary twilight zone between the nonliving and the living. Moelling and Broecker believe that the very structure of viruses maybe what was required as an evolutionary precursor to self-replicating organisms.[25] Virology is undergoing a profound move away from the simplistic disease paradigm and today, much the same way that bacteria were once demonised, the *'nasty viruses'* are now being rehabilitated as an essential part of our existence and survival. According to Moelling in his book :

"Everything in virology is new. Unnoticed, a paradigm shift has taken place: the focus is no longer upon diseases, but rather on the positive side of viruses: viruses as drivers of evolution, viruses leading to innovation, viruses at the origin of life — or at least their presence from the very beginning."[26]

Viruses are responsible for all immune systems in living organisms, and in fact, provide us with antiviral defence mechanisms to prevent novel viruses taking over our cells.

24. Moelling K and Broecker F (2019) Viruses and Evolution – Viruses First? A Personal Perspective. Front. Microbiol. 10:523. doi: 10.3389/fmicb.2019.00523

25. Ibid.

26. Karin Moelling. Viruses: More Friends Than Foes (p. 1). World Scientific Publishing Company. Kindle Edition.

If it was not for a retrovirus humans would have to lay eggs to have children. The virologist Moelling wrote:

"Retroviruses can induce immune deficiency, a property, which caused one of the biggest catastrophes of mankind, the AIDS epidemic, caused by HIV. The ability of a virus related to HIV to suppress the immune system allows an embryo to grow within the womb of the mother without rejection by her immune system."[27] About 40 million years ago a retrovirus enabled mammalian species to evolve generating a placenta to provide nutrients within the womb as an alternative to laying eggs.

It is only in the past 100 years that we have been able to differentiate viruses from bacteria. The viral Spanish Flu in 1918 was mistakenly believed to be bacterial. It is now known that these simple viruses perform complex processes within our bodies in order to prevent diseases. They are also our canary in the coal mine alerting us to a number of the problems that have been caused by humanity such as GHGs, overcrowding and deforestation. (Dis)ease is both a result of cell stress, a state of mind, and the first step in our evolutionary adaptation to change due to that environmental distress. COVID 19 is simply one of many diseases to warn us and to help us to adapt to a changing geological planet. If we were to incorporate its DNA, or some other virus, into our DNA it might be a fast track to evolutionary survival on a hotter planet caused by us burning fossil fuels. Viruses are responsible for *all* immunity systems in living organisms and so are our survival life raft. Most vaccines are discovered or created from viruses.[28] Moelling points out that:

27. Moelling, K. (2017) Viruses: More Friends Than Foes (p. 130). World Scientific Publishing Company. Kindle Edition.

28. According to the Simpson Report that reviewed the New Zealand

"Diseases occur when a balance is disturbed and changes of environmental conditions occur through poor hygiene, travelling, overpopulated cities, disappearing forests, water reservoirs, pollution or close contact with other species that carry viruses unfamiliar to us (zoonosis). Microbes not known to an organism may cause diseases without affecting those who are used to them. Most of our human diseases are self-made a strong statement!"[29] Moelling comments that there were two important scientific publications that changed our world view.

"One showed that viruses make up half of our genetic material, our genome, all of our genes, and the other revealed to us the dominance of microorganisms in our body and around us."[30]

All around this planet various governments rely on modelling, statistics and simulations in an attempt to predict and prevent the spread of the COVID-19 virus. However, there is far more to the origin, the spread, and the ultimate co-existence of the virus in our worlds than is discussed by the scientists, policymakers, or politicians. According to Professor, Devi Sridhar of the medical school at Edinburgh University, models have well documented limitations and it is the responsibility of the modelling communities to communicate that to politicians and the public. No one expert, species or thing has the answers to this pandemic, and philosophy is largely missing from the conversation. Sridhar pointed out that modelling is only part of a much bigger puzzle – in principle we need to appreciate modelling is a gross simplification of reality.

---

healthcare system "Over time it is estimated that 90% of cervical cancers could be prevented by human papillomavirus (HPV) immunisation alone."

29.   ibid, p.4.
30.   ibid, pp. 4-5.

He wrote: "Another good principle is one of humility. No discipline has all the answers, and the only way to avoid "groupthink" and blind spots is to ensure representatives with diverse backgrounds and expertise are at the table when major decisions are made. Finally, mathematical models do not include value systems or morals so their outputs must be used cautiously, and with attention to ethics."[31]

We Are C

C could mean COVID-19, carbon, climate, contagion, culture, collective, creativity, or consciousness. Right now it means all of those things, but C ultimately stands for community. I will use an unusual assemblage to illustrate how all things exist within a context or community of relationships with the focus on the COVID-19 assemblage of things. One of the legitimate criticisms of the philosophy that introduced the scientific methodology is reductionism. This powerful thinking tool was a means to take big problems and keep breaking it down into smaller and smaller bite-sized morsels thereby reducing the complexity of the problem and allowing the scientist to solve the component parts. This had massive success and was so popular that it has remained in the scientists' toolbox for 500 years. However, what it cannot explain is how things communicate and interact in a community of things. Simply put you can not hope to understand the Anthropocene by focusing purely on carbon, especially a

31.   Sridhar, D. & Majumder, M.S. (Published 21 April 2020) Modelling the pandemic: Over-reliance on modelling leads to missteps and blind spots in our response. BMJ 2020;369:m1567 doi: 10.1136/bmj.m1567

single carbon atom, let alone the components of carbon, its electrons, protons, or subatomic particle family.

Ecological philosophy introduced an entirely new way to practice science using systems. This new thinking tool allows scientists to consider extremely complex communities of agents and their motivations (human and nonhuman) as holistic systems. Earth System science has revealed interactions and casualties that were hidden from view when only the individual agents, events or resources were examined by themselves, putting them together is also known as agent-based modelling. When we look at the world from this perspective it reveals a multiverse of realities. Every agent, whether it is a carbon atom or a climate pattern, 'observes' their own reality or world. For example, in the case of a carbon atom, it is attracted and repulsed by other elements in the periodic table. The world of carbon is very different to that of hydrogen or a cow yet when they all come together in a community with the help of a grass variety such as clover, a primitive organism known as Archaea, known collectively as methanogens, a powerful GHG, methane is released into the atmosphere helping to heat the planet. Also, the fermentation process, cows, carbon, and hydrogen are just one small part of a much bigger community of agents that produce methane that helps heat the climate. Now you might consider carbon, grass and hydrogen to be resources but they can also be considered agents that act. There is always some activity, but their action is what causes events to occur. Our anthropocentric, or self-obsessed human worldview is often blind to events that are too small, too large, or too fast, and too slow for human sensory perception, and so they end up out of sight and out of mind. Many of the wicked problems we currently face as a planetary community have stemmed from human ignorance and arrogance towards networks. The world of fungi, like viruses, is beyond our comprehension even though we might study them closely, as the mycologist Merlin Sheldrake pointed out:

"Biological realities are never black and white. Why should the stories and metaphors we use to make sense of the world – our investigative tools – be so? Might we be able to expand some of our concepts, such that speaking might not always require a mouth, hearing might not always require ears, and interpreting might not always require a nervous system?" [32]

The philosophy that preceded the empirical scientific method imagined a mechanical universe made up of components that could be mentally removed from the master machine, examined, explained and put back. However, our technological confidence that any problem can be explained as a machine, and fixed like a machine, has led to a wickedly complex network of overlapping worlds, and agents that impact on human worlds, but at scales of time and space that we are only beginning to imagine, and indirectly see, using tools that can map to our human perceptions. Much of the damage humanity has caused to other worlds, including our own, are because we have not had the thinking tools to imagine these other worlds and how they overlap. Simple questions not only lead to simple answers but to technological fixes commissioned and designed by simpletons that cause catastrophic disasters.

In this one chapter I can't possibly hope to give a comprehensive description of the massive community of

---

32.   See Sheldrake, M. (2020) Entangled Life: How Fungi Make Our Worlds, Change Our Minds and Shape Our Futures.

C. What I hope to do is give you a very quick sketch to illustrate, in this cartoon of C, how the philosophy of the OOO or *thing related reality*, can change the way we think and help us to co-exist and co-evolve with the nonhuman majority. At the very least we might lessen the damage we have done to date, and remediate and partially restore their worlds. If 'We are C' then, to borrow the phrase of the New Zealand Ministry of Health, *'we need to check who is in our bubble?'* The Big C that has got our attention at the moment is COVID 19. It is a pandemic that continues to evolve even as the world undergoes mass vaccination, and it has already forced us to adapt with it. We are currently in an uncomfortable co-existence with a coronavirus, but it has always been part of our community, we just couldn't see who it was and understand that we can't contain it.

Therefore if COVID 19 has managed to get inside our bubble, or more accurately we have joined its bubble, then how might we try and understand or describe their world and the intersection with ours? Moelling corrects our impression of viruses, he writes: "We are the invaders in a world of microorganisms, and not the other way round."[33] We will pick up the story around 4 billion years ago. The reason I am starting here is that there is some evidence that this was when life first began on this planet.[34] Ironically, as I have discussed earlier a virus is considered to be a transitional thing between the nonliving and living. According to Moelling there is good evidence because of their zombie-like or dead/undead existence, viruses were the beginning of all life on Earth, and our evolutionary ancestors. Viruses most likely began life as extremophiles, hidden from the energy of the Sun, much like those that

33.   Moelling, K. (2017) ibid.
34.   See chapter 2, The Story of Carbon

can exist 200 meters deep, within hydrothermal vents at 400° C. They are the oldest living things on our planet, yet while they cannot replicate, there is evidence that they once were capable of it. Moelling wrote: "That is almost generally accepted today. In the oceans, 200 meters deep and beyond, there is no sunlight any more, so it is not the sun that supplied the energy: rather, it was chemical reactions that provided the energy for early life. The beginning of life, without sunlight was based on chemical energy production. That was the motor for life. In the opinion of many scientists, it was there that the first biomolecules, such as RNA, arose."[35]

So rather than considering viruses as our enemies, we should consider them as our friends. While viruses makeup half of our genetic material and have been found in every living organism on our planet, and we are all interrelated – this is a very Big C. Viruses if they could be isolated and removed from their hosts, would make up the biggest biomass on Earth. Moelling points out that "Humans are a superorganism, a complete ecosystem. Healthy humans comprise 10^{13} cells which are authentically human, our "self", and in addition we host about 10^{14} bacteria and, in addition, at least ten to a hundred times more viruses."[36]

Right now COVID 19 threatens human lives, and yet the question remains why do so few people die when it is so contagious ? The survival of the human species, from the pandemic, is not in question. Despite the growing mortality rate, it is considered highly unlikely that humanity will meet its end due to COVID 19. The discovery of the virus is so recent that scientists still don't have a definitive answer, however, research into its coronavirus

35.   Moelling, K. (2017) ibid.

36.   Moelling, ibid. Kindle Edition, Location 310.

relatives such as SARS, the common cold, and other viral diseases have given them clues.

I am going to attempt to track some of the actants, resources, and events of the pandemic, and the Anthropocene as an example of how agent-based modelling might provide a lens for understanding the wicked world of Zero Carbon. My analysis and narrative can only be cursory as it would struggle to be confined to several volumes many times the size of this intended book. You may find it a little strange but I will also weave myself into this narrative to try to illuminate how capitalism implicates us all. The Western tradition of stories, myths and legends typically assume a linear structure, although they can also jump around in time, focusing on different characters. It is, however, fair to assume that complexity increased with time and that older oral storytelling tended to be more simple, and more linear. There are striking similarities between myths and legends within Western culture that are also shared with Eastern and Oceanic traditions. This has been noted by a number of literary scholars, psychologists, and archaeologists of ancient cultures such as Sir James George Frazer, Joseph Campbell, Carl Jung, Sigmund Freud, and Stephen Booker to name some who might be familiar to a modern audience. According to the English scholar Brian Boyd, these stories served an evolutionary purpose in order to quickly and simply convey the wisdom of elders about their social and natural environment.[37] Dramatic, linear, and emotionally impactful stories educated and entertained younger people about the dos and don'ts of survival in a community. However, linear tales are not necessarily, or even at all

---

37. See Boyd, B. (2009). On the Origin of Stories: Evolution, cognition, and fiction. Cambridge, Mass.: Belknap Press of Harvard University Press.

likely, reflections of the complex inter-relationships of objects in the material world. Even an object that may seem basic such as a rock is in continual 'relations' with the atmosphere, the biosphere, the hydrosphere, and of course, its closest of kin the lithosphere. Even a rock buried deep beneath the earth is most likely home to some microbial family, deprived of oxygen this rock has other atmospheric gases it plays with, and while water may be scarce today it is highly likely that at one time in the rocks life it was either underwater, or guiding water to a lake or an ocean. The rocks around our rock will have been grinding and moving our friend the rock around over the course of millennia. All of these relationships take place in multidimensional spacetime beyond the full understanding, and certainly control, of even the most ecologically astute geologist. The COVID 19 pandemic has once again reminded us that we are not the 'masters of the universe' but bit part players in a tragic comedy. Thacker points out:

"Our very concepts regarding disasters generally betray a profound anxiety. That some disasters are "natural" while others are not implies a hypothetical line between the disaster that can be prevented (and thus controlled), and the disaster that cannot. The case of infectious diseases is similar, except that the agency or the activity of this "biological disaster" courses through human beings themselves – within bodies, between bodies, and through the networks of global transit and exchange that form bodies politic."[38]

So, I am going to tell you a short story about Wuhan that will hopefully show the ludicrous simplicity of political narratives that apply Simple Simon's logic. The Wuhan

38.   Thacker, E. ibid.

assemblage may seem absurd and yet if you engage your imagination you may also gain insights about the relationships of zero carbon, COVID 19, culture and the process of worldbending, breaking and building. The world that is constructed in the naming, and connecting of the assembled nodes of agents or actors, events, and resources will define the boundaries of that world. However, it will hopefully be obvious that no matter how tightly you try to define what, when, how, and who, the world is still full of porous holes and withdrawn things. The universe, and our planet are full of multiple worlds that are in flux and are not static or hermetically sealed. As Bogost wrote:

"The moment we try to arrest a thing, we turn it into a world with edges and boundaries. To the hammer everything looks like a nail. To the human animal, the soybean and the gasoline look inert, safe, innocuous. But to the soil, to the piston? Ethical judgment itself proves a metaphorism, an attempt to reconcile the being of one unit in terms of another. We mistake it for the object's withdrawn essence."[39] The way we sense phenomena, animal, plant, fungi, or mineral, and the way we communicate with them are hopelessly inadequate. Sheldrake, after many years of studying fungi wrote: "Our descriptions warp and deform the phenomena we describe, but sometimes this is the only way to talk about features of the world: to say what they are like, but are not."[40]

Wuhan Fun

Let us begin by considering some of the things that we

---

39. Bogost, I. (2012) Alien Phenomenology, or What It's Like to Be a Thing. Posthumanities Book 20

40. Sheldrake. M. (2020) ibid.

can know about the COVID 19 pandemic such as where it was first discovered, in the city of Wuhan. The Chinese city of Wuhan became internationally famous after the Wuhan Municipal Health Commission in China reported a cluster of cases of pneumonia in Wuhan, Hubei Province, on the 31st December 2019. A novel coronavirus was eventually identified. On the 4th of January 2020 the WHO (World Health Organisation) reported on social media that there was 'a cluster of pneumonia cases – with no deaths – in Wuhan, Hubei province.'[41] On the 5th January 2020 the WHO published its first Disease Outbreak News on COVID 19.[42] However, despite this apparently speedy reaction to the Chinese reports, it has emerged that China was attempting to cover-up the outbreak. According to the Pulitzer prize winner, Laurie Garrett: "Nothing is known about the earliest case later discovered in Wuhan—the individual who was hospitalized on November 17—but it's now clear that 266 Chinese were suffering from the coronavirus before the Western New Year's Eve." The first officially reported case was a patient who experienced symptoms on December the 1st 2019 and another 13 patients in this first group had not visited the official epicentre of the outbreak the Hua'nan seafood and poultry market. Garrett surmises that: "This suggests that there was already widespread community transmission inside Wuhan well before the Christmas season."[43] China reported its first death to COVID 19 on the 11th January 2020. The

41. https://www.who.int/news-room/detail/27-04-2020-who-timeline---covid-19

42. https://www.who.int/news-room/detail/27-04-2020-who-timeline---covid-19

43. See Garrett, L. (2020) online at New Republic. https://newrepublic.com/article/157118/trump-xi-jinping-america-china-blame-coronavirus-pandemic

61-year-old man who died was a regular customer at the Hua'nan market. The report came just before one of China's biggest holidays when hundreds of millions travel around China. Wuhan, a city of more than 11 million, is a major travel hub for flying and rapid train travel. By the 23rd of January, the whole of Wuhan was cut off from the rest of China with a travel ban in and out.

Patient Zero

There has been some urgency and intense interest to try and find what is known as *'patient zero'*. However, the situation in Wuhan illustrates the complexity and difficulty of trying to track down the original cause of a pandemic. Politics, social behaviour, ecology, zoology, microbiology and economics are just some of the epidemiological *actants* in a sprawling network of relationships that have spread throughout the world. The original source of the Spanish Flu is still not definitively known, and while the hypothesis that the COVID 19 outbreak was 'transmitted from living animal to a human host' is considered by the WHO to be the most likely, it has not been proven. The 2014 to 2016 Ebola outbreak in West Africa killed in excess of 11,000 people and spread to 10 countries including the US, Spain, the UK and Italy. It was thought to have originated with a two-year-old boy from Guinea who 'may have been infected by playing in a hollow tree housing a colony of bats.' The stigma of *'patient zero'* was felt keenly by Gaetan Dugas, a Canadian gay flight attendant and is regarded as 'one of the most demonised patients in history'. He was blamed for spreading HIV to the US in the 1980s but a 2016 study showed the pandemic originated in the Caribbean and spread to the US in the 1970s. In 1906, Mary Mallon, a cook for wealthy families in New York, became well known as *Typhoid Mary*, and was blamed for spreading typhoid fever amongst the rich of New York. Interestingly, it is the very origins of the phrase itself, *Patient Zero*, that signals a red flag amongst all reports of linear casualties describing events, resources, and agents both human or nonhuman. *Patient Zero* was itself the result of data corruption as in the classic Terry Gilliam movie, *Brazil* (1985), in which the data was corrupted when a fly was splattered by the teletype machine thus turning a T into a B changing the fate of the hapless Mr Buttle. This was no doubt based on the true story of the first computer bug, commonly thought of as a glitch caused by a coder.

On September 9, 1947 a group of computer engineers opened up the Mark II computer to ascertain what was causing its errors – they discovered a trapped moth that had caused the machine errors.[44]

Patient Zero was created by investigators who were searching for the origins of HIV in San Francisco and Los Angeles in the early 80s, the CDC, (Centers for Disease Control and Prevention), used the letter O to refer to cases *'Outside of the state of California.'* However, the letter O was misinterpreted to be the number o or to further disambiguate *zero*, 'and so the concept of patient zero was born.'[45] The example of the moth and patient zero illustrates how random data errors can result from very material relationships and even a name can be misspelled

---

44.   Sep 9, 1947 CE: World's First Computer Bug.
      https://www.nationalgeographic.org/thisday/sep9/worlds-first-computer-bug/#:~:text=On%20September%209%2C%201947%2C%20a,bug%20id
      entified%20in%20a%20computer.

45.   Duarte, F. 24th February 2020. Who is 'patient zero' in the coronavirus outbreak? BBC Future. https://www.bbc.com/future/article/
      20200221-coronavirus-the-harmful-hunt-for-covid-19s-patient-zero

with unintended consequences. The worlds of things includes bugs, computers, and names and they all interact even if no one is looking.

What's in a Name?

Incorrect data can propagate *'alternative facts'* that define new worlds. Social media and populist leaders have conspired with an assemblage of many other things to create a crisis of health, hate, and truth.[46] In a *'post-truth'* world, President Trump has done as much as anyone in an attempt to politicise and weaponise the name of the pandemic. Trump has called COVID 19 the *'China Virus'* and *'Kung Flu'*, he also used the hashtag *'#WuhanVirus'* which became popular in the US.

We have not only the pandemic to contend with but also an *'infodemic'*. According to a research paper on unreliable and untrustworthy information in Aotearoa: "Local, regional, and global infodemics map to the spread of the pandemic (Islam et al., 2020). An infodemic is "an over -abundance of information—some accurate and some not—that makes it hard for people to find trustworthy sources and reliable guidance when they need it" (World Health Organisation, 2020a).[47]

Historically, it has often been the location of the first known cases of the pandemic that has given it its title.

---

46.   (8/06/2020) Prince Harry: Social Media is dividing us. Together, we can redesign it. Fast Company. https://www.fastcompany.com/90537682/prince-harry-social-media-is-dividing-us-together-we-can-redesign-it

47.   N:B This paper has not been peer reviewed. Max Soar, Victoria Louise Smith, M.R.X. Dentith , Daniel Barnett, Kate Hannah, Giulio Valentino Dalla Riva , Andrew Sporle. (6th Sept. 2020) Evaluating the infodemic: assessing the prevalence and nature of COVID- 19 unreliable and untrustworthy information in Aotearoa New Zealand's social media, January-August 2020. Te Pūnaha Matatini: Centre of Research Excellence for Complex Systems and Networks, New Zealand

Often the location has been wrong and the reason was either a mistake or often, political. *The Spanish Flu* most likely originated in a Kansas military camp then spread to Europe. The Spanish blamed the Italians and called it the *Naples Solider* after a popular operetta show tune by the same name because the song was said to be as 'catchy as the flu'.

The name given to an event, such as a pandemic are often designed to deflect and target an alien agent to rally political supporters behind governments who want to avoid responsibility for their own disease mismanagement and or channel xenophobic hatred as a powerful weapon to rally unity behind those in power in their own country.

According to Spinney, who wrote about the Spanish Flu, it most likely began in the US and then travelled to the trenches of France during WWI. Then, with a little nudging from their governments, the French, British and Americans started calling it the Spanish Flu. Other countries blamed others, as Spinney wrote:

"In Senegal it was the Brazilian flu and in Brazil the German flu, while the Danes thought it 'came from the south'. The Poles called it the Bolshevik disease, the Persians blamed the British, and the Japanese blamed their wrestlers: after it first broke out at a sumo tournament, they dubbed it 'sumo flu'."[48]

Spinney went on to write that naming is the first priority to controlling the menace of contagious disease. When it is incorrectly labelled it can lead to stigmas such as AIDS that was incorrectly named the *gay plague* and stigmatised the homosexual community. *Ebola* came from the River Ebola in Central Africa, but an epidemic outbreak happened in

---

48.   Spinney, Laura. Pale Rider (Kindle Locations 836-838). Random House. Kindle Edition.

West Africa in 2014. *Zika* was first named after a forest in Uganda where it was first isolated in 1947 but spread around the world and in 2017 was considered a major threat in the Americas.

To try to avoid the politicisation of diseases the WHO in 2015 issued guidelines stipulating that disease names should not make reference to specific places, people, animals, or food, and should not include words that engender fear, such as 'fatal' or 'unknown'. As I have earlier noted zero, the void, and the nothingness of the unknown is closely related to fear and the horror of the abyss – at least in Western culture.

To attempt to avoid the politicisation of the virus the WHO named it COVID 19, previously know as 2019-nCoV, now otherwise known as SARS-CoV-2, it is a strain of coronavirus similar to SARS and was discovered in 2019. It gets the prefix *corona* because the virus has a crown-like structure. According to an article in MIT's *Technology Review*:

"A SARS-CoV-2 virion (a single virus particle) is about 80 nanometers in diameter. The pathogen is a member of the coronavirus family, which includes the viruses responsible for SARS and MERS infections. Each virion is a sphere of protein protecting a ball of RNA, the virus's genetic code. It's covered by spiky protrusions, which are in turn enveloped in a layer of fat (the reason soap does a good job of destroying the virus)."[49]

The size of a single virus particle is extremely small, 1 nanometer is 1 billionth, or 10^{-9} of a meter. The protein spikes on the virus attach to a protein on the surface of a

49.   Patel, N. (2020). How does the coronavirus work? Technology Review. MIT. https://www.technologyreview.com/2020/04/15/999476/explainer-how-does-the-coronavirus-work/

cell known as ACE2 usually used to regulate blood pressure. However, when COVID 19 binds to the ACE2 it starts a chemical chain of events that fuses the membranes around the cell with the virus forcing an entry into the cell which allows the injection of the virus's RNA genetic material. This is a typical parasitic viral approach to setting up a new home and replicating itself by using the cell's own protein-making process, thus creating tens of thousands of new virions in only hours of inhabiting the human cell. From there it can rapidly replicate itself to infect lots of other healthy cells. The host's body temperature rises as another line of defence in an attempt to kill off the virus. White blood cells rush to the scene to pursue the infection and some will ingest and destroy the infected cells while others create antibodies. In this way, they inhibit further infection by blocking the virions, and others go on to make toxic chemicals that kill off the infected cells.[50]

Those with existing respiratory or weakened immunity may suffer a life-threatening illness or if they only get an upper respiratory infection they may go on to develop the first stage of a co-existence with our new friend who could eventually contribute to a more resilient human population as we adapt to our rapidly changing environment.

COVID Kith and Kin

Our co-existence with our tiny family of viruses will also depend on how quickly the virus mutates and adapts to the changing environment. There is much hype around vaccines, but even if we are successfully vaccinated and the virus changes, we may lose immunity.

What is harder to anticipate is that because of our

---

50.   Patel, N. (2020) ibid.

lifestyle pandemics will follow other environmental disasters – co-existence is not easy when the network is fast-moving and dynamic, our kith and kin may become unrecognisable very quickly, or we could get a totally new crowd in from out of town.

The paradox that we face in naming and attempting to understand, or explain an other *thing* is that rather than *being* static all things are in a continual process of becoming which according to Whitehead is the creative evolution of the universe. This is something, neither as an individual or as a species, we are capable of stopping. As Donna Haraway points out:

"The task is to make kin in lines of inventive connection as a practice of learning to live and die well with each other in a thick present. Our task is to make trouble, to stir up potent response to devastating events, as well as to settle troubled waters and rebuild quiet places. In urgent times, many of us are tempted to address trouble in terms of making an imagined future safe, of stopping something from happening that looms in the future, of clearing away the present and the past in order to make futures for coming generations. Staying with the trouble does not require such a relationship to times called the future. In fact, staying with the trouble requires learning to be truly present, not as a vanishing pivot between awful or edenic pasts and apocalyptic or salvific futures, but as mortal critters entwined in myriad unfinished configurations of places, times, matters, meanings."[51]

We need to study the family tree of not just coronaviruses but a massive network of other agents,

---

51.   Haraway, Donna J.. Staying with the Trouble: Making Kin in the Chthulucene (Experimental Futures) (p. 1). Duke University Press. Kindle Edition.

events, and their resources (in the case of viruses otherwise known as reservoirs) all of which are in their own process of becoming, constantly in flux.

Like SARS, MERS, AIDS, and Ebola, COVID 19 is a zoonotic disease, that can jump from another species to human hosts. One theory is that this is what happened in Wuhan and is thought that it might have spread from bats, pangolin or maybe some other species. According to scientists studying COVID 19 it is unlikely that it spread from bats to humans, but rather like SARS it got there by way of intermediate species, in the case of SARS, the masked palm civet[52] before getting to us. This is what originally happened with SARS. Scientists using genome sequencing to track the lineage of the various strains of the disease believe that SARS, (that is 76 percent identical to SARS-CoV-2 or COVID 19) was spread by bats, and SARS-CoV-2 is the closest relative to the horseshoe bat virus that shares 96% of its genome with samples collected in 2013.

However, that small 4% variance represents decades of steady mutation and that the horseshoe bats from Yunnan province, almost 1,900 kilometers away from Wuhan, means that the origin of SARS is 50 years old. The next theory was that at 99% genetic concordance the evidence then pointed to the scaly pangolin, also known as the scaly anteater, although scientists believe that there may be others, and have ruled out the specific pangolin found in Wuhan wet markets. It is most likely, because of earlier human infections that it was passed from human to human

52.   Isolation and Characterization of Viruses Related to the SARS Coronavirus from Animals in Southern China, Science 10 Oct 2003: Vol. 302, Issue 5643, pp. 276-278 DOI: 10.1126/science.1087139

in Wuhan.[53] It is still not known what the original animal reservoir is.[54]

It is of some urgency because as noted by scientists: "The identity of the intermediate host of SARS-CoV-2 is therefore a mystery that many researchers hope to solve, as knowing the intermediate host is very helpful for the prevention of further spread of the epidemic."[55]

The extraordinary thing about the civets, bats and the pangolin is that they have already adapted and evolved to happily host this virus – they are its kith and kin. Bats, civets, and pangolin are like humans, they are mammals, making it easier to jump to our species, the pangolin scales are made from keratin, the same protein that makes human fingernails.

The pangolin, the only mammal with scales, and found in Asia and Africa, is killed to harvest its scales to make traditional Chinese medicines to boost fertility, while the meat is considered a delicacy. Pangolin is the most trafficked mammal in the world. While in May 2020, the pangolin, that is threatened with extinction, was removed from China's officially approved ingredients for traditional medicines, such prohibitions have been shown to be ineffective in combating the illegal trade in medicines using banned species in China. The Chinese pangolin is almost non-existent in China. As animals like the pangolin

---

53. See Daalder, M. (Oct. 5, 2020) 'COVID 19 - The NZ strains: Our second wave.' Newsroom. https://www.newsroom.co.nz/the-nz-strains-our-second-wave

54. Smriti Mallapaty. May 18, 2020. Animal source of the coronavirus continues to elude scientists. Nature.https://www.nature.com/articles/d41586-020-01449-8

55. Yang Zhang, Chengxin Zhang & Wei Zheng, THE CONVERSATION (11 APRIL 2020) More Evidence Suggests Pangolins May Have Passed Coronavirus From Bats to Humans https://www.sciencealert.com/more-evidence-suggests-pangolins-may-have-passed-coronavirus-from-bats-to-humans

are often hunted in remote areas where there are few humans there has been little opportunity for the virus to *'get to know us'*.

What virologists have found is that with other viruses they have, over the course of many generations, become endemic and endogenous within humans. Endogenization is a process whereby a viral sequence becomes part of the genome of the host animal and therefore no longer causes illness. Thanks to reverse transcriptase endogenous viruses make up 50% of our genome and are an important part of our immunity to exogenous viruses that they keep out. In other words, viruses who have already shared their genetic material with us have become permanent, and therefore, harmless in humans. We have adapted to their presence, yet there are still new opportunities for other viruses that we have not yet adapted to. According to Fiona Gordon an environmental policy analyst and an ambassador at the Jane Goodall Institute in New Zealand: "Humankind has now poked so many holes in nature's firewalls that coronaviruses can now just walk right on through," her report says. "From the destruction of forests to our relentless encroachment into pristine habitats – in our failure to conserve the natural world, we have failed ourselves as well."[56]

Gordon warns that our inability, even in remote New Zealand, to prevent the illegal trade of *'wildlife'* has enabled COVID 19 and could well open our borders to the next pandemic after this. She explains that there are literally hundreds of endogenous coronaviruses circulating in carrier animals such as pigs, camels, bats, [birds] and cats.

---

56.   Gordon, F. (10 June, 2020) Covid-19 and New Zealand's Role in the Global Illegal Wildlife Trade https://africanelephantjournal.com/covid-19-and-new-zealands-role-in-the-global-illegal-wildlife-trade/

If humans have had limited contact with carrier animals then there is a greater chance for the transmission of a novel coronavirus that our body is not yet immune to. Thanks to the speed with which we move around the planet, our proximity to the wilderness, and our compulsive sociability, we are the ultimate super-spreaders of viral diseases.

What it's like to be COVID 19?

The question of what it's like to be this particular coronavirus sounds absurd but if we are going to adapt to its world we need to at least appreciate that like us COVID 19 faces the question of how to survive? The best that we have achieved so far in getting to know it is that we have some educated guesses around how it slipped into our bloodstream. We know a little about how it might enjoy a symbiotic relationship with a bat (but not so much), and how it might have jumped species to live with pangolin, but we don't have definitive proof of this. Philosophically this should alert us to our insurmountable ignorance about any *thing* else, including inanimate objects, in which we come into contact with.

In the documentary, *My Teacher Octopus*, (2020), Craig Foster, a diver and filmmaker followed an octopus for a year and became astounded by how little was known about how they think, and their enormous intelligence. As he described his experience, '*she had ignited my imagination*' and he began '*thinking like an octopus*'. However, he acknowledged how challenging it was, and how incomplete his experience was because her mind was so different. According to Foster, two thirds of an octopus' brain is not centralised but is distributed outside of her brain and in her arms, with two thousand suckers attached that taste, touch and probe their environment. Foster found his 'teacher', the octopus, to be playful, curious, friendly, social, and incredibly creative. After millions of years of evolution the octopus has only 1 year to live and must use all of the lessons from its evolution and adaptation to survive.

Like an octopus, a horseshoe bat, that might be infected with SARS CoV-2 is ultimately unknowable to us and to itself. Thomas Nagel wrote a famous philosophical paper, *What it's like to be a bat?* (1974) that outlines the difficulty of knowing a bat. Surely, to get to know COVID 19 we might have to also learn what it's like to be a pangolin? To be a pangolin scale? To be the medium that transmitted COVID 19 from a bat to a pangolin and then perhaps to a human? What is it like to be COVID 19? Nagle still approached a bat from an anthropocentric point of view but he did highlight the difficulty, if not the impossibility for us to know what it is like to sense our environment like a bat, and to be infected with COVID like a bat?

As Ian Bogost wrote in his book, *Alien Phenomenology, or What it's Like to Be a Thing*, it may seem to be objective and easy to measure, record or otherwise externally observe phenomena but: "such an observer cannot have the experience that corresponds with those phenomena, no matter how much evidence he or she might collect from its event horizon."[57] The scientists may succeed in finding the superficial aetiology of COVID 19 but there is always evidence withheld or withdrawn by the virus, by the environment, by events missed and agents unmet. Reductionism can only ever approach the thing, and apply a metaphor to the experience i.e. to 'see' *like* a bat or for

---

57.   See Bogost, I. (2012) Alien Phenomenology, or What it's Like to Be a Thing

a bat to 'hear' *like* a human is never exact, but rather an approximation using a measure that is nothing like the *reality* experienced by the bat or COVID 19. Their world is very different, and they do not even speak our language, or have the same sensory perceptions – they do not share our means of communication. As cited by Sheldrake: "By contrast, in English, writes Kimmerer, there is no way to recognise the 'simple existence of another living being'. If you're not a human subject, by default you're an inanimate object: an 'it', a 'mere thing'. If you repurpose a human concept to help make sense of the life of a non-human organism, you've tumbled into the trap of anthropomorphism. Use 'it', and you've objectified the organism, and fallen into a different kind of trap."[58]

China's Demons

Between 2007 and 2015 I travelled to China many times venturing beyond Beijing, Shanghai and Guangzhou, visiting some more out of the way places that were still home to millions of people such as Wuhan. I also travelled by train from Beijing to Shanghai and saw a countryside that seemed empty of so-called wildlife. I had attended numerous official banquets as leader of a number of film and mayoral trade missions. As the President of a Chinese animation company, Mili Pictures Worldwide, I would often dine with the executive team and we would entertain business partners in lavish style. You don't have to do this very often before you realise with a shock that China must consume vast volumes of food that is bought, harvested, slaughtered and consumed with huge gusto. Anyone that

58. Sheldrake, M. (2020) Entangled Life: How Fungi Make Our Worlds, Change Our Minds and Shape Our Futures.

has observed the super cities of China and extrapolated the daily appetite of 1.4 billion people can understand why the countryside seems eerily empty of any nonhuman majority. Once I travelled to the largest outdoor studio in the world to shoot a promotional video for my partners, Hengdian World Studios. It was a 2-hour high-speed train ride from Shanghai then I travelled in a bumpy van for an hour and a half into the country to arrive at my destination. It may have been a mistake or a bit of fun at the expense of the 'foreign devils' known as gwailo (ghost person) in Guangdong Wah, or Cantonese, but my friends and I were served a steaming hot glass bowl of pick and mix fried crickets, bugs, and pupae. It was then that I began to understand that in China anything that could be eaten was.

I had visited a number of animal markets around China and like some other Asian countries there is a premium placed on fresh, and often live animals to be sold for eating. China has suffered massive famines such as one from 1959 – 1962 that killed 30 million people by starvation. Once I visited the pet market in Shanghai, with the production designer, Grant Major; it was a massive assault on the senses. We heard the high pitched yelping of puppies, crickets trapped and chirping in intricate bamboo cages, scorpions danced and silent snakes slithered peering out from their terrariums. I was therefore somewhat surprised when I first heard about the pangolins in the wet markets of Wuhan, this was something I had not yet seen. I assume that they were not domesticated because feeding them an endless supply of ants would test anyone's patience but I admit I am largely ignorant of their habits. I did know that the Chinese were well-practised in animal domestication. I had once asked why the ducks on Chinese farms don't fly away and was told 'because they have been bred so long in captivity they have forgotten how to fly'. It turns out that the pangolins that were tested for the virus had come from Malaysia and were probably not the origin of the disease. Just tracing the etiological origins of a pandemic is fraught but to get the full picture of any phenomena is exponentially complex if not wholly impossible.

It has been at least six years since I met the Mayor of Wuhan in Beijing and was invited to visit his city. If you watched the news on TV about the COVID 19 outbreak occurring in Wuhan, and saw the less than flattering news footage of the wet markets, you could be excused for thinking this was some country backwater. It is regarded as anything but that in China. The capital of the famous Hubei province, is known as the 'cradle of Chinese culture' it is an ancient place that is home to over 11 million residents; the major research and production centre for fibre optics; some of the top science and technology university research facilities; Mao's summerhouse was located on Wuhan's wonderful East Lake; and an incredible museum that houses the famous sword of Goujian that is still razor sharp and untarnished after two and half thousand years.

Despite Mao's ruthless Cultural Revolution (1966-76), that attempted to stamp out 'the old ways, the four olds: old thinking, old culture, old customs, and old habits,'[59] China has seen a resurgence of interest in TCM, (traditional Chinese medicine), mythology, and superstitions. While Wuhan may not have been widely known in the West before the pandemic, in many ways it is a good place to examine the *actants* of COVID 19 because of the strange assemblage of things that span spacetime, COVID 19 is a hyperobject.

---

59. S. A. Smith, Talking Toads and Chinless Ghosts: The Politics of "Superstitious" Rumors in the People's Republic of China, 1961–1965, The American Historical Review, Volume 111, Issue 2, April 2006, Pages 405–427, https://doi.org/10.1086/ahr.111.2.405

I had met with the Mayor of Wuhan as a favour to the Chairman of Mili Pictures, Jack Zhang. I had met a couple of Jack's production team in Qingdao at a large animation school with Grant Major. Grant had won an academy award for his production design of *Lord of the Rings* and we were conducting a workshop with my production partners, Phoenixland Studios, to develop characters for my fantasy movie, *Two Dragons*. I had become aware that Chinese audiences, like many throughout the world, had a rekindled fascination for fantasy in the guise of movies such as *Lord of the Rings*, directed by New Zealander, Peter Jackson. It seemed the Chinese were longing to see their own myths and legends writ large on the silver screen. Following the workshop Grant and I were invited to Suzhou to speak at a visual effects symposium hosted by Mili Pictures which is where I met the Chairman of Mili, Jack Zhang. Short, chubby and boyish, Jack, was an ambitious member of the Chinese Communist Party, he would become excited and animated when he talked about the superheroes of Chinese legends. Somehow he managed to shake off the dusty party ideology that warned of the old ways, and just like a *Transformer* toy, Marx's *scientific materialism* was switched out for a teenage fantasy version of the Chinese classic, *Creation of the Gods*, with Jack's superheroes throwing thunderbolts and running across the clouds. He would excitedly puff on his perpetual cigarettes as he accompanied his confusing story of how the gods, that I could not pronounce, came to be. This was all accompanied by his sound effects 'piow piow bam bam bam!' predictably ending with a cataclysmic explosion and the hero's triumph as he giggled with delight and pride. Surely, he told me, this was a story that the world would love to see as a movie. I was intrigued and excited to accept his offer of employment, but thought better of the position of Chairman, and instead suggested President of Mili Pictures with the responsibility to transform the company into an international entertainment business to be called, Mili Pictures Worldwide.

What does this have to do with COVID 19 and what Bruno Latour might well have called the *Pathologisation of China?* In *Worldbending* I had argued that our ancient myths and legends shadow the algorithms, common sense, and ideologies of our capitalist world and that apocalyptic fantasies lurk around the backstreets of our mind ready to jump out and mug us. It is important that we pay attention to the narratives and language used when we discuss wicked problems such as the pandemic. Capitalism and liberal individualism continue to shape our conversations and even our science and how we think. In the paper on infodemics it was noted:

"Recent Aotearoa New Zealand research on effective policy communication highlights the challenges for communicating effectively about COVID-19 within these contexts, as audiences bring individualistic personal values or frames to bear on the overwhelming plethora of information presented in social and mainstream media (Berentson-Shaw, 2020). Concerns around effectiveness of medical interventions, particularly those promoting herd immunity or 'secret' cures, are representative of an individualistic mindset, and an us-vs-them mentality, rather than a cooperative community approach."[60]

While no *sensible* adult will come out and seriously discuss ghosts, demons, and gods with superpowers, it is remarkable how often these archetypes leach out in news

---

60.   Soar, M. et. al (6th Sept. 2020) ibid.

reports, political rhetoric, economic theory, and even scientific papers. Mao's China had a really good go at purging these superstitions using brutal murders, torture, and imprisonment for those who had lapsed into the 'old ways'. Yet despite being a fascist dictator, Mao Zedong, who would have denied the boyish, Jack Zhang, his fantasies, was someone that Jack, like a majority of CCP, still thanked Mao for his materialist revolution that had pulled China out of the mud of the mid-20th century. So why was Jack comfortable to be talking openly about ancient myths and legends that would have had him imprisoned, reeducated or shot as a counterrevolutionary during the Cultural Revolution?

The demons have risen from the dead, and I can imagine Mao is probably turning in his grave. Philosophically, and economically China has developed a new historical materialism. In the 21st century, the spectre of Marx's, 'Communism', has been banished in China, replaced by the material trinkets and shiny toys of state capitalism. Ironically, today I also heard a commentator for the capitalist mouthpiece *Bloomberg* refer to the US economy as state capitalism due to the injection of cash into the sagging American economy. For China state capitalism is not a paradox widely acknowledged by the many Chinese entrepreneurs I met. This was a new *thing related reality*. In ontological terms, it would be what Eugene Thacker has dubbed *demontology*. A demonic view of reality in which demons are not just some evil spirits but represent an ontological or philosophical horror of the unknown, a fear of an uncertain future that could suck China back into the black hole of the humiliating past two hundred years. Thacker's book explores the relationship between philosophy and horror, through the motif of the 'unthinkable world'. Movies can create 'what-if' worlds by creating speculative realities and yet that also poses a risk for Chinese technocrats who prefer a singular reality, even if it is President Xi's *China Dream*. The Chinese movie industry still struggles to understand its own audience, and while the CCP's hatred for America grows to an ideological fever pitch, Trump's government whipped up a reciprocal animosity, and yet at the same time Hollywood movies continue to dominate the Chinese box office. Archetypal myths and legends have become more and more popular around the world from the *Marvel* comic universe to the ongoing *Stars Wars* franchise that follows the plot of Joseph Campbell's *Hero's Journey*. For the Chinese film officials that approve movie releases today Chinese mythology may seem a harmless indulgence popular with the people who have forgotten the strictures of the Cultural Revolution. What they do not seem to notice is the similarities in the subconscious purpose of the genre based on universal archetypes. To them this is simply a way to promote merchandise and imitate the capitalist objectives of Hollywood propaganda chained to the Chinese ideology but repurposed for the Chinese people. Strangely, this was something that the Disney corporation had taught them. As Thacker has pointed out in Hollywood:

"the mythological has become the stuff of the culture industries, spinning out big-budget, computer-generated films and merchandise; the theological has diffused into political ideology and the fanaticism of religious conflict; and the existential has been re-purposed into self-help and the therapeutics of consumerism."[61] China is beginning to study the ancillary markets of the Hollywood movie

---

61.   Thacker, Eugene. In the Dust of This Planet: Horror of Philosophy vol. 1 (p. 3). John Hunt Publishing. Kindle Edition.

industry and despite some residual ideological struggles are in the process of creating a replica of the US consumer *'dream'* using marketing and brand building.

Mao Zedong had banned many traditional Chinese operas, and in particular what was known as *Ghost Dramas*, in which the victims of murder took revenge on the living that had killed them. In March 1963 Mao had banned these ghostly stories. According to their biography of Mao, Chang and Halliday, wrote that Mao: "having just been the agent of tens of millions of deaths, he regarded these on-stage avengers as uncomfortably close to reality."[62] By the end of 1963 Mao accused all art forms, including cinema, of being *'feudal, capitalist* and *very murky'*. [63] Mao feared the past and it was not just human ghosts he feared but the nonhuman majority that lurked in the shadows of ancient buildings, monuments, walls and temples that fell to his destructive whims. For Mao they were the hiding place of the material and immaterial spectres of things from the past. In trying to make fantasy movies in China I discovered that forty years after Mao's death the Communist Party of China was still frightened of Mao and the ghostly things that might seek revenge upon them. Not only were ghost stories still unofficially banned but time travel was also a no no.[64]

When I visited Wuhan in 2013 all official co-production movies had to be approved by the CFCC, (China Film Co-production Corporation) which was a government organisation run by the massive and very powerful

---

62.   Chang, J., Halliday, J., & Vintage (Londyn). (2007). Mao: The unknown story. Vintage Books. p.593

63.   Chang, J., Halliday, J., (2007) ibid.

64.   You may know of the exception to this unwritten censorship, the movie, *Looper* (2012) which was co-produced by DMG Entertainment had time travel as a central theme. I suspect Mao would not have been pleased.

department, known at the time as, SARFT, (State Authority of Radio, Film, and Television). It could be argued that beyond the military and the economy, no other area of China was so tightly controlled as the vehicles of state propaganda and culture. There was no official censorship in China, or rather no official rating system, unlike the self-regulated Hollywood system run by the Motion Picture Association representing the five major film studios in America. As a result in China the rules were never written down, such as no ghosts, or time travel. I once spoke to one of the biggest independent movie producers in China, a founder of the Huayi Brothers Media Corporation and they told me that they would love to have a rating system in China similar to the US. One of the brothers told me that before going to see a movie the audience had no idea about the suitability of a movie for younger audiences. He had taken his five year old son to a Huayi movie that his company had produced and was shocked to see a blood bath of rampant violence. Getting approval for a co-production was notoriously difficult and so I spent a great deal of time testing ideas and concepts for my movies with Chinese film officials. However, there was little understanding of Western movie development and the ultimate approval for a movie could go right to the top of SARFT if not high ranking party members who could be capricious and mercurial. At the time I believed that China had a special affection for New Zealand because of the historically close relationship the kiwi, Rewi Alley, had with Mao. New Zealand was the first to recognise China's rule of Hong Kong and backed China's entry into the WTO when America opposed it.

Our purpose for visiting Wuhan was to negotiate the construction of a Mili Pictures film studio and theme park to be called the *Miliverse*. I had introduced our company to the Disney concept of *family entertainment* and transmedia production. Our presentation to the Wuhan East Lake High Tech Development Zone described our Sino New Zealand co-production eco-cultural film park. I had invited a team to Wuhan including Mili executives, Grant Major, and John McConnell of the McConnell construction group from New Zealand, and Chris Edwards of the pre-visualisation company, Third Floor in LA. If I was to apply Latour's Actor Network Theory these wonderful people were only a few of the actants in this strange assemblage I am referring to as *Wuhan Fun*. We all arrived at Wuhan's international airport at 3am in the morning and were hosted by the city in a 5 star hotel. However, our pitch to the city was scheduled for 10 or 11am the same morning of the same day so we didn't get a great night's sleep.

I presented a vision of a sustainable, ecologically friendly studio and theme park. Living in China for weeks at a time and experiencing the horrific and maleficent black mist of Beijing and Shanghai I was convinced that the only business I could be involved in had to be ecologically sustainable. After our presentation our team was bused to what appeared to be the countryside on the remote edge of East Lake looking back to the Wuhan metropolis, it was easy to see why some had dubbed Wuhan the Chicago of China. Grant, John and I wandered around the shoreline through the long rushes with our guides and a map imagining where our studio and theme park would sit. We were excited if not a little drunk on the scale and ambition of the Wuhan development and to be frank we parked any critical speculation as to whether such a massive project could avoid a negative impact on this relatively untouched wetland environment. On reflection we were lost in the fantasy of an anthropocentric construction of a world-for-us and ignored the complex assemblage of the nonhuman majority all around us. Grant had told me about when he first visited Shanghai in the 1980's and stayed at the only international hotel for foreigners, The Peace Hotel. He said that when he looked across the river from the Bund side all he saw back then was countryside. Today, that side of the river is dominated by skyscrapers peaking out from the murk of a polluted city, while millions of tons of sewage and toxic waste run through the river to the sea.

In Wuhan, I now realise, it was exactly the sort of development that we had planned that might give a partial explanation for the COVID 19 pandemic. The environmental distress that millions of tons of concrete, the multitudes of human tourists and their waste dumped on the East Lake ecology of Wuhan reminds me of the dangerous development of what happened in Flushing Meadows in New York,[65] or the construction of Disneyland in Orlando, Florida. It is when cells in a body, human or nonhuman are put under novel stress that mutations arise and disease has its opportunity. It was fortunate for the planet, and many other actants, that our *Miliverse* project did not happen in Wuhan, or in any of about five other cities in China that we visited. We had even visited a pristine nature reserve just outside of Shanghai city, another possible location for our theme park, where ironically the Godwits nest before they begin their 12,000km migratory flight to New Zealand. I realise today that humanity's inability to truly consider the ecology and world building of the nonhuman majority was leading to the anthropogenic harm to the environment, and this in

---

65.   See chapter , Worldbending for a discussion of the anthropogenic history
       of Flushing Meadows.

turn was hurting our species as well. It is highly unlikely that the pangolin that may have passed on COVID 19 from a bat came from the surrounding countryside of Wuhan. But I can see how the pressures of capitalism and immaterial dreams become the material nightmares for all things that get dragged into its voracious vortex of consumer desires.

China's Dream

In the early 80s, around 1981, I had studied Asian development, and China. At the time I was informed that a vegetarian diet, and plant farming would feed many times more people than could be achieved with meat based agriculture and is part of the reason I stopped eating meat in 1984. But China was 'modernising' under Deng Xiaoping and as the decades slipped by, the burgeoning middle class began eating more and more meat. Meat has become a symbol of consumer affluence and Deng had managed, with the assistance of the ever-helpful Milton Friedman and company, to use shock treatment to transform the stayed Maoist leadership into free market ideologues. This resulted in mass unemployment and rampant party corruption as a capitalist lolly scramble offered up the State's jewels. It made communist party leaders and high placed bureaucrats very rich and created a new generation of 'princelings'. The party elite handed down to their descendants ill gotten profits from their neoliberal revolution. It was much like what happened in Russia under Gorbachov and a similar mass transfer of wealth to a new class of oligarchs. Large numbers of students, and workers demonstrated in Tiananmen Square to protest for greater democratic freedoms but just as in latin America democracy was not part of the economic agenda and

Friedman turned a blind eye to what has been called a massacre. The number of victims from the brutal crackdown by 300,000 troops varies from hundreds to several thousand. According to Naomi Klein's sources this version of state capitalism was violently opposed by the demonstrators who saw this as the brutal imposition of capitalism with corruption, yet without democracy. This was the groundwork for the future of *'disaster capitalism'* in China and I realise now I could have been a part of it. Greed on the part of Mili executives and Shenzhen venture capitalists narrowly averted a disaster in Wuhan but no doubt they did not give up trying.[66]

This whole book could have focused on *China's Dream* and the destructive potential of Xi Jingping's state capitalism but I just wanted to give you a taste of how speculative realism can provide a tool to possibly avoid the bifurcation of the human and the nonhuman worlds by telling a personal story. It is fitting that the Wuhan wet market was the suspected origin of the transmission of COVID 19. However, while we can look at food as both an actant and a resource in the distributed network of the pandemic, it should be obvious that this is a multi-dimensional network without any true epicentre or singular historical cause. The reality of the COVID 19 pandemic is that it is a massive spatiotemporal hyperobject and the connection between the living and the nonliving, the food and the concrete is not just blurry but murky. *China's Dream* may well replicate the *American Dream* and become the world's nightmare.

---

66.   See Klein, N. (2007) The Shock Doctrine: The Rise of Disaster Capitalism.

6

Food for Thought

There is nothing quite like the cooking and eating of exotic animals, especially if they are endangered, to grab the headlines. For most Westerners, reporters, and social media influencers a shopping list that included a scaly anteater or pangolin, followed by a bat intended for a soup, or deep fried in a wok, would trigger revulsion. The idea that these feral animals could be human food has stuck in the Western brain and become ground zero for the transmission of the coronavirus from animal to human. But what if we were forensic detectives, where would we pick up the story of the pandemic? The OOO version would treat all the objects of the disease as equal. The pandemic is a hyperobject that spans spacetime beyond the human scale. It has no beginning and no end but is a complex web of interconnected nodes in a distributed network, from the distant past of the big bang, to the unimaginable future that predicts the cold demise of our galaxy and beyond. To understand the actors, events, and resources in a network requires a transdisciplinary sensitivity that is missing from most scientific, engineering, design, and art forms. *Reality is stranger than fiction*, so most narrators ignore it. The litany of COVID 19 might include these strange bedfellows:

- a demon vixen from the *Creation of the Gods*
- a Boeing 737
- the sword of Guolin
- a Third Floor previzualization
- an illegal pangolin smuggled into China
- an Oscar for *Lord of the Rings*
- level 4 lab containment near the wet market in Wuhan
- East Lake
- a whole bat soup recipe
- the Miliverse
- late night KTV deals
- Mao's Ghost Dramas
- Gwyneth Paltrow in *Contagion*
- baijiu
- the 3 Gorges
- 1 in 100 year floods.

I know only a little about how these events, resources and agents have interacted and can only hint at their relevance albeit to say that given the chaotic reality blender of our universe, they may or may not have significance, but they are all part of the mix that was a world I co-inhabited. This slice of reality is mine, yet all reality is subjective and relative, but it is also very real, material, and standoffish. Food is an essential, sometimes dangerous, metaphorical act, event, and resource, all at once. It is relative to the context of the agent or actant, the time and the space, and has a queer meaning to all who encounter it.

The truism *you are what you eat*, can be extended to *you think what you eat*. Various scientists, nutritionists, and philosophers have challenged the common view that the brain is the singular location of thought and one thing's food is another one's poison. As a psychologist who studies creativity, Antonio Damasio has pointed out our feelings, emotions and thoughts are part of a complex assemblage of muscles, neural sensors, food resources, biochemical and electrical reactions, viruses, bacteria, nerves and synapses within the interior body that then reaches out and extends into the vast assemblage of things exterior to our body that comes into contact with our skin, our optical nerves, our auditory senses, proprioception, taste sensors, into a physical and immaterial cosmos, a web of imaginary metaphors beyond our comprehension. Damasio writes:

"The circumstances, actual or recalled from memory, that can cause feelings are infinite. By contrast, the list of elementary contents of feelings is restricted, confined to only one class of object: the living organism of their owner, by which I mean components of the body itself and their current state. But let us dig deeper in this idea, and note that the reference to the organism is dominated by one sector of the body: the old interior world of the viscera that are located in the abdomen, thorax, and thick of the skin, along with the attendant chemical processes. The contents of feelings that dominate our conscious mind correspond largely to the ongoing actions of viscera, for example, the degree of contraction or relaxation of the smooth muscles that form the walls of tubular organs such as the trachea, bronchi, and gut, as well as countless blood vessels in the skin and visceral cavities. Equally prominent among the contents is the state of the mucosae—think of your throat, dry, moist, or just plain sore, or of your esophagus or stomach when you eat too much or are famished. The typical content of our feelings is governed by the degree to which the operations of the viscera listed above are smooth and uncomplicated or else labored and erratic. To make matters more complex, all of these varied organ states are the result of the action of chemical molecules—circulating in the blood or arising in nerve terminals distributed throughout the viscera—for example, cortisol, serotonin, dopamine, endogenous opioids, oxytocin. Some of these potions and elixirs are so powerful that their results are instantaneous. Last, the degree of tension or relaxation of the voluntary muscles (which, as noted, are part of the newer interior world of the body frame) also contributes to the content of feelings."[1]

Damasio has reprioritized our world building process by ordering our cognition to take place after feelings, *'from pain and suffering to well-being and pleasure.'* Feelings are the precursor to thoughts and to determine where a feeling starts and stops is a liminal exercise in futility. A feeling is itself a hyperobject and our definition of them highlights the limitations of our knowledge. In science in particular we have an epistemological bias to only define two sorts of knowledge about a thing i.e. science wants to know what it is made from and what it does? That knowledge defines what it is but a thing is much more than the sum of

---

1.   Damasio, Antonio. (2018). The Strange Order of Things (p. 103). Knopf Doubleday Publishing Group. Kindle Edition.

knowing those things. Poetry is another form of knowledge that is not defined by what it does or what it is made from but rather it is a metaphorical thing – a poetic object is like something else. We learn from poetry about things by looking at some thing comparing its commonality and difference with some thing else. We can not help but to anthropomorphize other things that are not human. We *wander lonely as a cloud*; we are not a cloud but we imagine a solitary cloud in an empty blue sky and think *'how melancholy'*. Damasio's description of how feelings occur is also deceptive as it is primarily biochemical and mechanical and yet food is also part of a complex web of connections that also has a political, cultural and spiritual dimension.

As Jessica Hutchings points out in her book, *Te Mahi Māra Hua Parakore – A Māori Food Sovereignty Handbook*: "It is important to think about the wider context in which we grow and acquire kai to feed our whanau. Acting locally has global implications and this has never been more important than it is now. the drive to achieve Māori food sovereignty is part of a complex political landscape that is shaped by factors such as climate change, peak oil, GE and the tension between multinational and local; food production."[2]

Hutchings outlines six pillars of food security:

1. Focus on food for people

2. Value food providers

3. Localise food systems

4. Put control locally

2. Hutchings, J. (2015) Te Mahi Māra Hua Parakore - A Māori Food Sovereignty Handbook. p.30

5. Build knowledge and skills

6. Work with nature

"It seeks to heal the planet so that the the planet can heal us. It rejects methods that harm beneficial ecosystems' functions or that depend on energy intensive monocultures and livestock factories, destructive fishing practices and other industrialised production methods that damage the environment and contribute to global warming."[3]

Food is relative and to explain food we often use metaphor. Food is a feeling and a metaphor as well as a scientific reality that is bound to the internal reality of the organism that *imagines* then *consumes* it. My delicious soup of pangolin and bat might be a transportation hub for a virus who ultimately hopes to jump from one animal to another and ultimately be digested as food by my cells by disguising itself as nutrients or food. If the virus's crown or corona of protein was really food and not 'fake food' it would provide my cells with energy. Instead it pretends to be food and my cells would then be sucked in or rather they would suck in the virus. We could call this infection *eating*. The virus then uses my cell's replication factory to multiply its RNA or DNA. The purpose of eating food is to gather energy to be able to reproduce and so perpetuate life. Viruses use cells to reproduce and *steal* energy. What a thing is, and what a thing does is shaped by spacetime scales that will change their definitions according to context and perspective. Infection vs eating; food vs. virus; life vs. non-life are all relative – so you can see that the food concept is a messy metaphor.

---

3. Hutchings, J. (2015) ibid. p.43

When and what do you call food?

Just as philosophers have struggled with the definition of life or non-life, we can see there is a real problem with defining food because it is so relative and human relativity is challenged by the multiple dimensions of spacetime. According to Grant Maxwell in his book, *The Dynamics of Transformation: Tracing an Emerging World View*, humans understand and can operate in approximately three and a half dimensions of conscious reality, i.e. three dimensions of space, and about half a dimension of time. To illustrate this I will examine the NZ Zero Carbon Act through the lens of what is food? For those of you not living in Aotearoa this may seem like a diversion yet throughout most of the *'overdeveloped'* world the farming and agricultural lobbies have been successful in protecting this *'sacred cow'*, defended by politicians and subsidised through the effective work of professional lobbyists. The Paris Agreement is testament to how many countries have their own protected industries and many countries have not even addressed agricultural emissions in their Nationally Determined Contributions, NDC. The so-called *'Green Revolution'* producing an abundance of food, and that swept across the *'overdeveloped'* world since the 60s, now looks decidedly black as we have come to realise it was impossible without fossil fuels, and massive clouds of GHG emissions. As noted by New York chef and ecologist, Mark Bittman: "With the machinery and chemical industries both highly dependent on fossil fuel, and with human labor increasingly displaced, the heartland became a producer of what might as well be called "petrofoods." As Wendell Berry wrote in 1977, "That we should have an agriculture based as much on petroleum as on the soil—that we need petroleum exactly as much as we need food and must have it before we can eat—may seem absurd. It is absurd. It is nevertheless true."[4]

When we discuss the ontology, the being, or world of the New Zealand legislation popularly called, *The Zero Carbon Act*, it is apparent that the law is only aware of this half dimension of time. Admittedly it is a politically designed piece of legislation, even though it claims that it is taking a *'long-term approach that endures political cycles'*. It is unmistakably anthropocentric and founded on political compromise. Was it ever capable of anything else? Unfortunately, while it is better than nothing, by pretending to be non-political it obscures our view of some uncomfortable scientific facts and unknown uncertainties. I say this because in the Zero Carbon Bill's 2018 discussion document it outlined a timetable based on three five year carbon budgets starting in 2020 and take us through to 2035 – this was created by humans, for humans, and follows New Zealand's Emissions Trading Scheme and its political creation and evolution according to the political cycles of this country and the IPCC (Inter-governmental Panel on Climate Change). In the Summary of the Zero Carbon Bill, the NZ Ministry of the Environment, explained how the budget works:

"For each five-year period, the emissions budget will state the quantity of greenhouse gas emissions permitted in CO_2 equivalent. It will cover all greenhouse gases – CO_2, methane, nitrous oxide, hydrofluorocarbons, perfluorocarbons, sulphur hexafluoride, and nitrogen trifluoride."[5] In the Zero Carbon Act agriculture and its

---

4.   Bittman, M. (2021) Animal, Vegetable, Junk: A History of Food, from Sustainable to Suicidal. Boston. Houghton Mifflin Harcourt.

5.   See the Climate Change Response (Zero Carbon) Amendment Bill:

carbon dioxide, methane and nitrous oxide emissions will not be included in the ETS until 2025. However, the next review of the sector will be in 2022 and the government can put the sector into the ETS sooner if they have not made progress. If that did happen farmers would have to pay for their emissions but would receive a 95% discount as part of the coalition agreement between Labour and New Zealand First. The recent election in 2020 has returned Labour and removed New Zealand First with a weakened Green party giving Labour the ability to make policy on their own but while maintaining a strong tie with the Greens. James Shaw retains the position of the Climate Change Minister outside of cabinet. The very temporal sensitivity of politics shows again how human spacetime has a very limited context when dealing with geological time scales. Farmers will begin voluntarily reporting emissions in 2023 and will be mandated to do this by 2024. Supposing that farmers make some progress and they enter the ETS in 2025 it is still a five year delay and a lot can happen in five years, including unexpected governmental instability. In 2019 a high ranking member of the NZ Green party, Jack McDonald resigned. He said, "When the IPCC says we have 12 years to save the world from climate catastrophe, we simply don't have time for centrism, moderation or fiscal austerity."[6]

Why should any of this matter? Surely it is just an unavoidable political reality. I want to zero in here on a nonhuman component of the Earth System and discuss

Summary, https://www.mfe.govt.nz/sites/default/files/media/ Climate%20Change/climate-change-response-zero-carbon-amendment-bill-summary.pdf

6. Radio NZ. (4 August 2019) High ranking Greens member pulls pin before election. https://www.rnz.co.nz/news/political/395936/high-ranking-greens-member-pulls-pin-before-election

how our half dimensional temporal consciousness has made us myopic about future World Building. We cannot think clearly about time and space beyond the human scale. The bottom line is that our limited understanding of time is luring us into a false reality that could mean the physical end of this world while we try and sort out a political compromise. The end of human time is sneaking up on us and will most likely take us all by surprise.

This existential threat is most obvious if we consider the 3 big bad boys of global warming and Earth System change: carbon dioxide; nitrous oxide; and methane. Carbon dioxide is the poster child for the Anthropocene. However, these 3 gases are only part of the massive nonhuman majority that shapes our reality, and a mass of others, and yet we barely know them. They are according to the ecological philosopher, Timothy Morton, *hyperobjects*, that create their own worlds beyond our human spacetime comprehension. Their age stretches back to the beginning of the universe, and while their atoms and subatomic particles are constantly changing their home address, the components of the gases are immortal, and they are also part of us. Earth System Science is an attempt to understand and explain their infinite life cycles. So I am going to have a go at thinking aloud about how these three gases intersect with our *Zero Carbon Action* and hopefully alert you to the potential risks and uncertainties that have been embedded in the NZ law, and the resulting protocols and algorithms that control our lives. There are ways in which the CCC (Climate Change Commission) can advise the government about more stringent means of controlling emissions but it is advice only and will not be mandatory.

What is food?

To make this personal for you I thought I would begin with food to start unwinding the life cycle of the three gases we are most concerned about. Food glorious food – what is it? Well, that simple question has a complicated answer or rather starts a complex discussion. If we slice and dice this word what do we learn about what it is we are doing and the World we co-create with other humans and the massive nonhuman majority, including our three gaseous buddies? An Oxford dictionary definition will probably not cause too many complaints:

"food (mass noun) – any nutritious substance that people or animals eat or drink or that plants absorb in order to maintain life and growth" – Oxford Dictionary.

But then there is also the phrase *"food for thought"* meaning 'something that warrants serious consideration'. And surely food does just that! But look at what is hidden in this meaning, beyond our common-sense view of the world. We assume it is a thing that is eaten or drunk by living beings to keep them living, growing, reproducing. Eventually when the consumer of food dies the food nutrients are recycled by the life process. And, that is exactly the life cycle that our 3 gases participate in. Without them, we would all be dead. But those gases are actually compounds or molecules made from chemical reactions – they are the products of intimate relationships between so-called nonliving chemicals – carbon, oxygen, nitrogen, and hydrogen. From these building blocks – life as we know it is somehow created? So, as we love life one would think that lots of these chemicals would be food for lots of lovely beings. It turns out that the difference between life and death, the living or nonliving, is far from black and white.

As Vladimir Vernadsky author of *The Biosphere* (1926) once said '*humans are walking talking minerals*'. Of course, all of these three gases help to build our bones, skins, nails, blood, and other bits and pieces, and these go into creating our food – whether it is plant or animal. Truly they are our food, and in turn, we are food for others! Therefore, more food, more gas, more life!

We now know that this utilitarian argument is wrong and that more of these three gases now threatens our planet with death and the acceleration of the sixth mass extinction. How is it that if some of these gases are good for life, why is it not better to have more of them? The obvious answer relates to the proportionate amount of any one thing. Too much can lead to toxicity. According to the Wikipedia entry on toxicity:

"A central concept of toxicology is that the effects of a toxicant are dose-dependent; even water can lead to water intoxication when taken in too high a dose, whereas for even a very toxic substance such as snake venom there is a dose below which there is no detectable toxic effect."[7]

Even the most delicious food, the best wine, or the purest water can become toxic if the dose overwhelms a system, whether it is a population, a single being, or a stable Earth System. For the past 12,000 years of the Holocene we have co-existed with other '*foods*' enjoying a world that was born out of five previous mass extinctions[8] caused by an overabundance of certain foods, that became toxic to some, and a banquet for others. The first mass extinction was probably caused by too much oxygen. However, rather than

---

7. https://en.wikipedia.org/wiki/Toxicity
8. Brannen, P. (2018). Ends Of The World: Volcanic apocalypses, lethal oceans and our quest to understand earth's past mass extinctions. S.l.: Oneworld Publications.

understanding that the Earth System is a temperamental species killer we came to think of the Earth as our reliable and unbreakable World that would obediently supply us with every thing that we needed to not only survive but to thrive and indulge in the aesthetic and joyous pleasures of life on Earth.

The beginning of civilisation was thought to have begun at this time when the chaos of swirling gases of oxygen, carbon dioxide, methane and nitrous gas settled down to maintain temperatures within a mean range of only about 1 degree Celsius in variation. This provided humanity with the confidence to also settle down and start growing local things to eat. The four gas giants, oxygen, carbon dioxide, methane and nitrous oxide cycled in happy chemical equilibrium and our weather was relatively pleasant and procreative – our population exploded. This lasted for around 12,000 years, up until now, but the geologic World is beginning to wake up.[9] Our government believes that equilibrium is the usual state of our climate and atmosphere but there is scant evidence for this belief. The chemical interaction of just four of those gases is unimaginably complex but what if their assumptions about just one of them, methane, are wrong? As was outlined in their Summary of the Zero Carbon Bill they believe: "Unlike carbon dioxide (CO_2), methane is a short-lived greenhouse gas. It degrades in the atmosphere over decades. Once in equilibrium, it can continue to be emitted at a stable rate without increasing its concentration in the atmosphere."[10] What we don't remember is that climatic chaos was previously normal, and fluctuations in greenhouse gas

---

9.   See Bill McKibben. (2013) Waking the Sleeping Giant

10.   See the Climate Change Response (Zero Carbon) Amendment Bill: Summary, ibid.

concentrations, and therefore extreme climatic conditions, were normal in the past when considered within the inhuman timescale of geology. It has been around 800,000 years since we have seen the same concentration of methane in the Earth's atmosphere – there has not been so much food in the air for ages.

Food – it's common sense

I find that the saying '*It's common sense*' is one of the most worrying phrases I know. It assumes that the majority knows best and implies a non-scientific methodology. The statement is politically loaded and is designed to shut down enquiry or debate. It kills creativity and has an inbuilt conservatism that reinforces the status quo and the vested interests of the day. Before Copernicus it was simply '*common sense*' that the Sun revolved around the Earth – anything else was heresy and the Church and State responded with violent opposition.

Today agriculture, as we know it, or as Timothy Morton in his book, *Dark Ecology*, calls it *agrilogistics*, is modern day '*common sense*'. In New Zealand and many other countries, agriculture is the religion, or ideology, that defines the '*World as we know it*'. Citing Morton, the blogger, Kirsten Everson explains, "agrilogistics; [is] 'a technical, planned, and perfectly logical approach to built space' that arose when our ancestors figured out a way to assure their future food supply – agriculture. Agrilogistics by necessity caused humans to shift their way of looking at the world from one of fear, wonderment and magic to one of rational, linear organization and control."

Yet, the common sense of agrilogistics is more ideological than logical. Try to explain the logic of food production that is such a significant contribution to global warming and the *'end of the world as we know it'*. For those who are concerned about the survival of the species, known as humanity, or the sixth mass extinction of an even larger subset, humans and nonhuman, the conclusion must be that agriculture does not equate with food. In other words, it is not *'maintaining life or growth'*. Driven by capitalism it is no longer concerned with the aesthetic pleasure of co-mingling with the biome of our guts or the life-giving qualities of their nonhuman companions.

The marketing of desirable food has been reduced to an emphasis on 4 essential things – sugar, fat, acid, and salt. Too much of these and not enough essential nutrients will kill you but the pleasure senses in your brain keep begging for more, which is something that advertising understands all too well. Appearance is only surface deep – who is tracking the biography of the less compatible nonhuman co-inhabitants in the body. We no longer know or totally trust the origins, and personalities of the food we eat. It is about productivity, efficiency and profitability – the accumulation of capital at all cost delivers food that has become cheaper, and cheaper, but less and less nutritious, and greater negative impact on the environment.

We have to move beyond the stupid economic modelling that is focused on *'growth'* and GDP – capitalism is founded on extractive industries – agriculture is the oldest. What is more, industrial agriculture, at the current scale, would be impossible today without fossil fuels, including the farming of livestock, which is the most very inefficient use of that energy. On average it takes ten calories of fossil-fuel energy to produce one calorie of food energy that we call food. And so we get back to our favourite foods – the three gases. Meat consumption around the world continues to rise as incomes in LDCs increase so have their diets become more calorific and their daily meat intake has risen causing personal health problems and more global heating. The editors of the book, *Drawdown*, have estimated that if we were all to switch to a plant-rich diet that by 2050 we would reduce CO_2 emissions by 66.11 gigatons. They go on to say that not only do livestock account for 18 to 20 percent of greenhouse gases but that globally they are only second to fossil fuels. "If you add to livestock all other food -related emissions – from farming to deforestation to food waste – what we eat turns out to be the number one of the greatest causes of global warming along with the energy supply sector."[11]

Methane – more than a bad smell

I am going to concentrate on methane, CH_4. Methane, over a 20 year period is between 80 and 104 times stronger at heating the planet than carbon dioxide, and 28 times stronger over a 100 year period. The IPCC and the NZ government have used the longer period of a century to calculate methane's carbon dioxide equivalence and add it to the budget, but there are many assumptions and uncertainties that this 100-year formula or algorithm hides. In the latest report by the IPCC, they wrote "Eliminating emissions of these [Short Lived Carbon Forcers, such as methane] results in an immediate cooling relative to the present." (p.65) The experts concluded that if we could reduce all anthropogenic (human-caused) emissions to zero immediately then it is likely that there would only be 0.5°C heating over the next 2-3 decades and

---

11.   Hawken, P. (Ed.). (2017). Drawdown: The most comprehensive plan ever proposed to reverse global warming.

it is also likely it would be less than 0.5°C over the next 100 years. Of course, they noted that an instantaneous reduction to zero was not feasible because of 'techno-economic inertia'. In other words we know what we need to do but we just pretend we don't.

A well-respected climate scientist Dr Wolfgang Knorr said in an interview, "What we should really ask is: has there been any discernible impact of climate policy on emission rates of greenhouse gases? ... However difficult climate diplomacy might be, we should not shy away from clearly stating that nothing has been achieved as long as we cannot see an effect on global measurements. Therefore, the political response has been inadequate." Current politics has not been up to the task and Knorr says that scientists, including the IPCC, have let us down. He says it is time for us all to take action at the political and decision-making level. "What is happening now is that Extinction Rebellion and similar protest movements increasingly seize the public discourse on climate change and leave much of the scientific community behind, in particular, the IPCC. And that is a good thing because we need to move on from the current paradigm... We need a sober, grown-up look at all the risks climate change entails. No doomsaying, no preaching, no exaggeration in order to convince others, but also not shying away from speaking out things that are painful."

The Tipping Point

It is all about the time left before a temperature threshold is exceeded. The important thing to understand is that we may be approaching a tipping point and that it is only by attacking SLCFs (Short Lived Carbon Forcers) like methane, and Long-Lived Carbon Forcers, like nitrous oxide, as well as carbon dioxide, that we might be able to quickly reduce the global heating effect of GHGs. So back to the Zero Carbon Act and the assumptions, and political compromise built into it. According to Manaaki Whenua – Landcare Research NZ: "The inventory of NZ's greenhouse gas emissions is calculated annually by the Ministry for the Environment and reported internationally. According to this inventory, methane emissions from ruminants have increased by 10% since 1990. (Over the same period, carbon dioxide emissions from road transport have grown by 62%, and nitrous oxide emissions from agricultural soils by 25%.) There is considerable variability between individual ruminants so measurements need to be expanded to paddock or farm scale. Readings by Landcare research in Canterbury with herds of 270 to 550 cows showed the "seasonal methane emission rates ranged from 284 to 427 g/day per cow." Landcare Research in "Collaboration with NIWA and AgResearch led to comparative micro-meteorological ('top down') and sheep and soil ('bottom up') measurements being made during Autumn 1997. There was good agreement by the two methods with an average net methane emission rate from sheep populated lowland pasture of 160 kg/ha per year."

So from these calculations, we can guesstimate the number of GHG emitting ruminants based on 2017 statistics. The total dairy cattle in New Zealand = 2.6 million in the South Island and 4 million in the North Island. Beef cattle have decreased to 3.6 million. Sheep have decreased also to 27.5 million. By my calculations that is 10,573,320 million tons of methane a year for cattle alone.

"Per capita, New Zealand has the largest methane

emission rate (0.6 t per person per year)—six times the global average. "

The carbon Emissions Trading Scheme with its exemption of agriculture has resulted in a 19.6% increase in CO_2 equivalence from 1990 until 2016. "The main drivers for this change in emissions are an increase in the application of synthetic nitrogen fertiliser of about 650 per cent since 1990 and an 89.6 per cent increase in the dairy herd population."[12] However, when trees are subtracted (because of more harvesting) the increase in GHGs between 1990 and 2016 is 54.2%. One report that measured the change in atmospheric methane between 2014 and 2017, and published in February 2019, described it as very strong growth. "The rise in atmospheric methane (CH_4), which began in 2007, accelerated in the past 4 years. The growth has been worldwide, especially in the tropics and northern midlatitudes. With the rise has come a shift in the carbon isotope ratio of the methane. The causes of the rise are not fully understood and may include increased emissions and perhaps a decline in the destruction of methane in the air. Methane increase since 2007 was not expected in future greenhouse gas scenarios compliant with the targets of the Paris Agreement, and if the increase continues at the same rates it may become very difficult to meet the Paris goals. There is now urgent need to reduce methane emissions, especially from the fossil fuel industry."[13]

12.   Ministry for the Environment.(2019) New Zealand's Greenhouse Gas Inventory 1990-2017. Vol.1, Chps 1-15. Fulfilling reporting requirements under the United Nations Framework Convention on Climate Change and the Kyoto Protocol.

13.   Nisbet, E. G., Manning, M. R.,Dlugokencky, E. J., Fisher, R. E.,Lowry, D., Michel, S. E., et al. (2019).Very strong atmospheric methane growth in the 4 years 2014–2017: Implications for the Paris Agreement. Global

The uncertainty and lack of data surrounding methane emissions does not justify an argument that New Zealand should make little or no change in trying to mitigate those emissions. There are a number of possibilities that scientists are considering including the possibility that methane may be already reaching a tipping point and that the hydroxyl that converts methane to carbon dioxide and water may have become saturated and unable to absorb any more. It is important to remember that while methane is a short-lived GHG it is both far more potent and also eventually converts to carbon dioxide so that global heating continues but just more slowly. According to the Climate Action Tracker website: "New Zealand is one of the few countries to have a zero emissions goal enshrined in law, its Zero Carbon Act, but short-term policies cannot yet keep up with that ambition. Noteworthy is the fact that the 2030 greenhouse target was not updated in the newly-submitted NDC; only a hint was included that this could happen in 2021. Methane from agriculture and waste (over 40% of New Zealand's emissions) is exempt from the zero emissions goal, and has a separate target not yet covered with significant policies.... The adoption of the Zero Carbon Act in 2019 was a step forward, but implementation is key and the methane exemption weakens the target considerably. The Act aims to achieve net zero emissions of all greenhouse gases, except for methane emissions from agriculture and waste, by 2050...The lion's share of these methane emissions is from agriculture. A true commitment to net-zero would require a further reduction of 18-25 $MtCO_2e$ in 2050, corresponding to the residual methane emissions that would remain unmitigated by the

Biogeochemical Cycles, 33 ,318–342. https://doi.org/10.1029/2018GB006009

2050 target. These would need to be achieved in other sectors, in particular through full decarbonisation of energy and industry emissions by 2050."[14] We have to think about the scale and the impact of current agriculture; the climate tipping point; and why we eat? Food for Thought?

Plague Politics

The anthropogenic harm that has already begun to affect food supply, the amount of arable land, water resources, and land that has reached optimal productivity and is increasingly suffering from over-fertilisation. In the South Island of New Zealand the once pristine fresh water river ways have become toxic to life due to intensive dairy farms.[15] In New Zealand, (with some regional variability) according to AgResearch dairy requires approximately 1000 litres of water to produce 1 litre of milk. What needs to be also accounted for is the waste or effluent produced as a result of this milk production. It has been calculated that 1 cow produces as much effluent as 14 humans, and as we have 10 million cows New Zealand farms produce the equivalent of 140 million people urinating and defecating untreated straight onto the ground and finally ending in the streams and the sea.[16] Dr Mike Joy and a team of researchers has estimated that it will take NZD\$15 billion to repair the ecological damage caused by dairy farming. "Dairying's impacts are chiefly: the amount of water used to create milk; the effluent that pollutes the soil, aquifers

---

14. See https://climateactiontracker.org/countries/new-zealand/

15. See 'Behind New Zealand's '100% Pure' Image lies a Dirty Truth.' https://youtu.be/a_mrSrvlFlQ' ABC, Foreign Correspondent

16. Gerard Hutching. (Aug 25 2018) Milking it: The true cost of dairy on the environment. https://www.stuff.co.nz/business/farming/106546688/milking-it-the-true-cost-of-dairy-on-the-environment. Stuff.co.nz

and waterways; the way in which soil is compacted by heavy animals; and the greenhouse gases that cattle emit.In addition dairy processors are significant energy users and greenhouse gas emitters. Fonterra burns about 410,000 tonnes of coal to turn liquid milk into powder. Based on one tonne of coal producing 2.86 tonnes of carbon dioxide, Fonterra's factories pump out 1.17 million tonnes of the climate warming gas, making it one of New Zealand's top greenhouse gas polluters."[17]

Around the world the degradation of land encourages farmers to move to greener fields and encroach further into jungles, forests, and results in more infertile land that deepens the harm. All of this is a recipe for pandemics, pest infestations, poverty, famine, fires and floods – what the *Bible* referred to as *The Plagues*. Meanwhile successful farming lobbies convince politicians to give them more time and more money. To me this is *Plague Politics*. Successive governments continue to listen to the elites who benefit from a lenient regulatory approach to the agricultural sector while we are running out of time to implement effective land use reform and tighter controls on pollution, over stocking, and GHG emissions.

Next year I turn sixty which may give me some perspective on at least the life I have seen to date. I studied a Master in Political Studies in the 1980s and was fortunate to be taught about political theory and history by Professor Andrew Sharp. When I was in my early twenties I was as idealistic as any of my peers so it is with some shock that I re-read one of Professor Sharp's prescribed text about Jeremy Bentham's reformist ideals by James Steintrager who wrote:

---

17.   Gerard Hutching. (Aug 25 2018) ibid

"Only after those sixty years of disappointment did he [Bentham] realise that 'man from the very constitution of his nature, prefers his own happiness to that of all other sensitive beings put together."[18]

Peter Turchin in his book, *Ages of Discord*, has warned of the increasing societal instability and conflict that is likely to come from hungry populations, competing elites, and increasing numbers of epidemics and pandemics. The ruling elites will continue to resist change and land use reform while those who are looking to replace them will agitate for change. Turchin's Structural-Demographic Theory is founded on agrarian historical data which he explains:

"According to this theory, population growth in excess of the productivity gains of the land has several effects on social institutions. First, it leads to persistent price inflation, falling real wages, rural misery, urban migration, and increased frequency of food riots and wage protests. Second, rapid expansion of population results in elite overproduction—an increased number of aspirants for the limited supply of elite positions. Increased intra-elite competition leads to the formation of rival patronage networks vying for state rewards. As a result, elites become riven by increasing rivalry and factionalism. Third, population growth leads to expansion of the army and the bureaucracy and to rising real costs. States have no choice but to seek to expand taxation, despite resistance from the elites and the general populace. Yet, attempts to increase revenues cannot offset the spiralling state expenses. Thus, even if the state succeeds in raising taxes, it is still headed for a fiscal crisis. As all these trends intensify, the end result

---

18.   Steintrager, J. (2011). Bentham. Routledge. p.44

is state bankruptcy and consequent loss of military control; elite movements of regional and national rebellion; and a combination of elite-mobilized and popular uprisings that expose the breakdown of central authority."[19] While Turchin is writing about pre-modern agrarian societies it is not hard to imagine this happening in the future, and may already be starting to happen in America.

To date New Zealand has been fortunate enough to contain the COVID 19 pandemic with a small number of deaths and very low community transmission. For various reasons New Zealanders have been returning home from living overseas because it is safe, or they have lost their jobs, or they want to be with their families in this time of crisis. Those arriving are first accommodated in what is called, managed isolation and quarantine. Turchin continued to layout his theory of what happened during times such as the Black Plague as he explains:

"Sociopolitical instability resulting from state collapse feeds back on population growth via depressed birth rates and elevated mortality and emigration. Additionally, increased migration and vagrancy spread the disease by connecting areas that would have stayed isolated during better times. As a result, epidemics and even pandemics strike disproportionately often during the disintegrative phases of secular cycles."[20] So, it would seem that the degradation of arable land, extreme weather, forest fires and population density presents a perfect storm for future plagues and the breakdown of our complex societies.[21]

---

19. Turchin, Peter. (2016) Ages of Discord: A Structural-Demographic Analysis of American History (p. 11). Beresta Books LLC. Kindle Edition.

20. Turchin, P. (2016) ibid (p. 11)

21. See Tainter, J. A. (2017). The collapse of complex societies. Cambridge University Press.

Two days before the Zero Carbon Bill passed into law in New Zealand a Guardian article stated: "The world's people face "untold suffering due to the climate crisis" unless there are major transformations to global society, according to a stark warning from more than 11,000 scientists."

"We declare clearly and unequivocally that planet Earth is facing a climate emergency," it states. "To secure a sustainable future, we must change how we live. [This] entails major transformations in the ways our global society functions and interacts with natural ecosystems."

There is no time to lose, the scientists say: "The climate crisis has arrived and is accelerating faster than most scientists expected. It is more severe than anticipated, threatening natural ecosystems and the fate of humanity."[22]

Two days later on the 7th Nov. 2019, the New Zealand government passed into law the Zero Carbon Bill, more formally known as the Climate Change Response Amendment Bill. It passed by an almost unanimous vote of 119 to 120 MPs and is a credit to political negotiations. According to the news website, *Spinoff*, James Shaw the leader of Labour's coalition partner, The Green Party, and Climate Change Minister had this to say in parliament.

"Some things are too big for politics, and the biggest of them all is climate change. The intent of the Zero Carbon Bill was, is, and always should be to elevate climate change policy beyond petty politics and partisanship, to transcend and transform a problem so wicked and so stuck that we have made virtually no progress on it in the 30 years [sic] we have been aware of it, in spite of the very best efforts of

22. Carrington, D. (5 Nov. 2019) Climate crisis: 11,000 scientists warn of 'untold suffering': Statement sets out 'vital signs' as indicators of magnitude of the climate emergency https://www.theguardian.com/environment/2019/nov/05/climate-crisis-11000-scientists-warn-of-untold-suffering

many, many good people. Climate change policy has been a political football kicked up and down the field, and frequently into touch, by changes of government and, in fact, changes within governments. This unstable policy environment has prevented progress and sent contradictory signals, which has stymied decisive action until this, the 11th hour and 55th minute before midnight."[23]

The legislation is undoubtedly a significant achievement and milestone but before we congratulate ourselves it should still provide us with more food for thought. Since 2009 the National government and the farming lobby successfully delayed action on agricultural GHG emissions for ten years. The new coalition government of the Greens, Labour, and NZ First promised to do something about agriculture and climate change. However, the passing of the Zero Carbon Bill has come to mean there is zero carbon consequences for agricultural GHG emissions as the agricultural lobby is awarded a hiatus of another 2-5 years. This is despite the fact that the sector contributes the largest single percentage of New Zealand's GHGs at 48%, and the next closest is Transport at a distant 20%.[24]

The political compromise in the Zero Carbon Act sets the methane interim requirement to reduce emissions to 10% below 2017 levels by 2030, and reduce gross emissions of biogenic methane within the range of 24% to 47% below 2017 levels by 2050; all other GHGs including nitrous oxide

---

23. Shaw, James. (7 Nov. 2019) Hansard Climate Change Response (Zero Carbon) Amendment Bill — Third Reading. https://www.parliament.nz/en/pb/hansard-debates/rhr/document/HansS_20191107_054000000/shaw-james

24. The declaration that New Zealand is facing a 'climate change emergency' maybe just symbolic or a represent a serious commitment in doing something about it now Labour has a strong majority in parliament. See The Independet article https://www.independent.co.uk/environment/new-zealand-climate-emergency-jacinda-ardern-b1762221.html

would have a net emission reduction to zero by 2050. According to the website, *MyClimate* "Net zero emission means that all man-made greenhouse gas emissions must be removed from the atmosphere through reduction measures, thus reducing the Earth's net climate balance, after removal via natural and artificial sink to zero. This way humankind would be carbon neutral and global temperature would stabilise."[25]

When the NZ government was asked why it had not gone further on setting more aggressive targets for methane they replied because methane (they don't mention nitrous oxide) is a short lived GHG and within 50 years it will disappear and so the amount in the atmosphere will have stabilised. However, that only applies if methane levels no longer continue to rise. It also ignores the fact that methane converts to CO_2 when it degrades so it continues to contribute to the cumulative heating caused by carbon dioxide.

The NZ Interim Climate Change Commission report makes an important point. "The warming caused by methane is not as short-lived. The warming today will still be felt several centuries from now as the climate absorbs and redistributes the heat trapped while the methane is in the atmosphere."[26]

To stress this point, methane may only have an estimated mean half life of 9.1 years, but within the first 9 years it has a Global Warming Potential (GWP) of 80-104 times that of carbon dioxide and in 100 years a GWP of between 28-32 times, and remember that CO_2 is not

25.	What does "net zero emissions" mean? https://www.myclimate.org/information/faq/faq-detail/what-does-net-zero-emissions-mean/

26.	ICCC (30 April 2019) Action on Agricultural Emissions

nothing but is recognised as the major cause of most of the global heating of the planet.

A Catastrophic Update

In October 2018 the IPCC shocked the world with its updated report for policy makers. "One of the key messages that comes out very strongly from this report is that we are already seeing the consequences of 1°C of global warming through more extreme weather, rising sea levels and diminishing Arctic sea ice, among other changes," said Panmao Zhai, Co-Chair of IPCC Working Group I."[27]

The report stressed the urgency of action to reduce all GHGs and that included land use. "The report finds that limiting global warming to 1.5°C would require "rapid and far-reaching" transitions in land, energy, industry, buildings, transport, and cities. Global net human-caused emissions of carbon dioxide (CO_2) would need to fall by about 45 percent from 2010 levels by 2030, reaching 'net zero' around 2050. This means that any remaining emissions would need to be balanced by removing CO_2 from the air."[28]

Every little bit counts and that certainly means a reduction in powerful GHGs that come from our food production such as methane, and nitrous oxide. This is all a question of time and urgency, therefore, it is hard to reconcile kicking agricultural emissions down the road when immediate action will have immediate results at a time when it is most urgent and needed. According to New Zealand's Interim Climate Change Commission "One thing

---

27. IPCC (18 Oct. 2018) Summary for Policymakers of IPCC Special Report on Global Warming of 1.5°C approved by governments

28. IPCC (18 Oct. 2018). ibid.

is clear – New Zealand must take action to reduce agricultural methane and nitrous oxide because these gases form such a large proportion of our national greenhouse gas profile. There is often less focus put on nitrous oxide – but this is a potent and long- lived gas and must be a part of efforts to achieve a net zero target."[29] According to the EPA in America the GWP of nitrous oxide is 265-298 times that of CO_2 for a 100 year timescale and N_2O emitted today remains in the atmosphere for more than 100 years on average.[30]

The major cause of these GHGs is our farming systems that includes monocultural animal husbandry and cropping, intensive water use in dairy; heavy fertiliser use; water quality degradation; and concentrated effluent with methane and nitrous oxide consequences. It was the 'father of the fertiliser industry', Justus Liebig in the 19th century who explained the importance of balanced soil nutrition that plants needed to grow including nitrogen, phosphorus, and potassium. He recognised that a shortage of nitrogen in the soil would hinder plant growth, but he also recognised that animal dung could be used instead of synthetic fertilisers and promoted guano. However, for all Liebig's brilliance as an organic chemist and his understanding of the carbon cycle, metabolic systems, and the chemical nutrients necessary for healthy animals and humans, his understanding was rudimentary. While his research informed generations of farmers who came to understand soil degradation and the necessity of cycling new resources back into the soil the dangers of over-fertilisation were not appreciated until later in the 20th

---

29.   ICCC. (30 April 2019) ibid.

30.   See https://www.epa.gov/ghgemissions/understanding-global-warming-potentials

century along with their contribution to global heating. Marx who read Liebig came to foresee the dangers that extractive farming could cause a *'metabolic rift'* that could not be fixed with synthetic fertilisers. Regenerative farming, organics, and biodynamic farming have all grown in recent popularity, with a political and spiritual dimension that challenges a strict materialist view of Earth System science. These have formed an alliance with indigenous cultures that have pushed back against the harm caused by anthropocentric supremacy and arrogance. The Māori food and soil sovereignty movement, Hua Parakore is using maramataka, the Māori calendar and Māoritanga, or knowledge, to inform a more mindful land use that insists that humans are merely the guardians, or kaitiaki of the Earth Mother, Papatūānuku. The arrogant dismissal of *'non-scientific'* beliefs and philosophies ignore the uncomfortable reality that capitalism, science and technology have created more harm than traditional land use, and indigenous wisdom is better at mitigating GHGs than the market pricing of carbon and the preservation of profits.[31]

The ICCC stressed that agricultural GHGs cannot be exempt and "Whatever the target relating to methane ends up being, we know that we need to reduce emissions. It is time to get on with the job... Continued delay is not an option. It is critical that we get started now."[32] Unfortunately, political compromise has done just that and caused a delay.

The IPCC is warning that even limiting global warming

---

31.   See Hutichings, J. Te Mahi Māra Hua Parakore: a Māori Food Sovereignty Handbook, and Te Mahi Oneone Hua Parakore: a Māori Soil Sovereignty and Wellbeing Handbook.

32.   ICCC. (30 April 2019) ibid.

to 1.5°C will cause catastrophic death of coral reefs, forest fires; and sea level rises that will inundate low lying islands and coastal communities. The impact of just 1°C of heating has already resulted in the death of coral; ice cap melting; sea-level rise; extreme weather; floods; droughts; heat death; increased poverty and starvation; species extinction threats and mass migration by climate refugees.

The Independent reported on the secretive planning by Shell and BP: "Oil giants Shell and BP are planning for global temperatures to rise as much as 5°C by the middle of the century. The level is more than double the upper limit committed to by most countries in the world under the Paris Climate Agreement, which both companies publicly support."

"The discrepancy demonstrates that the companies are keeping shareholders in the dark about the risks posed to their businesses by climate change, according to two new reports published by investment campaign group Share Action. Many climate scientists say that a temperature rise of 5°C would be catastrophic for the planet."[33] What is more, for every 1°C the amount of methane released by freshwater microbes could be several times greater, and possibly even outpace CO_2 emissions as the atmosphere moves past the tipping point and goes into a positive feedback loop that keeps heating beyond our control.[34]

New Zealand may only contribute 0.17% of CO_2 equivalent emissions to the global total, however, most of

33.   Chapman, B. (Oct. 2017) 'BP and Shell planning for catastrophic 5°C global warming despite publicly backing Paris climate agreement.' Independent. https://www.independent.co.uk/news/business/news/bp-shell-oil-global-warming-5-degree-paris-climate-agreement-fossil-fuels-temperature-rise-a8022511.html

34.   See Princeton University (2014) A more potent greenhouse gas than carbon dioxide, methane emissions will leap as Earth warms. Science Daily. https://www.sciencedaily.com/releases/2014/03/140327111724.htm

our citizens expect to make a contribution to climate change mitigation and help improve our stewardship of the environment. New Zealand is, however, surprisingly in the top 5 countries of total GHG emissions per capita in the OECD (2014). Here is the table of the others:

1. Australia – 22.4 tonnes per capita

2. USA – 21.55 tonnes per capita

3. Canada – 20.51 tonnes per capita

4. Luxembourg – 19.59 tonnes per capita

5. New Zealand – 17.98 tonnes per capita [35]

Catastrophic Carbon Bombs

Catastrophe has already hit as the job is made much harder by a perfect feedback loop. Cattle ranchers and farmers have been given a green light by the Brazilian government to increase the slashing and burning of the Amazon to make way for cattle grazing. As a result not only does the Earth lose three football fields a minute of oxygen-producing forest but the carbon that was sequestered in the trees is released in a gaseous explosion of carbon dioxide, and methane from the biomass. What is more, deforestation leads to lower rainfalls and places like the Amazon may reach a tipping point where the ecosystem will no longer support regrowth.

New South Wales in Australia is today bracing itself for fires that are officially designated a 'catastrophic fire rating'. One observer who witnessed the dark red and smokey skies called the scene apocalyptic. As I have pointed out in my

---

35.   Source: OECD – https://figure.nz/chart/jMoS5wjQpAHSYx33

book *Worldbending*: "We live in apocalyptic times, not in any religious sense, but rather of our own making because we have physically and irretrievably damaged our world. John Hall explains that our apocalyptic outlook is not new, it is a belief that has a long history. In our past, the coming of the apocalypse may have caused us to look to the skies for the signs of the end, but in recorded history, despite catastrophic natural disasters, our planet, and our species have survived. Hall explains the historic meaning of the apocalypse is rather than the actual end of the world, the apocalypse is typically *"the end of the world as we know it,"* an extreme social and cultural disjuncture in which dramatic events reshape the relations of many individuals at once to history."[36]

Some have questioned the Australian government's commitment to climate change mitigation. They signed a target of 26-28% reductions in emissions by 2030 which some have criticised as insufficient for the world's largest coal exporter. The UN has reported that Australia was not on track to meet its commitment. According to the BBC, the PM of Australia, "Mr Morrison told the UN last year that Australia was doing its bit to address climate change, and "balancing our global responsibilities with sensible and practical policies to secure our environmental and our economic future". However, the Deputy PM referred to those who claim the NSW fires were caused by human industry as the *'ravings of inner city lunatics'.*[37]

And there you have it, the arguments given by just about every politician worldwide, and repeated again in New

36.   Hall, John R.. Apocalypse: From Antiquity to the Empire of Modernity (Kindle Locations 182-185)

37.   See The Guardian. https://www.theguardian.com/australia-news/2019/nov/11/dear-michael-mccormack-the-only-raving-lunatics-are-those-not-worrying-about-climate-change

Zealand. Our political system suffers from an addiction to growth economics and GDP modelling that even calculates disaster recovery as a positive uplift in GDP. And yet, it is growth economics based on fossil fuels and faster consumption of resources that has contributed to the parlous state we are currently in.

The Australian fires are a catastrophic lesson for us here in New Zealand. According to the BBC the fires killed or harmed three billion animals in Australia.[38] Fires will follow the droughts brought on by global warming and are becoming increasingly the *new normal* from California, to Siberia, Greenland and New Zealand. Following the droughts and the fires that remove the trees come the floods that then strip the top soil and prevent reforestation. Recent fires in the US were caused by powerlines and resulted in massive power outages as the power was cut to prevent more fires. This resulted in the bankruptcy of the utility giant, Pacific Gas & Electric, and is one example of the unintended and unexpected consequences of the Anthropocene and global heating. Another example is the Dunedin fires in NZ that could have caused three-quarters of the city's main water supply to be unfit for human consumption due to ash and toxic fire retardant. Toxic and carcinogenic pollutants have poisoned the US water table and the extent of it in New Zealand is as yet unknown.[39] What this indicates is the systemic fragility of our infrastructure and our lack of imagination when it comes to planning for what promises to be a very uncertain

38. Australia's fires 'killed or harmed three billion animals' (28 July 2020) https://www.bbc.com/news/world-australia-53549936

39. Felton & Kendall (31 March 2021) We sampled tap water across the US – and found arsenic, lead and toxic chemicals. https://www.theguardian.com/us-news/2021/mar/31/americas-tap-water-samples-forever-chemicals

future. Our survival is contingent on our ability to adapt and change to unprecedented scenarios.

Farmers for Climate Action

As climate change is now happening right in their backyard some conservative Australian farmers have become climate change activists. *Farmers for Climate Action* are a lobby group that wants action from a government lead by climate change deniers. They reached out to their own farming network to find out where farmer's stand on climate change? 80% of 1,300 farmers expressed their concern about climate change. Another farmer pressure group against coal miners and gas drillers, *Lock the Gate*, which formed in 2010, 'felt they had no legal rights, so they decided to lock their gates to coal and gas companies' from entering their properties.

New Zealand politicians and co-operative agriculture leaders such as Federated Farmers could well find themselves out of step with more and more farmers who are suffering from droughts, floods, water access, crop decline and stressed and dying animals due to heat exhaustion.

Increasingly consumers are moving away from buying products that are produced with poor farm practices that exacerbate global heating through GHGs emissions and environmental pollution from effluent and fertiliser runoff. There was a recent announcement that America's largest milk producer, Deans, has filed for bankruptcy due to declining consumer demand for milk, and as reported by Stuff.co.nz "health and animal-welfare concerns have also contributed, as more shoppers seek out non-dairy alternatives."

"Oat milk, for example, saw US sales rise 636 per cent to more than US$52 million (NZ$82 million) over the past year, according to Nielsen data. Sales of cow's milk dropped 2.4 per cent in that same time frame."[40]

Political and economic excuses for doing little with respect to agricultural GHGs are short sighted as it is quite clear we have a decade or less to avert disaster. Despite the fact that the two bad boys, methane and nitrous oxide, that have a GWP (Global Warming Potential) from 80 to 300 times more potent than CO_2 at heating the atmosphere; agricultural emissions have been left alone until 2025 thanks to political compromise. Both methane and nitrous oxide could, in the near future, come under the microscope, as human activities that increase the natural baseline of these GHGs becomes more critical. There are 50 gigatons of methane trapped in the permafrost in the Arctic that could explode into the atmosphere as the ice continues to melt at a faster than expected rate.[41] What that will mean is that methane could become a significant problem for global heating. Nitrous oxide has also been identified as more important than previously thought. At the end of 2012, it was calculated that nitrous oxide and methane may contribute as much as 25% of California's GHG emissions. The point is that these very potent GHGs, that have been downplayed by the Zero Carbon Act, may suddenly headline and that could force sudden change in the IPCC reporting and subsequent advice from our own Climate Change Commission. The result could be shock legislation and more extreme pain for farmers. As I write

40.   Chapman, M. (13 Nov. 2019) The biggest milk company in the US declares bankruptcy https://www.stuff.co.nz/business/farming/117380386/the-biggest-milk-company-in-the-us-declares-bankruptcy

41.   See this story Fire and Ice on the show Sunday on TVNZ https://www.tvnz.co.nz/shows/sunday/clips/fire-and-ice

this COP26 is underway in Glasgow and already New Zealand is looking increasingly isolated in its decision to procrastinate about methane emissions. Under the leadership of President Biden the US, and EU have introduced a pact to reduce methane emissions by 30 percent of 2020 emissions by 2030. In total over 100 countries, including New Zealand, have signed the pact at COP26. According to the US Special Presidential Envoy on Climate Change, John Kerry, citing Dr Fatih Birol of the International Energy Agency, the 30 percent reduction in global methane would equate to all the transportation emissions on the planet, i.e. the GHG emissions by all the cars, all the planes, all the ships – the entire global transportation sector being reduced to zero by 2030. "Cutting back on methane emissions is one of the most effective things we can do to reduce near-term global warming and keep to 1.5°C," said Ursula von der Leyen, the European Commission president, referring to the 2015 Paris Agreement's toughest climate goal.[42]

However, the New Zealand government has said it will stick with its pedestrian reductions that will only see a 10 percent reduction below 2017 levels (not 2020) by 2030 and 24-47% by 2050. According to Radio NZ: "...there has been a growing focus on methane as a way of buying extra time to tackle climate change. Although there's more CO_2 in the atmosphere and it sticks around for longer, individual methane molecules have a more powerful warming effect on the atmosphere than single CO_2 molecules. "We cannot wait for 2050," EU Commission chief Ursula von der Leyen told the summit. "We have to cut emissions fast." She said

---

42. Read more: https://www.newscientist.com/article/2295810-cop26-105-countries-pledge-to-cut-methane-emissions-by-30-per-cent/

cutting methane was "one of the most effective things we can do to reduce near-term global warming", calling it "the lowest hanging fruit"."The pledge covers countries which emit nearly half of all methane, and make up 70 percent of global GDP, the US president [Biden] said.[43] Journalist, Rod Oram, has reported that the New Zealand farming lobby is out of step with the rest of the world, the IPCC and the general tenor of COP26.[44]

Delay in dealing with changes to New Zealand's farming system will simply mean that more needs to be done in less time – in other words, the crisis will become more extreme. Meanwhile, there are more and more examples of changes to the farm system internationally and New Zealand is also at risk of looking out of touch with global concerns on mitigating agricultural emissions. Just recently, Nestle the biggest food group in the world has committed to invest USD$3.58 billion over the next five years to progress towards net zero emissions by 2050. The company has vowed to halve its 92 million tons of GHGs by 2030 and to switch to 100% renewables across 800 sites by 2030. "The company said it would work with farmers and suppliers to promote regenerative agriculture practices – such as restoring soil health – saying it expects to source 50% of its key ingredients from farmers using these techniques by 2030, and will scale up its reforestation programme."[45] We may have even just reached that tipping point as the scientists continue to argue about the validity of data that

---

43. COP26: US and EU announce global pledge to slash methane https://www.rnz.co.nz/news/world/454806/cop26-us-and-eu-announce-global-pledge-to-slash-methane

44. Listen to his interview here: 'Rod Oram at COP26 summit.' From Nine To Noon, 9:09 am on 1 November 2021 https://www.rnz.co.nz/national/programmes/ninetonoon/audio/2018818620/rod-oram-at-cop26-summit

45. See Reuters as cited by Carbon Pulse https://carbon-pulse.com/116368/

affects the certainty of their computer simulations. The Arctic is heating faster than the rest of the planet and scientists have just announced their findings that up to 400 times the usual amounts of methane have been detected at one of the locations in the Laptev Sea near Eastern Siberia. According to one of the Swedish scientists, Örjan Gustafsson, of Stockholm University: "At this moment, there is unlikely to be any major impact on global warming, but the point is that this process has now been triggered. This East Siberian slope methane hydrate system has been perturbed and the process will be ongoing." While there is still scientific uncertainty, and the paper that reported this has not yet been peer reviewed, still the precautionary principle would recommend that everything that could be done now to prevent more methane emissions should be. As The Guardian's Global Environmental journalist wrote: "Temperatures in Siberia were 5°C higher than average from January to June this year, an anomaly that was made at least 600 times more likely by human-caused emissions of carbon dioxide and methane. Last winter's sea ice melted unusually early. This winter's freeze has yet to begin, already a later start than at any time on record."[46]

In a statement published in the journal of *BioScience*, (and reported in the *Guardian*), on the 40th anniversary of the first world climate conference, 11,000 scientists from 153 nations endorsed a statement written by dozens of scientists. The lead author, Professor William Ripple of Oregon State University said he was motivated to write it because of the extreme weather he had observed. "The scientists say the urgent changes needed include ending

46.   Watts, J. (27th Oct. 2020). 'Sleeping giant' Arctic methane deposits starting to release, scientists find. The Guardian.
       https://www.theguardian.com/science/2020/oct/27/sleeping-giant-arctic-methane-deposits-starting-to-release-scientists-find

population growth, leaving fossil fuels in the ground, halting forest destruction and slashing meat-eating." Why we must be focused on the total picture and not just carbon dioxide is that we simply do not know enough about Earth System Science and that we could see a sudden phase change forced by a relatively small addition of methane, nitrous oxide, hydrofluorocarbons, carbon dioxide or all of the above interacting with the hydrosphere, biosphere, atmosphere and lithosphere. As the scientists pointed out: "Despite 40 years of global climate negotiations, with few exceptions, we have largely failed to address this predicament. Especially worrisome are potential irreversible climate tipping points. These climate chain reactions could cause significant disruptions to ecosystems, society, and economies, potentially making large areas of Earth uninhabitable." A domino effect could take any hope of mitigation out of our hands at any point.

On the 31st January, 2021, the independent Climate Change Commission published its draft advice for consultation[47] awaiting feedback from the public and the government. It's chief recommendations reported by the New Zealand Herald were:

• Winding down imports of fossil fuel light vehicles with internal combustion engines – or conventional cars – by 2032.

• Slashing livestock numbers by around 15 per cent by 2030.

• Planting 380,000 hectares of new exotic forestry by 2035.

• Cutting all greenhouse gas emissions by 36 per cent –

---

47.   See He Pou a Rangi Climate Change Commission Draft Advice for Consultation (2021).https://ccc-production-media.s3.ap-southeast-2.amazonaws.com/public/evidence/advice-report-DRAFT-1ST-FEB/ADVICE/CCC-ADVICE-TO-GOVT-31-JAN-2021-pdf.pdf

or from an annual average 69.2 megatonnes (Mt) carbon dioxide equivalent (CO2-e) of long-lived gases in 2018 to 44.6Mt – by 2035.

• Reducing gross carbon dioxide. This would need to fall 35.1 Mt CO2-e in 2018 to 22.2 Mt CO2e in 2035 – a 36.8 per cent reduction.

• Bringing down biogenic methane by 1.32Mt CH_4 (methane) in 2018 to 1.11 Mt – a near-16 per cent reduction – by 2035.[48]

The CCC and IPCC will wait to hear what the NZ government decides to do before it evaluates the new commitments, or NDC (Nationally Determined Contribution) from New Zealand. Without these new targets the CCC advised in their draft: "It is our assessment that current policy settings do not put Aotearoa on track to meet these targets. To do so, Aotearoa must accelerate action on climate change."[49]

In the final report to the government, after public consultation, the CCC doubled down on using the ETS 'by boosting the price signal and incentivising businesses and individuals to make choices that lower emissions." They seem to think that they can still tweak the ETS by measuring the system level 'alongside the ETS to help overcome market problems and to spur innovation."[50] The CCC acknowledged that there was strong business support for the ETS, but they also received submissions that thought the ETS should be repealed. While the CCC said

---

48. Fewer cows, mass EVs: what NZ must do to hit climate targets. (31 Jan, 2021) https://www.nzherald.co.nz/nz/fewer-cows-mass-evs-what-nz-must-do-to-hit-climate-targets/

49. He Pou a Rangi Climate Change Commission Draft Advice for Consultation (2021) ibid.

50. See He Pou a Rangi the Climate Change Commission (2021). Ināia tonu nei: a low emissions future for Aotearoa. p.238

that the ETS will remain the key policy tool for reducing emissions, it also believed that it needed support from a 'comprehensive suite of climate policies' such as 'taxation, electricity pricing and grants or subsidies.' However, carbon pricing and a carbon market remains the central mechanism for the CCC and 'that pricing works best when decisions about emitting activities are made based on optimising costs.'[51]

Yet, the CCC admits that policy makers have played with carbon pricing to combat perceived defects in the ETS with large corporations sitting on unused carbon credits. They wrote: "Every functioning ETS in the world today contains market stability mechanisms that alter the number of units available depending on the market price or other factors."[52] It would seem that while the CCC still believes that the markets are the best option, in this report they have backed away from a blind faith in the 'invisible hand' and accept that the ETS cannot totally address updates to scientific data, market failure, or increasing carbon emissions but requires a 'balanced' mix of market mechanisms and policy regulation. It remains to be seen if this advice goes far enough and whether the mechanisms for regulation through the ETS will work; by the time 2035 comes around it maybe too late to find out.

We have to move on from the political election cycle and think about what we can personally do – lobby the government, the Federated Farmers, and dramatically reduce, if not stop eating dairy and meat altogether. This may sound radical but time is literally running out and it is not good enough just because we like to eat certain food and consume an unsustainable lifestyle. Food is, after

51.   He Pou a Rangi the Climate Change Commission (2021).ibid p.214
52.   He Pou a Rangi the Climate Change Commission (2021).ibid p.214

all, a cultural ritual and the one certainty we have about the future is that our culture is likely to change rapidly and without any sense of predictability. We are about to all become firefighters, but what will we do if we run out of water, oxygen, and a cool summer breeze? It is food for thought.

7

Ground Zero

I have been exploring the origins of 'zero carbon' and the awkward truth, that for sixty years we have known that there is something wrong with the way we have been living, especially in the W.E.I.R.D world, (Western, Educated, Industrial, Rich, Democratic). The etymology or origins of the word weird also hints at the origins of our weirdness. According to the Oxford Dictionary, weird is an adjective meaning: "something supernatural; unearthly: weird, inhuman sounds. Informal very strange; bizarre: a weird coincidence – all sorts of weird and wonderful characters." And has an archaic meaning, mainly from an old Scottish noun meaning a person's destiny, but also as a verb in an informal North American usage 'weird someone out' meaning to 'induce a sense of disbelief or alienation in someone'. The Oxford Dictionary describes its origins as: "Old English wyrd 'destiny', of Germanic origin. The adjective (late Middle English) originally meant 'having the power to control destiny', and was used especially in the

Weird Sisters, originally referring to the Fates, later the witches in Shakespeare's Macbeth; the latter use gave rise to the sense 'unearthly' (early 19th century)."

But really, what is our 'ground zero' for our weird behaviour, our self-destructive, cognitive dissonance that ignores the facts; that twists our beliefs to support a deathstyle and ignores an authentic lifestyle – a mad anthropogenic head long rush towards oblivion – is that the destiny for humanity? Retracing our steps we might look at what the psychologist Arnett pointed out in 2008 that most articles in American Psychological Association journals were fixated on the U.S population yet ignored 95% of the global population.[1] The very term 'ground zero' has an American military origin referring to the concentric circles that radiate out from a nuclear explosion, also known as 'surface zero' and has also come to be associated with earthquakes, pandemics, and disasters. It was first used to describe the Trinity test in Jornada del Muerto desert, New Mexico and the atomic bombings of Hiroshima and Nagasaki. The United States Strategic Bombing Survey of those devastating nuclear attacks wrote in June 1946: "For convenience, the term 'ground zero' will be used to designate the point on the ground directly beneath the point of detonation, or 'air zero.'"[2] And herein lies the problem.

The very language we use has been shaped by what is commonly called cultural imperialism or cultural colonisation, in which cultural symbols morph – beginning

---

1. Arnett, J. J. (2008). "The neglected 95%: Why American psychology needs to become less American". American Psychologist. 63 (7): 602–614. doi:10.1037/0003-066X.63.7.602. PMID 18855491

2. U.S. Strategic Bombing Survey: The Effects of the Atomic Bombings of Hiroshima and Nagasaki. June 19, 1946. President's Secretary's File, Truman Papers. Page 5.

life by being imposed by foreign invaders, to finally becoming the *'natural'* everyday turn-of-phrase used by indigenous people in someone else's political, social, and environmental context. It begins with the traders and the merchants; then backed up by invading military forces; followed by the socialisation of politicians and vested interests. Today, on the dark side of the Earth, as the sun sets on the conquistadors of the Northern Hemisphere, Aotearoa, the New World, is seeing the dawn of a new day. Yet, while I type this the for ever helpful algorithmic paperclip[3] no longer pops up to annoy us with a goofy bit of tech advice, but instead, one colonial language is seamlessly supplanted by that of the next invader. My use of English is covertly modified into North American English by the assumption that if I am writing this with U.S. technology then I must want to spell the same way, no matter what my *'color'*. *'Colour'* is not black and white and neither is history.

Language and history are technological constructs that belong to the victors. Cultural imperialism is not purely symbolic but inherently kinetic, or in other words, language and the rewriting of history in our own cultural image have real material consequences for those cultures we supplant. The atomic tests of the 40s, Ground Zero, may appear to be an appropriate place to start our investigation of Zero Carbon but history did not end and begin again in the desert that was once a home to indigenous people. That was before the Spanish invaded Mexico, once upon a time known as *New Spain* (1598), and before the U.S. invaded *New Mexico* (1848).

---

3. A loathed and often mocked symbol of Microsoft's past and the cultural imperialism imposed on 95% of the world's population by U.S. information technology.

It was in 1840, Aotearoa, that the British Empire did an unusual thing that could suggest that the *New Zealand* treaty was a new shift of attitude towards colonisation. While English colonialists did declare sovereignty over these South Pacific islands, they did also sign Te Tiriti o Waitangi with Māori, the indigenous people, the tangata whenua, of Aotearoa. This was not a total colonial erasure of Māori tikanga; Māori culture and mātauranga Māori,[4] or knowledge. However, the cult of the new was embedded in this history and this New World was typically and deliberately overwriting the previous indigenous culture. Ironically, *Zeeland* was Dutch and it was Dutch cartographers that named their so-called '*discovery*', '*Nova Zeelandia*'. It was ironic that while the Dutch did not invade Aotearoa they used the Latin language to name New Zealand. The Dutch tried to impose a new history on an occupied land, but the Roman Empire that had occupied Europe still left traces of their own culture long after invasion and occupation. The cultures that wiped away the Roman Empire did not succeed in erasing the last traces of Rome, although many, including the British, and later the Americans, would claim their own *ground zero* moment in history. The Latin phase *terra nullius* or nobody's land was applied to Australia, despite the aboriginal occupation of the continent for over 50,000 years. *The End of History* is a repeated concept that is used by invading cultures, their ground zero, it follows in the aftermath of colonisation, the figurative and literal uprooting of previous cultures consisting of plants, animals, people and minerals. In New Mexico the surface of the ground was weeded free of the past in order to transplant a new monoculture, defined by

---

4.   See Mātauranga Māori and Museum Practice by Te Ahukaramū Charles Royal (2007)

oil drilling, mineral extraction, cattle ranching, forestry, military and nuclear weapons testing. The colonisers, in many lands have characterised indigenous agriculture as *'unproductive'* and were often wilfully blind to the light touch of *'first nations'*. Colonisers often described the land they wished to take, and enclose, before they claimed it as their own, as native *'wastelands'*. Political economy and the history of colonial property rights are deeply implicated in the forgotten history of invasion and confiscation that preceded neoliberal cultural imperialism, and the so-called *'free market'*. This short list of New Mexico's economic activities are all responsible for some contribution to anthropogenic harm, and indeed following July 16, 1945 and the Trinity nuclear bomb test, technological contributions to GHG emissions have sharply accelerated.

As of the beginning of this year we can examine how the international community is progressing towards *'net zero'*. As Simon Lewis, the Professor of Global Change Science at University College London and University of Leeds, has bluntly explained: "The science of net zero is simple: every sector of every country in the world needs to be, on average, zero emissions."[5] Lewis notes that while we might know how to do this in certain industrial sectors such as electricity, cars, and buildings, politicians have been blithely ignoring air travel, shipping logistics, and agricultural emissions, and they are being assisted by accounting tricks used by big business lobbies. The ambiguity of *'net zero'* is turning an ambitious goal to save us into a novel weaponised future, our new *'ground zero'*, our point of no return. Lewis writes: "Net zero increasingly

---

5. Lewis, S. (March, 2021) " The climate crisis can't be solved by carbon accounting tricks." The Guardian. https://www.theguardian.com/ commentisfree/2021/mar/03/climate-crisis-carbon-accounting-tricks-big-finance

involves highly questionable carbon accounting. As a result, the new politics swirling around net zero targets is rapidly becoming a confusing and dangerous mix of pragmatism, self-delusion and weapons-grade greenwash."[6] 'Net zero' is probably not understood by most because the industries, dates and percentages are in constant negotiation and as a result we are not getting much closer. Lewis notes: "The world is on track for emissions to be just 0.5% below 2010 levels by 2030, compared with the 45% needed on the road to net zero by 2050."[7] 2030 is shaping up instead to be our 'ground zero'. It really could be our 'nuclear moment'. This is not what our PM, Jacinda Adhern intended to imply.[8] Rather our creative accounting tools, and market led economics, are disguising the historical past that remains interlocked with nuclear weapons baring down on us. Meteorology and the study of climate patterns, required to understand climate change, was developed to support nuclear weapons testing and the need to predict where nuclear fallout would spread according to wind directions and weather patterns.[9]

Wai – an Economic Metaphor

Economists who still believe that pricing signals will magically reduce carbon to a level that will prevent ground zero also believe that our Emissions Trading Scheme can be tweaked to better inform the market. As Raworth has

---

6. Lewis, S. (March, 2021). ibid.

7. Lewis, S. (March, 2021). ibid.

8. Adhern attempted to rally the Labour party and public around 'our nuclear moment' implying this was a moment in NZ's history similar to the ban of nuclear ships in the 80s.

9. See Worldbending. The End is Nigh for a brief history of atmospheric testing and meteorology.

pointed out public policy has been in the mental vice of economics for seventy years. Our ability to think about policy in new and innovative ways will be determined by the extent we can identify the way international economics continues to set the agenda for our public debate. After Keynes this country has been largely led by American neoliberal theories. Raworth writes that economic theory has been supported by the power of graphic visualisation, bar charts, and piecharts of economic data. In 1948 Paul Samuelson, who has been called the *father of modern economics*' wrote an introductory economics text book that was used by returned servicemen to re-educate them for new jobs. Many studied engineering, and economics was taught as a minor course in the curricula. Samuelson found that pictures were very helpful in explaining economics and he included one of the first diagrams explaining macroeconomics using water flow and plumbing as a metaphor for the flow of money, complete with a hand water pump and tap. [10] He was so successful that his text book sold four million copies in forty languages around the world and dominated for thirty years. Samuelson introduced a circular closed flow diagram based on the water metaphor. The following year, 1949, a kiwi engineer and economist, Bill Phillips, designed and built the MONIAC, (Monetary National Income Analogue Computer) as a teaching device for his classes at the London School of Economics. It was the world's first hydraulic water computer. While the device was genius and well before its time, eclipsing the complex mathematical algorithms that became the staple of economic computerisation, it was still deeply flawed. The system was

---

10. See Raworth, K. (2017) ibid. p.17

closed and ignored energy flows, raw resources, and social impact. Yet, this circular flow model has survived and even lives on as a metaphor in the refined design of Raworth's Doughnut Economics. What is more today when we stare into the watery abyss contemplating both droughts and floods we have all been re-educated to be neoliberal economists. It is no longer sufficient for a scientist to point out what is going wrong with our most precious resource, water, they also feel compelled to offer solutions.

Scott Jasechko and Debra Perrone have recently published an alarming study of close to 40 million water wells around the world. It shows that globally 40% of our water needs, including agricultural irrigation, and rural and urban drinking water, are supplied by shallow groundwater wells that are threatened by climate change. One in five could run dry if underground water levels fall by only a couple of metres. Around the world, including New Zealand, agriculture is one of the biggest extractive industries of water. According to Jasechko and Perrone pumping water is the most extractive mining activity carried out by humanity. Jasechko is not an economist, but he described groundwater flow in monetary terms: "One can think of groundwater reserves kind of like a bank account so there is income, that is what we call recharge as groundwater scientists and that might take the form of rainfall, or snowmelt trickling through soil profiles to enter the aquifers and replenish it – so that's the income. The losses or withdrawals from the bank account can occur either through natural flows of groundwater that just exit and maybe replenish the flow of water in rivers but it can also take the form of pumping through wells."[11] He went

---

11. Radio NZ. 27th May, 2021. Groundwater wells at risk of running dry:

on to offer two solutions to the decline in groundwater: "There are solutions, should we choose to enact them...In places where groundwaters are declining there's a number of options one could be developing and enforcing policies that limit groundwater depletion. Another may take the form of market forces we could develop groundwater markets to develop trading."[12] While neoliberals would be horrified by policy intervention the ability for a water trading market is likely to be as successful as the ETS, in other words a failure.

It is one of the pillars of neoliberalism that any government intervention is suspect and will distort the 'natural' order and that the free market is part of that natural order. It is why economists continue to have faith in markets and carbon pricing because capitalism is the most 'natural' and most efficient mechanism to deliver almost everything.

Hayek, Friedman and those of the Mont Pelerin group believed that Keynes was wrong and that government intervention in almost any area would create market distortions. For the neoliberal Fukuyama, who pronounced the 'end of history' in 1989, this was the end of Communism and the 'final liberation from Keynesianism'.[13] Today, there is almost a consensus amongst economists that economics, as they preach it, will continue to provide solutions to wicked problems like climate change. However, once again language, paradox, and metaphor have obfuscated the past and history has been rewritten by the victors. Economists have attempted to create their own ground zero moment.

---

research. https://www.rnz.co.nz/national/programmes/ninetonoon/audio/2018797188/groundwater-wells-at-risk-of-running-dry-research

12.   Ibid

13.   Klein, N. (2007). ibid.

Just as there have been a series of catastrophic mass extinctions, prior to the current sixth mass extinction we are just now beginning, each crisis appeared to wipe the slate clean. There have been many attempts by cultures to reset the clock, to give their own cosmological and historical explanation for their presence and their self-proclaimed success – their own ground zero moment in history as if they too had wiped the slate clean.

Tabula Rasa

In property development a greenfield project is one described as 'undeveloped' land.[14] Yet this terminology is weird as it often applies to land that has been previously used for agriculture or landscaped and is subsequently seen as an opportunity by a property developer. How can it be undeveloped if it has been touched by agriculture or a landscape designer? The use of language is important with respect to the land or whenua. It was climate change that originally enabled humans to stay in one place for long periods of time. The Holocene is an unusually stable period of warm weather that has lasted 12,000 years and made it possible for civilisation to establish itself. But philosophically, from a geologic, biologic, and hydrologic perspective was there ever a place or time when the whenua was not occupied by some thing? Not only was the colonial term *terra nullius* – land of nobody or land of none, non sense, but by association the Latin legal phrase 'res nullius' or 'nobody's thing' probably more accurate because the land does not belong to anybody, but to **every thing**. Ever since

14.   Bradley, Naima; Harrison, Henrietta; Hodgson, Greg; Kamanyire, Robie; Kibble, Andrew; Murray, Virginia (2014). Essentials of Environmental Public Health Science: A Handbook for Field Professionals. OUP Oxford. p. 101. ISBN 978-0-19-150540-9.

agriculture first took hold of our imagination humanity began to assume some sort of control over the resources that led to property boundaries and assumptions about human exclusivity. It was not just a succession of cultures that believed that their occupation of the land was superior to those cultures that came before but many cultures privileged human culture over the culture of other things that were already in place. Certainly, Greek-Judaeo-Christian culture assumed a Great Chain of Being that descended from an anthropomorphic God down through angels, men, then women, in a strict order that valued a rock as the lowest of the low.

It was Aristotle who is first credited with the phrase *'unscribed tablet'* to describe the human mind that was blank until it had a thought, just as a blank stone tablet, a wax tablet, a chalk board, or a computer tablet were blank before anything was written on them. *Tabula rasa* or the erased tablet was a mind-like metaphor. It suggested that just like a wax tablet that could be heated and then smoothed blank, or a chalk board that could be erased, the mind could also be wiped clean.

Once it was handed down to the 20th century, and filtered through two thousand years of cultural imperialism, behavioural psychology picked up the concept of *'tabula rasa'*. It was a concept that has underpinned the frightening and barbarous use of shock therapy. Put simply, B.K. Skinner believed that the human mind could be wiped clean and reprogrammed using rewards and punishments. It was taken up by an ambitious, Dr Ewen Cameron in the 50s, and whose cruel experiments using very high voltage shock treatments, combined with high dosage medications of LSD and other mind altering drugs, became the basis for the CIA's modern interrogation techniques used from Chile to Iraq and Guantanamo Bay.[15]

Shock and Awe Downunder

Why did present day New Zealand economists hold fast to the faith and conviction that the market is the most natural way to control climate change? Given the global influence of neoliberal Friedmanite economics the answer should be obvious, but to give a more detailed explanation I will quickly review the shocking history of New Zealand's bloodless 80s coup. I am only half joking when I call *Rogernomics*, a movement named after Roger Douglas, an undemocratic economic insurrection. I admit I have had a blindspot about the mid-to-late 80s in New Zealand as I was then living in Australia with my wife-to-be, Sarah. I had earlier been studying New Zealand politics, completing my Masters thesis in 1984. I had been researching the deregulation of Thatcher's economy and warning of the possible consequences here in New Zealand, and yet, when I look back 35 years later, it is as if the whole country has suffered from mass amnesia. Bruce Jesson's book, *Only Their Purpose is Mad*, published in 1999, wrote:
"New Zealand was a hollow society, a society without texture, a society without centres of resistance, as 1984 and its aftermath have demonstrated. New Zealand had such a shallow culture that most New Zealanders knew little about their country's history. Amnesia is not a recent development but is part of the colonial condition."[16] I had to go back in time to figure out who, and how this coup had

---

15.   See Klein, N. (2007) ibid.

16.   Jesson, B. (1999). Only Their Purpose is Mad. The Dunmore Press. New Zealand. pp.70-71.

happened, and when I read about our *state of amnesia* in Jesson's book it began to make sense. He was right, when I studied New Zealand politics and history as an undergrad many of my friends and family considered them unusual subjects. At the time Keith Sinclair's slim, *History of New Zealand*, (1959) was considered an academic book read by few. New Zealand had to wait until 2003 when Michael King's best selling, *The Penguin History of New Zealand*, was published. It marked a watershed moment in our history when we began to take a popular interest in our past. Even today, Aotearoa has been slow to introduce our history into the school syllabus.[17] I was certainly aware at the time that neoliberalism was leaching into the country from the UK and US, but it was not until the 21st century did I realise how profound a revolution had taken place while I was away. The *free market* algorithm had been installed and the *culture machine* was left running, thus producing an invisible logic that has been reinforced by 21st century gizmos. A number of economists, and political commentators such as Evans, Grimes, Wilkinson & Teece made claims that the 80s economic reforms were an inevitable *pursuit of efficiency* following the almost Eastern Bloc communist control of the Muldoon years. According to Evans et. al, these rapid reforms were: "Triggered by a constitutional and foreign exchange crisis in July 1984." I recall a drunken Prime Minister, Robert Muldoon, or Piggy, as he was called by those who disliked him, ignoring an historic convention and refusing to comply with the wishes of the incoming government. The Treasury, economists from the Reserve Bank, and Roger Douglas had for a number of years called for a 20 percent devaluation in the

---

17.   See https://www.beehive.govt.nz/release/nz-history-be-taught-all-schools

New Zealand dollar. High unemployment and the loss of New Zealand's long-standing triple-A credit rating all led to a snap-election and a sense of an impending disaster. However, as Easton and Jesson have both pointed out these were poor excuses for a so-called crisis. Evans et. al went on to explain: "New Zealand launched into a sequence of economic reforms which David Henderson (1995), an experienced OECD observer, has called: "one of the most notable episodes of liberalization that history has to offer."[18] Despite, the high drama surrounding the election of the Fourth Labour government, Jesson wrote: "Yet the collapse of the regulated economy in 1984 cannot be ascribed to an economic crisis."[19] Jesson argues that the traditional business and bureaucratic elite lost confidence in themselves, leaving a vacuum for the New Right to fill. Brian Easton called the speed of the economic change a 'blitzkrieg' and Jesson observed: "Subsequently, the coup of 1984 was justified with the chant, borrowed from Margaret Thatcher. 'There is no alternative,' but this was not the case."[20] According to Goldfinch: "Nor was the New Zealand experience of economic restructuring simply a copy of what was happening in other nations throughout the nineteen-eighties; liberalization in New Zealand was carried out more extensively, more quickly and showed a degree of theoretical purity that was probably unparalleled anywhere in the world until, arguably, the liberalization of eastern Europe."[21]

18. Evans, L., Grimes, A., Wilkinson, B., Teece, D. (Dec. 1996) "Economic Reform in New Zealand 1984-95: The Pursuit of Efficiency". Journal of Economic Literature. Vol. XXXIV. No.4 pp. 19856-1902. American Economic Association.

19. Jesson, B. (1999). ibid. p76.

20. Jesson, B. (1999). ibid. p77.

21. Goldfinch, S.(1998). "Ranking New Zealand's Economic Policy: Institutional

Roger Douglas was part of a Troika that included Richard Prebble, and David Caygill. But despite their introduction of neoliberal *'bullshit bingo'*, a game in which they bandied around words like *freedom*, and *progress*, scholars have not credited them with any deep seated theoretical understanding of what they did. Yet, they were praised and recognised around the world by none other than the Trump ally, and neocon Republican, Newt Gingrich, who wrote the following in praise of Richard Prebble's thin book, *I've Been Thinking* (1996):

"The popular revolution for lower taxes, less government spending, and more individual freedom is sweeping the globe. New Zealand has been riding point on this revolution, and Richard Prebble's book explains the recent past and near future of our movement."[22]

The Troika catalysed an ideology that decimated unions and collective bargaining, and hailed the kiwi entrepreneur's myopic vision of *'two cars, a house, a bach, and a boat'*. This was once the *'Kiwi Dream'* also known as *'God Zone'* or *'God's Own'*, made famous by the British MP, Austin Mitchell, *The Half Gallon Quarter Acre Pavlova Paradise*, it was a country where *everyone* owned their own house on a quarter acre piece of property. This was a compelling dream that Thatcher and Gingrich also used to enlist political support under the moniker of the *'ownership society'*. Thatcher focused her attention on the British public council estates which she believed were anathema to a *'free'* capitalist society. Not only did the government have no business being in housing, but more importantly to

Elites as Radical Innovators 1984-1993". Governance: An International Journal of Policy and Administration. Vol. 11. No.2. p.177

22.   See Prebble, R. (1996). I've Been Thinking. Seaview Publishing. New Zealand

Thatcher, the government was killing the entrepreneurial spirit of the nation. Klein wrote that Thatcher believed that if those impoverished council residents could be induced to buy their own flat they would come to associate with the wealthy ownership class and resist the concept of wealth redistribution.[23] This also remains the bedrock of the *Kiwi Dream* and is the blockade that prevents a just and equitable life for an increasing number of New Zealanders. Subsequent to a negative public opinion poll the Labour government has abandoned a capital gains tax. Debt and property prices have soared as the government has propped up property prices during the pandemic, pitching renters against property investors. While the Reserve Bank continues to tweak the margins of economic policy, by withdrawing interest rate deductions for landlords, the economy is still managed with a free market mentality. During the 80s the lack of tariff protection, free trade, and a floating currency made the New Zealand dollar a favourite of *'Belgium dentists, and Japanese housewives'*[24] exposing our manufacturing businesses to the financial logic of Friedman's supply side economics, high interest rates, a strong dollar and cheap imports. Once the finance markets were deregulated it opened up New Zealand to currency exchange speculators, both within and outside the country. The pain that followed the devaluation of the dollar and the rapid legislative changes that shocked the country into submission opened our economy to exploitation by the few over the many. While there was almost unanimous disapproval of Muldoon's tight fisted control, many have

---

23. See Klein, N. (2007) Shock Doctrine. p.135

24. Currency trading was no longer connected to the productive side of the NZ economy. See https://www.stuff.co.nz/business/blogs/nick-smith/2414081/The-Dr-Dolittle-currency

come to assess the reforms by Douglas and then Ruth Richardson, in the Jim Bolger National party administration, as going too far. When Bolger came to power in 1991 there were 70,000 state rentals with a cap on the rent of no more than 25 percent of the tenant's income. National set about getting out of the business of looking after low income accomodation by selling off state houses. The 1991 housing stock for low income families was 6000 more than today despite a population surge of 46 percent. In her article on the parlous state of New Zealand housing, Rebecca Macfie wrote: "The housing changes went hand in-hand with the Bolger government's headline reforms: deep cuts to welfare benefits and radical labour market deregulation that forced down wages for many."[25] The poverty that has besieged New Zealand should be no mystery and yet the neoliberal response is that the poor have done it to themselves. However, the invasion of the private buccaneers and speculators who have turned property and accommodation into a get-rich-quick game of monopoly was politically devised and executed. Before the 90s around one third of all houses built were low-cost houses. After the reforms, carried out by both parties, that had declined to 15 percent in 1996 and then by 2014 is only 5 percent. Today, our housing is the most overpriced in the OECD, and the money pumped into the economy since the pandemic has only helped to fuel another spike in the average housing prices in almost every region of the country.[26]

John Key, who would go on to lead the National party and New Zealand as PM was once a currency speculator

---

25. Macfie, R. (Sept. 2021) 'The Great Divide: how the property market is destroying our social fabric. North & South. Issue 419. p.33

26. Macfie, R. (Sept. 2021) ibid.

who had enjoyed the cortisol rush of playing with currency exchange and the future of countries that have become disenfranchised by globalisation. New Zealanders have lost almost all control over their banks and have had to sit by while multinationals continue to repatriate profits with little to show for it except increased carbon emissions, lower wages, and becoming tenants in their own country. It is the root cause of poverty in New Zealand and was justified by Friedman's 'trickle down' theory, (another algorithm) that did not deliver. Instead, public wealth was swapped for private property accumulation. We got absentee landlords and the most unaffordable housing market in the world. The New Zealand experiment has failed to deliver an equitable outcome for many and the gap continues to widen.[27]

Jesson wrote that what Douglas understood was politics and not economics. The real credit for the corporatisation, deregulation and privatisation of the New Zealand economy goes to the shadowy bureaucrats and business elite, Jesson named the 'oligarchs' of Treasury, the State Services Commission, the Reserve Bank, the IMF, World Bank, OECD, and New Zealand Business Roundtable. According to, Alan Bollard, the economist and former Governor of the Reserve Bank of New Zealand (2002-2012), during the late 70s and early 80s the views of Muldoon as the PM and Minister of Finance, "increasingly conflicted with the treasury's advice, taking some pleasure in appealing to the general public over the Treasury economists' head."[28] The Treasury 'eventually came up with

---

27.   See https://www.milestonedirect.co.nz/articles/why-do-cities-become-unaffordable

28.   Bollard, A. 'Australasia' p.90 in Williamson, J (ed.) (1995). The Political Economy of Policy Reform. Institute for International Economics.

their own views about New Zealand's economic requirements'.[29]

In 1982 Friedman wrote: "Only a crisis—actual or perceived—produces real change. When that crisis occurs, the actions that are taken depend on the ideas that are lying around. That, I believe, is our basic function: to develop alternatives to existing policies, to keep them alive and available until the politically impossible becomes politically inevitable"[30] And this is precisely what happened with the so-called crisis that enabled Roger Douglas to begin his shock tactics. Douglas had fallen foul of the leader of the Labour Party in opposition, Bill Rowling, by publishing an unauthorised, *Alternative Budget*, that argued for a 20 percent devaluation of the New Zealand dollar. According to Chris Trotter, Douglas had *'inadvertently'* released a single sheet of paper during the early part of the campaign in the run up to the 1984 election. As the shadow Minister of Finance, with the National party under siege, when Douglas mentioned a 20 percent devaluation of the dollar, this was seen by many currency speculators to be the inevitable future policy, and an attack on the dollar began.

A Treasury official reported to Muldoon the country was facing a *'currency crisis'*. The IMF, that had been infiltrated by economists known as the Chicago Boys, had produced a report on Muldoon's economic interventions that was unflattering. It was well recognised around the world that both the World Bank and IMF made their loans conditional on neoliberal reforms. It has not been found who leaked this IMF report, but the leak undoubtedly exacerbated the sense of a crisis as speculators attacked the New Zealand currency. This was certainly in line with similar crises

29.   Bollard, A. (1995) ibid p.90
30.   Klein, N. (2007). The Shock Doctrine: The Rise of Disaster Capitalism.

orchestrated by Friedman's Chicago Boys in other countries. According to Klein: "The colonization of the World Bank and the IMF by the Chicago School was a largely unspoken process, but it became official in 1989 when John Williamson unveiled what he called "the Washington Consensus." It was a list of economic policies that he said both institutions now considered the bare minimum for economic health—"the common core of wisdom embraced by all serious economists." These policies, masquerading as technical and uncontentious, included such bald ideological claims as all "state enterprises should be privatized" and "barriers impeding the entry of foreign firms should be abolished." When the list was complete, it made up nothing less than Friedman's neoliberal triumvirate of privatization, deregulation/free trade and drastic cuts to government spending."[31] Democratic debate and decision making may have been the usual route for economic policy, "However, if an economic crisis hits and is severe enough—a currency meltdown, a market crash, a major recession—it blows everything else out of the water, and leaders are liberated to do whatever is necessary (or said to be necessary) in the name of responding to a national emergency. Crises are, in a way, democracy-free zones—gaps in politics as usual when the need for consent and consensus do not seem to apply."[32] Trotter asked, was the New Zealand crisis a setup?[33] Certainly, there was secrecy that surrounded much of the strategic discussions around how to get 'reluctant politicians

31.   Klein, N. (2007) ibid.

32.   Klein, N. (2007) ibid.

33.   Trotter, C. (2007) No Left Turn: the distortion of New Zealand's history by greed, bigotry and right wing politics. Random House. pp.285-6

to embrace policies that are unpopular with voters'. [34] The invitation only conference, sponsored by the Institute for International Economics in Washington, D.C. on January 13th, 1993 was convened by the economist, John Williamson, and had the rather dull title of, *The Political Economy of Policy Reform*, which later became a book. What was discussed was acknowledged by the participants could be construed as 'Machiavellian' but they agreed it was necessary if the reforms that were thought to be in the 'public interest' were opposed by vested interests that might block them. The attendees were economic and financial 'technocrats' and 'technopoles', otherwise known as technocrats with political responsibilities. Much of the attention around the ten points of the neoliberal, 'Washington Consensus', has been focused on the LDCs, in Latin America, Asia, and Africa. As the economist Rodrik put it: "Stabilize, privatize, and liberalize" became the mantra of a generation of technocrats who cut their teeth in the developing world and of the political leaders they counselled."[35] Yet as pointed out by the Australian economist, Joseph Wallis, one important lesson that Williamson drew from the colloquium was that countries where their technopoles were burdened with the complexities of democracy, (like New Zealand and Australia), did not face an insurmountable obstacle to radical and comprehensive economic reforms (CRPs). Wallis wrote that: "[Williamson] could point to New Zealand and Australia as being examples of democracies which, since the mid-1980s, have unilaterally undertaken

---

34. Klein, N. (2007) ibid.

35. Rodrik, Dani (2006). "Goodbye Washington Consensus, Hello Washington Confusion? A Review of the World Bank's Economic Growth in the 1990s: Learning from a Decade of Reform" (PDF). Journal of Economic Literature. 44 (4): 973–987. doi:10.1257/jel.44.4.973. JSTOR 30032391

CRPs which clearly fall within the ambit of the Washington Consensus. "[36] In attendance were a number of technocrats who believed politics was too important to leave to the politicians. From New Zealand Dr. Alan Bollard, who would become Governor of the Reserve Bank summed up the economic reforms in NZ pointing to a number of attributes of the Washington Consensus. The currency crisis was one that created the opportunity for Douglas to begin his programme of microeconomic *structural adjustment.*' Klein wrote: "It was "Dani Rodrik, a renowned Columbia University economist who worked extensively with the World Bank, described the entire construct of "structural adjustment" as an ingenious marketing strategy. "The World Bank must be given credit," Rodrik wrote in 1994, "for having invented and successfully marketed the concept of 'structural adjustment,' a concept that packaged together microeconomic and macroeconomic reforms."[37]

Following the successful election of the Labour Party, the incoming Prime Minister, David Lange, said in an interview dated 11 July, 1986, "The circumstances of those first few days in government gave Roger [Douglas] the opportunity to do what he had always wanted to do anyway. But he wouldn't have been able to do that had we gone through the orthodox routine of an election in November, then a budget in June...When the crisis hit in July 1984 [when early elections were called] it was Roger Douglas who, above all, had thought through the economic issues

---

36.   Wallis, J.L. (Jan. 1997) Policy Conspiracies and Economic Reform Programs in Advanced Industrial Democracies: The Case of New Zealand. UNE Working Papers in Economics No. 3. Ed. Brian Dollery. Dept. of Economics. University of New England. Australia. ISBN 1 86389 405 5 ISBN 1 86389 405 5

37.   Klein, N. (2007) ibid.

– so when the Cabinet needed to fall back on economic philosophy, it was Douglas who had one."[38]

After the Bretton Woods conference, (1944), The International Monetary Fund (IMF) was originally set up as a multilateral institution to manage the foreign exchange arrangements between trading nations.[39] Yet, after a few years of Friedman's economic evangelism in the 70s, boosted by Reagan and Thatcher, the IMF, The World Bank, and OECD had all become colonised by the ideology of the Chicago Boys. In many ways it was not so much Friedman's economic theories of supply side monetarism as his evangelical belief that State intervention must be reduced and the invisible hand of the market freed. There was a revolving door whereby New Zealand government officials, such as Roderick Deane, the head economist and Deputy Governor of the Reserve Bank, spent time working for the IMF. Deane had a senior position at the IMF in Washington as Alternative Executive Director, (1974-76) and their technical advice about the 'market' was shared with Treasury officials such as Roger Kerr, described by Deane as "the most outstanding economics student I ever had when I was teaching."

The 'free market' doctrine had first been tried in Chile and spread through the Southern Cone, and while some of Friedman's theories may have seemed to our Treasury to be inappropriate, (because of the take up of his concepts by Latin American dictatorships), there were still plenty of concepts that reduced the power of the Welfare State that

38.   Bollard, A. (1995) ibid p.99

39.   Krueger, A. O. (2003). IMF Stabilization Programs. In M. Feldstein (Ed.), Economic and Financial Crises in Emerging Market Economies. Chicago USA: Chicago University Press.

they agreed with.[40] Douglas appears to have been tutored by Treasury officials who were waiting for a crisis just like this, if in fact they did not precipitate this one. Friedman's shock doctrine spread rapidly around the world, yet New Zealand stood out in the early 80s as the only non-dictatorship that had willingly adopted a political economy that favoured capital accumulation in the hands of the few. The kiwi majority patiently waited for the *'natural'* redistribution of the wealth via the *'trickle down'* theory and the *'invisible hand of the market'*. Friedman and his economic disciples believed that this *'free market'* was the most democratic way to support public welfare, yet they shied away from public debate, and the kiwi Troika undertook their plot without public discussion. The result was a revolution that saw the biggest transfer of public wealth into private hands that the country has ever witnessed, (that is since the theft of Māori land), and yet it was a bloodless coup, because New Zealanders are, to quote Jesson, not only *'docile'* but *'anti-intellectual'*, and easily bored with history or politics. According to Klein: "In direct contradiction of Friedman's central claim, [Stephen] Haggard [a staunch neoliberal political scientist from the University of California] concluded that "good things—such as democracy and market-oriented economic policy—do not always go together." Indeed, in the early eighties, there was not a single case of a multiparty democracy going full-tilt free market."[41] That is, except for New Zealand.

Easton outlined how Douglas carried out the coup: "Like other successful generals, Douglas has written down his strategy:

---

40.   Easton (1997), Trotter (2007), Jesson (1999), Bollard (1995)

41.   Klein, N. (2007) ibid.

If a solution makes sense in the medium term, go for it without qualification or hesitation. Nothing else delivers a result which will truly satisfy the public.

[And] Do not try to advance a step at a time. Define your objectives clearly and move towards them in quantum leaps.

[And] It is uncertainty, not speed, that endangers the success of structural reform programmes. Speed is an essential ingredient in keeping uncertainty down to the lowest possible level.

[And] Once the programme begins to be implemented, don't stop until you have completed it. The fire of opponents is much less accurate if they have to shoot at a rapidly moving target."[42]

Easton noted that while Roger Douglas had begun to discuss neoliberal concepts in his book, *There has Got to be a Better Way!* in 1981, on the whole the book was a muddle of ideas that included state interventionism mixed with free market ideology. A helpful Treasury official, Doug Andrews, who was seconded to the Labour opposition, had helped Douglas to sharpen his neoliberal theories and, possibly with the help of others, developed his secretive strategy for rolling over his leftist Labour colleagues. This strategy was essentially one of '*shock and awe*' that had been successfully used elsewhere, and was what Naomi Klein called '*disaster capitalism*'. Treasury and Douglas jumped on the crisis to flip the switch on the way kiwis saw the State and the market and in accord with the Chicago School saw private enterprise as the paragon of efficiency that should be unimpeded by government regulation.

Jesson, Easton, and other critics such as Jane Kelsey,

---

42.   Easton, B. (1997). The Commercialisation of New Zealand. Auckland University Press. p.80

recognised the sticky fingers of Milton Friedman all over the Treasury papers briefing the incoming Labour government,[43] but it was Naomi Klein who laid it out for me, this strategy had been used before in other countries. The *blitzkrieg* that was part of Roger Douglas' 'military' strategy that could have been taken straight from the Chicago Boys and their playbook. Friedman had developed his rolling attack tactics, first in Latin America, in the late 50s, and then all around the world. In Chile, Pinochet's 1973 coup, that overthrew the democratically elected Salvador Allende, was not only a CIA plot, but was orchestrated and underpinned by economists trained by Milton Friedman at the Chicago School of Economics, disciples of Friedman, known as the Chicago Boys. Today, Milton Friedman's algorithms are still whirring away in the background. Meanwhile, following the social, economic, and cultural erasure by Rogernomics, the nuclear blast of Friedmanism created a ground zero in our psyche leaving a blank slate to rewrite our history. The New Zealand Welfare State had been weaponised and the relatively small equity gaps have subsequently become chasms in the 21st century. Disaster capitalism is an effective circuit breaker and Friedman has proven the shock doctrine works as a tactic for ramming home unpopular policies.

As we move into the 2020s a new breed of economists are aware that new shocks, disasters, and crises are beginning to suggest paradigm shifts away from the neoliberal market forces. The former Chief Economist for the

---

43.   While Easton makes a good argument for the influence of the Chicago School on NZ economists, there were other American influences, and the US was seen as the centre of all economic theory at the time. The OECD was a major influence and that in turn was influenced by American theory. Deane brought home from the IMF the US trend towards microeconomics and more market freedom. See A. Bollard, (ed.) (1989). Influence of American Economics on New Zealand.

Treasury, Dr Girol Karacaoglu, in his recent book, *Love You: Public policy for intergenerational wellbeing*, wrote that the COVID 19 pandemic is an opportunity to introduce a kinder, less fiscal and monetary approach to public policy based on wellbeing. Karacaoglu enthused: "Covid-19 provides a perfect platform and a great opportunity for implementing a public policy that is genuinely focused on enhancing intergenerational wellbeing."[44]

Uncle Milt's Shock Treatment

We are currently living in two states at once that includes neoliberal dogma mixed with the emerging ambition of *'intergenerational wellbeing'*. Yet, the Treasury's microeconomic market efficiencies still lurk in the background of many policy decisions made during the pandemic. While New Zealand may not have any collective memory of life before 1984 and the binary choice presented by Rogernomics,[45] yet we can still see the ideological breadcrumbs spread all over the neoliberal picnic blanket. 1984 was New Zealand's ground zero – the country was hit with such a nuclear blast that our memory and many other things were erased.

Milton Friedman's mentor and founder of the Mont Pelerin Society, Frederick Hayek, began to slice up the loaf and share his wisdom in 1944 when he published his anti-communist treatise, *The Road to Serfdom*. This was his rebuttal to Keynesian economics and the Welfare State. It took sometime for Hayek, and later Friedman, to shift the

---

44.   Dr Girol Karacaoglu. (2020) Love You: Public policy for intergenerational wellbeing. Apple Books. p.491

45.   Many still believe we have no choice but to either accept *communist* command and control, or a utopian *free market*

intellectual debate in economics away from the New Deal towards neoliberal freedom. However, as the Cold War developed the US government and universities under their sway started to bolster their economic theories with algorithmic gymnastics and mathematical displays of technical virtuosity. Economics, and particularly, neoliberal monetarism, laid claim to being a *'science'* that produced models and simulations of economic reality. By the late 1950's Friedman had found patronage with the Ford Foundation and together they convinced the Eisenhower administration that they could do their bit for the US economy and hold back the red tide of communism in Latin America. Klein explains: "The original plan was simple: the U.S. government would pay to send Chilean students to study economics at what pretty much everyone recognised was the most rabidly anti-'pink' school in the world – the University of Chicago."[46] In 1956 the indoctrination began with its first students arriving from Chile to be schooled in supply side economics, and in 1973, they were ready for the ideological *'blitzkrieg'* supported by two other forms of shock therapy, a political coup, and psychological torture developed by the CIA based on the shock therapy of Dr Ewen Cameron. Klein summarised: "These three forms of shock converged on the bodies of Latin America and the body politic of the region creating an unstoppable hurricane of mutually reinforcing destruction and reconstruction, erasure and creation. The shock of the coup prepared the ground for economic shock therapy; the shock of the torture chamber terrorized anyone thinking of standing in the way of the economic shocks. Out of this live laboratory emerged the first Chicago School state, and the

---

46.   Klein, N. (2007). The Shock Doctrine: The Rise of Disaster Capitalism. Penguin Books. Kindle p.59

first victory in its global counterrevolution."[47] The central plan of the Chicago School was to indoctrinate students who would later become leaders of their countries and while I have found no direct link between Friedman and the New Zealand Treasury in the early 80s, his ideas and strategies were definitely replicated in the land of *'milk and honey'* and his *'antistatist'* theories were evident in later years.[48] The Friedman's book, *Free to Choose*, (1980), has been described by one former Governor of the Reserve Bank, Dr Donald Brash, as *'enormously influential'* and Brash invited Milton and Rose Friedman to New Zealand where they joined him on a speaking tour of the country. Friedman shocked some kiwis when he declared that New Zealand should give up car manufacturing as he described it as a particularly *egregious* example of *protectionist* inefficiencies[49]

Goldfinch (1998) surveyed the players behind the scenes of the New Zealand ideological coup, conducting 86 interviews with 17 self completion questionnaires with cabinet ministers, CEOs of government departments and the top thirty company chief executives, along with various union leaders, political party leaders, media movers and academic supporters. What he revealed is the hidden influence of an elite that rallied behind the opportunity to build out the foundations of the *'new right'* that was constructing the edifice of 21st century neoliberalism. Just as the 80s property developers knocked down heritage buildings creating their architectural tabula rasa on which to erect ugly blue mirrored hulks – this was the ground

47.   Klein, N. (2007). ibid p.71.

48.   See Easton, B. 'From Reaganomics to Rogernomics' in *Influences of American Economics on NZ*. (1989) edited by Alan Bollard.

49.   Appelbaum, B. (2019) *The Economists' Hour: how the false prophets of free markets fractured our society*. Picador. p.87

zero for our recent economic history in New Zealand. I remember being shocked and angry seeing what had happened to my city as I woke up one morning to view the wholesale demolition of the CBD by the self-serving Chase Corporation, just one of many opportunists to dive into our country's lolly jar and steal our taonga.[50] One of the patron saints of this movement was the Secretary of the Treasury, Graham Cecil Scott, who held the position from 1986-1993, a one time member of the New Right, ACT party, and who travelled the world on behalf of the IMF and World Bank converting countries to neoliberal economics. According to Goldfinch, Scott's influence across economic policy decisions from 1984-93 ranked higher than Douglas, the Minister of Finance, Caygill, Prebble and David Lange, the P.M. Scott was found to be as influential in the economic structural reforms as Roger Kerr, ex-Treasury, and CEO of the right wing Business Roundtable with more than 6 major economic decisions that shaped this country.

Klein had found that as she began to dig deeper into the history of how Friedman's market model had swept the world, in the wake of the coup in Chile, she subsequently: "discovered that the idea of exploiting crisis and disaster has been the modus operandi of Milton Friedman's movement from the very beginning – this fundamentalist form of capitalism needed disasters to advance."[51]

Klein's argument is backed up by a recent form of

---

50. taonga (noun) treasure, anything prized - applied to anything considered to be of value including socially or culturally valuable objects, resources, phenomenon, ideas and techniques. Examples of the word's use in early texts show that this broad range of meanings is not recent, while a similar range of meanings from some other Eastern Polynesian languages support this (e.g. Tuamotuan). The first example sentence below was first published in a narrative in 1854 by Sir George Grey, but was probably written in 1849 or earlier. https://maoridictionary.co.nz

51. Klein, N. (2007). ibid p.9.

disaster capitalism in the US as poor, often undocumented, immigrants are either tricked or forced by poverty to join the ranks of disaster recovery workers in dangerous, underpaid work. These immigrants were often refugees from their own disasters and had suffered severe trauma and shock to the point that the threat of deportation by ICE, (U.S. Immigration and Customs Enforcement) often compounded their sense of *'learned helplessness'*. According to a journalist for *The New Yorker* magazine, Sarah Stillman, as climate change disasters become more and more frequent these vulnerable people are being exploited by the very global capitalist system that caused the anthropogenic harm in the first place. Stillman wrote: "In the past five years, private-equity firms have acquired dozens of disaster-restoration companies. In 2019, Blackstone, one of the world's largest private-equity firms, acquired a majority stake in Servpro Industries, reportedly for more than a billion dollars...Chasing disasters requires a labor force that is open to arduous work and is instantly mobile. Servpro promises to furnish workers to crisis sites within days, or even hours; one of its slogans is "Faster to any size disaster." To marshal this force, many companies turn to an ill-regulated group of subcontractors and labor brokers, which, in turn, cultivate social networks of migrants and other people seeking economic opportunity. As demand has grown, many of these workers have come to travel a yearly catastrophe circuit."[52]

Friedman and his followers knew that in order for his strategy to be effective, after the shock had hit, or applied, they had to move fast and in secret before the political

52.   Stillman, S. (Nov.1, 2021) The Migrant Workers Who Follow Climate Disasters. The New Yorker. https://www.newyorker.com/magazine/2021/11/08/the-migrant-workers-who-follow-climate-disasters

opposition had time to muster a defence. Klein argued that economic shock therapy shared a similar tactic borrowed from psychiatry – the objective was to use shock therapy to fix the patient, much like a computer, by wiping the disc clean so it could be re-programmed. The patient could be a highly regulated economy, sickened by the coddling of an over zealous Nanny state, or it could be an actual patient or political prisoner.

In the 50s the CIA had commissioned research from Dr Ewen Cameron at McGill University into the use of drugs, and shock treatment for psychiatric patients. The results may not have proven to cure patients but the data was still very intriguing to CIA interrogators who discovered that high voltages, mind altering drugs, and sensory deprivation resulted in a patient whose mind was virtually wiped clean, with almost no memories.

The founder of Behavioural Psychology, B.K. Skinner conducted many experiments electrocuting animals and analysing rewards and punishments. At the height of the Cold War in the 50s the Norwegian psychologist, Ivar Lovaas, developed his autistic therapy based on Skinner's behavioural psychology, known as Applied Behavioural Analysis or ABA, better known to the public as *conversion therapy*. Inhuman punishments such as electrocution that had been perfected on animals were modified for children and their rewards adapted to their personality. In 1965 Lovaas published a paper on his use of an experimental box he used to electrocute a couple of children he described as *schizophrenic autistic* twins.[53] Lovaas designed and built an horrific device with the polite name of the *Harvard*

---

53. Lovaas, O. I., Schaeffer, B., & Simmons, J. Q. (1965). 'Building social behavior in autistic children by use of electric shock.' Journal of Experimental Research in Personality, 1(2), 99–109.

Indocutorium'. Lovaas constructed the device using a Faraday coil, and metal strips that electrified the floor so the child could be administered a shock; if they did not respond to commands he applied *'aversive stimuli'* , otherwise known as punishment. According to Silberman in his book, with a forward by Oliver Sacks, Lovaas became something of a celebrity in the US in the 1960's. He wrote:

"Life magazine elevated the psychologist to international fame with a profile that ran under the memorable headline "Screams, Slaps, and Love." Praising Lovaas' work as "a surprising, shocking treatment that helps far-gone mental cripples," the article (and its photo spread, billed as "an appalling gallery of madness") shaped public perceptions of autism for decades to come."[54]

In 1967 Martin Seligman, an American psychologist, was searching for the cause of depression by using a similar device with dogs. The dogs had been taught that there was no escape from the shocks and when placed in a box dividing two floors, one that was electrified and the other not, the dogs displayed what he described as *'learned helplessness'*.

In country after country that applied Friedman's shock treatment the hapless populations soon exhibited *'learned helplessness'* as they lost all sense of agency as they felt incapable of escaping rapid economic reforms and inevitable shock.

After the introduction of Rogernomics the unions and traditional Labour supporters became paralysed by what happened next. Not only did the political economic landscape change irrevocably, but so did the ecological

---

54.   Steve Silberman, Oliver Sacks. (2015) NeuroTribes: The legacy of autism and how to think smarter about people who think differently. Allen & Unwin.

environment as the trajectory was set then for the future we have inherited today.

It was not the first time that shock treatment and erasure had been used in Aotearoa, New Zealand. As pointed out by the indigenous activist, Haylee Koroi, in *Climate Aotearoa*, environmental degradation went hand in hand with colonisation and industrial agriculture as capitalist ideologies tried to wipe out Māori collective ownership and the relationships of whakapapa that includes the non-human network of interconnections. She wrote: "These imported colonial ideologies justified the domination and exploitation of whānau, hapū and iwi, and by extension our more-than-human relations. This was indicative of broader global patterns of indigenous erasure and land and water exploitation by the British Crown. In more recent times, these colonial ideologies have become so deeply embedded in our collective psyche that they have become almost imperceptible. However, our local, national and global histories can return us to the truth of who we are, where we have come from and, more importantly, where we are going. As bell hooks says, 'At the heart of justice, is truth telling' (hooks, 2000, p. 33). Imagining futures beyond colonial domination is not only important but imperative, because it is there, and only there, that a just climate future is possible."[55]

According to Easton, up until the 80s the New Zealand Treasury officials shared knowledge through experience and precedence. However, a new school of economics dispensed with this quaint tradition and just like an electro-shock patient the New Zealand public and Treasury began with a blank slate – there was no recorded history

55.   Clark, H. (2021) Climate Aotearoa: What's happening & what we can do about it. Allen & Unwin. New Zealand.

that the patient could remember. The algorithm had to be installed on a clean disc, and because monetarism was mostly theoretical it required no previous experience, examples or memory for the algorithm to run – what came before was considered irrelevant, New Zealand was their tabula rasa. That is not to say that many of the patients that underwent Friedman's economic shock therapy weren't held up as shining examples, but the results were often fudged. Friedman economics was not just algorithmic, it was inoculated against history and real-world events because only the mathematical models of efficiency and optimisation really mattered.[56]

Today's New Zealand economists, who grew up during and after the free market insurrection of the 80s, believed that the *free market* algorithms were almost indisputable laws of nature, and that the only way to 'manage' an economy was to leave it to the market and tell the government to get out of the way. For them, and some of their elected ministerial students, the algorithms that drive the ETS machine have freed policymaking from some of the political interference of election cycles. At least, that is their argument. The reality is more sinister as the public are reassured that this hard to understand and technical mechanism is happily running itself and will save us from a fiery death, meanwhile it just keeps getting hotter and hotter.

Going Postal

In 1984, the Fourth Labour government was elected. This

---

56.   See Easton, B. (1997). The Commercialisation of New Zealand. Auckland University Press. Chp. 6 The Treasury: The Philosopher-Kings for Commercialisation.

was the year that I completed my Master's thesis in which I critiqued the proposed deregulation of the New Zealand Post Office, and telecommunications sector. The corporatisation, then the privatisation of the Post Office and Telecom meant that the New Zealand public handed over the state-run telecommunications monopoly to the share market, Fay Richwhite, Ameritech, and Bell Atlantic, who took the profits owed to New Zealanders and ran away to hid it in secret Cook Island bank accounts. This is exactly what happened in many Latin American countries. This was another ground zero, a place in time that was burnt by the nuclear blast of brutal economic reform – it was a devastating shock to the country from which we have still not recovered. History has been scrubbed clean, and many believe the narrative, often repeated by those who benefited from the reforms, that the majority is better off than we were before 1984. Vast amounts of public wealth were transferred to those who were put in charge of our assets without our agreement. The neoliberal ideology of the Treasury was so engrained that they were not even concerned about getting the best price for our forestries, telecommunications, and energy companies because they were convinced that private enterprise would be an improvement to the benefit of all.[57]

Ironically, it was made possible by legacy military technology that was invented to conduct kinetic wars. The bureaucrats, politicians, and business elite that cheered on the coup accused the operations of state owned companies of being inefficient. The Post Office and its services, that had a legacy stretching as far back as the Roman Empire,

---

57. See the documentary Someone Else's Country. (1996) Written and directed by Alister Barry. NZ on Screen. https://www.nzonscreen.com/title/someone-elses-country-1996

came under attack by Reagan and Thatcher, and so New Zealand's 'New Right' followed suit and sold them off. As a precursor to the Internet and modern telecommunications the Post Office was once considered a marvel of efficiency and speed. In Berlin, in the 19th century, a vast network of pneumatic tubes zipped letters and messages around the city, much like packet switching sends data around the Internet today.[58] The Internet shares the military technological legacy of the Post Office. They were both designed to afford military leaders intelligence and data about their troops and their enemies. Packet switching was invented during the Cold War to avoid a single point of failure should a central data collection location became ground zero in a nuclear attack by the Soviets. Yet, without computers and international data transmission the likes of currency traders, such as John Key (later PM of New Zealand) would not have been able to profit from the devaluation of the New Zealand dollar, and international share trading. Technology allowed the explosion of ground zero to happen in miliseconds. The shock led to a state of helpless resignation as the 1984 crisis gave the leaders of the coup an excuse to push through screeds of economic reform without significant resistance.[59]

Uncle Milt's medicine was swallowed by the bucketful and that is why we have no memory of how we got to this 'state of amnesia'?! What happened in 1984 left an indelible stain on our society and the New Zealand disciples that followed in the footsteps of Friedman ensured that the economic blast that created ground zero allowed the winners to steal the spoils of war and build mansions out

---

58.   See Graeber, D. The Utopia of Rules.

59.   Easton, (1997) *The Commercialisation of New Zealand.* p.82 Table 5.1
       Examples of reforms since 1984.

of the inequity of social, cultural and climatic injustice. Our treasury ensured that the vision of Rogernomics has lived on and those that were recruited and indoctrinated in this movement spread the word far and wide.

As a senior adviser within our Treasury, Graeme Wheeler, (1973-1984) could be assumed to have shared a similar economic philosophy with his boss, Cecil Scott, and he held the position of Director of Macroeconomic Policy in the Treasury under Scott (1990-1993) before leaving New Zealand to be appointed by the World Bank as Director of the Financial Products and Services Department. Wheeler, when he was advising New Zealand politicians at the time of the radical reforms of the 80s and 90s, obviously impressed the World Bank and was appointed as VP and treasurer of the the World Bank where he managed assets in a portfolio worth US$250b and an annual borrowing programme of US$10-20 billion. From 2006-2010 he eventually rose to the number two position as Managing Director of operations at the World Bank in Washington with 12,000 staff and an administration budget of US$1.7b. As recently as 2012 he was appointed by the NZ finance minister, Bill English, to be the Reserve Bank Governor. Wheeler's senior positions at the World Bank and in NZ strongly suggests that the World Bank and its ideologies still hold sway over our economic policy direction today.[60] Cyborg economists have become our new high priests who cast algorithmic incantations that their disciples, the policy wonks and politicians recite in their sleep.[61] Yet, despite their fanatical beliefs, these economic fundamentalists

---

60. See Scott's CV https://web.archive.org/web/20130208050445/
http://southerncrossadvisers.com/Downloads/
CV_Scott_Graham_Mar12.pdf

61. See Appelbaum, B. (2019) The Economists' Hour: how the false prophets of free markets fractured our society.

almost never get it right – not one of them predicted the Great Depression and only Nouriel Roubini predicted the GFC, earning him the title of Dr Doom. They still hold fast to the faith and point to the lines of code etched into the stones handed down from the Chicago School.

The original commandments are simple algorithms that Mirowski expands: "...its first commandment is that the market always 'works' in the sense that its unimpeded operation maximizes welfare. Its second commandment is that the government is always part of the problem, rather than part of the solution." The third commandment is that demand for a product, service, or commodity is the rock-bottom fundamental that will determine its price. [62] So, today the value of any thing is determined by its price tag, and if it has no price then it must be worthless. Ground Zero had erased all memory of what came before 1984.

The Carrot & the Stick

The shock treatment, that had erased everything, had proven to be effective in training dogs so that they learned to be helpless, and today there are countless examples of whole populations around the world who have been shocked into submission. One behavioural psychologist who went on to become the President of the American Psychological Association, Martin Seligman, came to realise that his shocking treatment of dogs was now out of favour and he no longer speaks about 'aversive stimuli', i.e. the stick. In 2002 Seligman published a book about a new fashion in psychology, *Authentic Happiness*. He would go on, a decade later, publishing another book, *Flourish*, (2011) that would define a new psycho-economic fashion,

---

62.   Mirowski, P. (2002) ibid. p.203

'well-being theory'. In that same year he would meet with the British PM, David Cameron, and encourage him to look into wellbeing moving beyond simple fiscal and monetary measures of prosperity.

The carrot and the stick of behavioural psychology have been disguised but still lives on in the assumptions behind neoliberal economics and the necessity to encourage growth as a material reward that promises happiness, even when material deprivation is the presumed punishment for the 'lazy and entitled poor'. It now seems that only if we put our minds to degrowth can we possibly pull back from the brink of this global ecological disaster. Wellbeing is a much needed correction to the economics that have focused on outmoded concepts such as GDP, and profit-before-people but growth still demands we keep refuelling the engine of capitalism. We are all participants in the final count down during the zero sum game in which capitalism wins and the planet loses. We may be exhorted by our Prime Minister to 'be kind' during the pandemic, however, if our economic system is unjust this kindness risks being performance theatre and will be but a sop that will not dry the tears of multiple generations of the dispossessed. A circular economic argument will not save us, and cost free, waste free, perpetual happiness engines are a fantasy.

8

Binary Worlds

This book is an exploration of worlds beyond the simplistic slogan of zero carbon, and how we construct those worlds with languages, ideologies, and technologies? But that is not to say that the fundamentals of the Anthropocene are not very real problems; we are facing the end of worlds as we know them, and have to adapt to new worlds that we co-design with other humans and the nonhuman majority.

It is now a truism to talk about the digitisation and virtualisation of every layer of the planet's stack. According to Benjamin Bratton in his book, *The Stack: On Software and Sovereignty*, The Stack can be defined as:

'six interdependent layers: Earth, Cloud, City, Address, Interface, User.' These six layers can be thought of as infected physical strata, both nonhuman and human, that intermix computational realities with physical worlds but include 'social, human and concrete forces' such as energy sources, gesture effects, 'hard and soft systems that intermingle and swap roles.'[1]

The Stack model is useful to understand how human culture and physical strata share a co-habitat. They not only mix together but threaten each other and just because we now have cloud computing this does not remove the physical infrastructure, and geological worlds that holds it up. The Anthropocene is comprised of rock, air, water, and biology, and these spheres are not hermetically sealed but are constantly mixing, morphing and becoming each other. The rock or lithosphere is washed by the hydrosphere releasing gases into the atmosphere and helping provide the building blocks of life into the biosphere.

When the world is digitised and virtualised this may obscure the physical assemblage of things that can no longer be seen, hence according to Taylor, the design is hiding in plain sight.[2] The circuitry that has enabled the miniaturisation of computing has retreated in scale to nanometers, a billionth of a meter. Designers have to work at layers of abstraction that prevent direct sensory perception of their designs.[3] When we use wireless computing and cloud based storage even the medium, the interface layer of the stack, disappears from our consciousness as the user concentrates on the message subconsciously pressing the hardware to their ear, or peering past the physical embodiment of computing through the 'window' of their screen into the virtual world of cyberspace. We no longer see the medium in front of

---

1. Bratton, B. H. (2016). The Stack: On Software and Sovereignty. Massachusetts: MIT Press.

2. Taylor, M. C. (1997). Hiding. Chicago, IL: University of Chicago Press.

3. Mau, B., Leonard, J., & Institute without Boundaries. (2004). Massive Change. London: Phaidon.

us, the deeper design is superficially encased in brushed silver metal or plastic that melds with our furniture both reflecting and hiding the minerals, gases, and microscopic fluids that are carrying bacteria, viruses, and other living beings beyond our consciousness. Decades before Marshall McLuhan told the world, 'the medium is the message' it was a Canadian theologian, Angus Maclean who wrote that 'the method is the message'. Yet we have become less and less interested in either the medium or the method and have instead become almost solely obsessed with the disembodied message, or content only. For example, the definition of social media does not so much refer to the medium, as the messaging, the platform, and what was said on what platform e.g. what did Trump, or someone else, say on Twitter, Facebook, WeChat, Instagram, or TicTok etc? In other words the physical medium could be the screen of a laptop, desktop, tablet, cellphone, Playstation, X Box, Smart TV, radio, or newspaper reporting. The message is seen as disembodied information, as per a simplified formula by Claude Shannon,[4] cybernetics and information theory. Cybernetics sought out and has largely delivered the dream of 'effective computation' in which the universal and disembodied algorithm promises us a solution to all of life's problems. Finn wrote this about algorithmic solutions: "From the beginning, then, algorithms have encoded a particular kind of abstraction, the abstraction of the desire

---

4.   Shannon, under the influence of Norbert Wiener, the father of cybernetics, developed information and communication theory. His algorithms have shaped digital communications and how we think about information. See Conway, F., & Siegelman, J. (2005). Dark hero of the information age: In search of Norbert Wiener, the father of cybernetics; also see Hayles, N. K. (1999). How we became Posthuman: Virtual bodies in cybernetics, literature, and informatics.

for an answer."[5] The game designer and philosopher, Ian Bogost says that we have fallen into a *'computational theocracy'* that replaces God with the algorithm. As Ed Finn argues: "Our supposedly algorithmic culture is not a material phenomenon so much as a devotional one, a supplication made to the computers people have allowed to replace gods in their minds, even as they simultaneously claim that science has made us impervious to religion."[6] However, a number of theorists and commentators have exposed the impact and outcome of different media and the way that our capitalist ideology is embedded in the software and hardware that in turn shapes our pervasive, and increasingly ubiquitous digital communications network and the content it mediates. The Internet has changed the way we write, read and think, much the same way that the printed book changed us around 600 years ago.[7] As early as Plato's *Phaedrus*, around 370 BCE, he warned us that a technology as simple as writing could weaken our cognitive abilities and memory.[8] Sean Cubitt in his book, *Finite Media*, exposed the very physical and geologic nature of what we have come to think of as media. He noted that: "When we speak of the media, we tend to refer to the technological media of the last two hundred years; but everything that mediates is a medium—light, molecules, energy. This flux of mediation is logically prior to communication and to the objects we have learned, through communication, to distinguish from the

---

5.   See Finn, Ed.(2017). What Algorithms Want: Imagination in the Age of Computing (MIT Press) . The MIT Press. Kindle Edition.

6.   Finn, Ed.(2017). ibid.

7.   Carr, N. G. (2010). The Shallows: how the Internet is changing the way we think, read and remember. London: Atlantic Books.

8.   Finn, E. (2017) ibid.

background hum. The flow of mediation precedes all separations, all distinctions, all thingliness, objects, and objectivity. It precedes the separation of the human and the environmental."[9] Cubitt pointed out that while computing is ostensibly said to be weightless and friction-free that by 2008 it had already outstripped the carbon emissions of the airline industry and was growing by at least 15 percent a year.[10]

Material Immaterialism

Long before we scurried down the digital rabbit hole Western culture was shaped by philosophical and religious beliefs that rejected the material world of flesh, blood, bones and minerals.[11] The religious doctrines of spirit and transcendence propagated a reality that was far more immaterial than physical. Over two thousand years of Western religion, philosophy, and emerging science shaped our worldview into a binary logic based on two states, true or false. The concepts of black and white, good and evil, female and male, zero and one, the creation and the abyss, darkness and light, a thing and no thing – all of these dichotomies have become *'common sense'* in Western culture and have laid the foundation for our digital world. These are, however, not universal truths, but co-exist amongst alternative worlds (or world-views), many of which are indigenous, that see these binary opposites as false dichotomies, false binary conceptions of reality,

9.   Cubitt, Sean. (2017). Finite Media: Environmental Implications of Digital Technologies (a Cultural Politics Book). Duke University Press. Kindle Edition.

10. Cubitt, S. (2017) ibid.

11. Wertheim, M. (1999). The Pearly Gates of Cyberspace: A history of space from Dante to the Internet. New York: W.W. Norton.

because not all worlds are defined by binary logic or on and off states. There are unknown states of being that cannot be defined, and may be always unknowable to the subject or thing. Typically, when this binary thinking is formalised it assumes a Boolean logic which means it has only a true or false state. However, human decision making, and physical reality are not that simplistic. In computing this binary logic has been made more sophisticated with Bayesian probability that applies a statistical analysis to a problem.

Norbert Wiener was a polymath who led a post-war academic movement known as cybernetics. It is from his concepts that we have inherited phrases such as cyberspace, cyborg, feedback loops, homeostasis, and transdisciplinary research. Following his very influential book, *Cybernetics; or Control and Communication in the Animal and the Machine* (1949), Wiener gathered a group of scientists, mathematicians, computer engineers, sociologists, anthropologists, biologists and medical clinicians, many of whom became authoritative in their fields. Cybernetics generated many academic papers that moulded the shape of modern thinking about computers and culture and the universal application of computer algorithms to almost any problem. Wiener embraced indeterminate uncertainty using statistics to predict the probability of an outcome. Even intractable problems such as shooting down a fast moving plane, or human decision making succumbed to the brute force of algorithmic number crunching. It was the vision of the mathematical philosopher, Gottfried Wilhelm Liebnitz, who in the 17th-18th century came up with concept of *matheisis universalis*. Today, many believe that there are no problems that we can not solve with the help of algorithms, software, and the '*big iron*' it is processed on. Computation and probability have subsequently become the universal solvent for *everything*, however, much has been lost in the process of disembodiment and context is the primary victim.[12] The application of a *universal* algorithm to anything from markets to carbon brutally simplifies reality and chisels the material context that surround the subject while looking for a solution. This has become a hegemonic ideology that has relentlessly attempted to supplant all other worldviews and has typically colonised indigenous cultures such as mātauranga Māori in Aotearoa.[13] This is not some recent attack but one that extends far back to Aristotelian philosophy and the later religious genocides of Christianity.[14] Yet, most of us are oblivious to the stealthy way that this immaterial faith has violently justified the exploitation of others, including: cultures, animals, plants and minerals. Algorithms hide and elide the violence of the past.

What is missing from binary algorithms is the messy reality, the fuzzy edges and data points that are not answered by true or false outputs. The world is mediated not just by material media that is inextricably woven into the fabric of reality but can be metaphorically true, and false, and sometimes neither. This is a known as a four state logic that allows for a result that could be 1) true 2)

---

12. See Hayles, N. K. (1999). How we became Posthuman: Virtual bodies in cybernetics, literature, and informatics. Also, Finn, E. (2017). What algorithms want: Imagination in the age of computing. Cubitt, S. (2017). Finite media: Environmental implications of digital technologies.

13. See Royal, Te Ahukaramū Charles. (2007) Mātauranga Māori and Museum Practice. Museum of New Zealand Te Papa Tongarewa. https://charles-royal.myshopify.com/products/matauranga-maori-and-museum-practice-a-discussion.

14. See Tarnas, R. (1991). The passion of the Western mind: Understanding the ideas that have shaped our world view (1st ed).

false 3) true and false 4) neither true nor false. Algorithms like any rules, or descriptions of the world are gross simplifications of insanely complicated relational networks of things that can't be fully explained by abstract descriptions but are deeply connected to their local context. This is not some binary reality but a multivalent logic, a fuzzy logic that acknowledges that reality is more like shades of grey than black and white. In his book, *Fuzzy Thinking*, Bart Kosko, introduces his skepticism about the mismatch between mathematics, Aristotelian logic, probability, and the description of reality. He wrote: "It shocked me to see Einstein doubt the very math framework of black and white science that he had helped build: 'So far as the laws of mathematics refer to reality, they are not certain. And so far as they are certain, they do not refer to reality.' So Einstein brooded about grayness too."[15] The irony is that the mathematical theorists, like Gödel, Church and Turing who helped build our binary worldview[16] first identified that the universe is ultimately non-computable, and that any logic must acknowledge the infinite, the unknown, and therefore uncertain reality of our existence. Why have we settled for a simplistic bivalent worldview that is encoded in computational algorithms? According to Finn the Western world has undergone something like a religious conversion – we have accepted the schism between the mechanical Newtonian world of deterministic physics and the probabilistic world of digital reality. The Internet of Things is the latest sect of a technological theocracy driven by simple rules and bivalent algorithms. Yet, as we build mirror worlds with invisible sensors and

15.   Kosko, B. (1993). Fuzzy Thinking: the new science of fuzzy logic. Flamingo. London. p.7

16.   See Finn, E. (2017) ibid.

software-instantiated simulations of reality we are moving further and further away from a material reality that physically, emotionally, and spiritually compliments the star dust we are made from. In settling for a simplistic explanation and ignoring the material metaphors that deny paradox and the multivalent truths about who, what, where and when we think we are, is to be happy with a basic colouring-in picture rather than to explore the rich creative freedom of multivalent logic in the context of an infinite multiworld view. Sure it is hard, but just because we are faced with wicked problems does not mean we have to dumb them down to simple answers – that is surely how we will unwittingly set the charges for a catastrophic end to this world.

Right now there is a doctrinal war brewing between the neoliberal titans of Silicon Valley and those philosophical heretics who claim that the universe, both beyond, and within us, contains non-computational realities. The documentary, *The Social Dilemma*, interviews those who have had the scales removed from their eyes, who once worked deep within the temple of tech, within the social media companies such as Facebook, and Twitter. These are the acolytes of new philosophers who are warning of our psychological imprisonment in the ubiquitous *'happiness machines'* – cultural algorithms that hide the material reality of our worlds. According to one outlier from the digital humanities, Ed Finn, deep in the circuits of this algorithmic universe, is an unwavering faith in *'computational ontology'*, an immaterial reality defined and created by logic, calculus, and reason. This liminal boundary between realities, in between worlds, are uncomfortable states of unknowing, states of horror and uncertainty about an algorithmic binary world that prefers probability, or risk management to the admission that we do not know, or can't calculate the probability of an event. There are multiple realities and worlds beyond the grasp of actuarial insurance calculators.

Aristotle's bivalent logic, that only sees reality as true or false has dominated for over two thousand years and has seeped into computer algorithms. The history of ideas is a complex negotiation of power, human culture, and the environment. According to Kosko, while our recent obsession has been shaped by post-war theories many theorists, technologists, coders and applied mathematicians have accepted binary logic because it mostly produced results and they found it: 1) easy and 2) it was a habit. They used the same artificial language to talk about matters of logic and matters of fact. They described math and the world with the same black-white 'symbolic logic' that Aristotle set in motion over two thousand years ago.'[17] In order to adapt to a more nuanced reality database technology added an extra state, NULL – a coding term for the unknown or in electronics it is the defined result of two signals that cancel each other out. Multivalent logic has been developed by mathematicians, computer scientists and philosophers who saw that yes/no answers are woefully inadequate to describe reality. In its most simple form multivalent logic added the term, NULL or *'unknown'*. While this also seems like a cartoon version of the world at least this three-state logic acknowledges the ambiguity and potential paradox of reality – the limits of what it is we supposedly know.[18]

---

17.   Kosko, B. (1994) ibid p.9

18.   S23M applies multivalent logic that allows domain experts to articulate the commonalities and variabilities of overlapping formal bodies of knowledge using a four-state Information Quality logic that supports the semantics of 'true' and 'false', as well as semantics of 'unknown' and 'not-

In 1931 Kurt Gödel published his Incompleteness Theorems that proved that: 1) that it is impossible for any algorithm or symbolic logical system, "to be internally consistent and provable using only statements within the system. The truth claim or validation of such a system would always depend on some external presumption or assertion of logical validity".[19] 2) the second theorem continues from the first, and proves that the system (or algorithm) cannot demonstrate its own consistency. Regarded by many as one of the most important foundations of computation, Alonzo Church and Alan Turing developed their own mathematical thesis about universal computation, however, rather than prove that algorithms could encapsulate all of reality, the so-called universal: "...Turing machine leaves open the question of what 'effectively computable' might really mean in material reality".[20] Simulations, computer models, and virtual worlds do not stand-in for that material reality – they are abstractions, distortions, and only ever partial truths. Yet, how often do we stop to question this logic or are even exposed to the assumptions made by a technological solution until we encounter a stupid auto-answer machine that only gives us the option of a yes/no answer, with the rare option of N/A, 'not applicable'.

In these immaterial worlds reality is defined by maths, biased concepts and ideas, according to the formula that

---

applicable'. In contrast to traditional databases and machine learning systems, Information Quality logic transcends the limitations of binary reasoning and Bayesian (probabilistic) reasoning. It has been specifically conceived for high-fidelity representations of tacit knowledge and cultural variability, including formal representations of unknowns and limits of [model] applicability.

19. Finn, E. (2017) ibid. Kindle. Location 519

20. Finn, E. (2017) ibid. Kindle. Location 532.

knowledge is justified belief.[21] According to the OOO philosopher, Graham Harman, many regard that there are only two basic kinds of knowledge about things: explanations about what they are made of, and what we can say about what they do? This knowledge system can be digitised, encoded by algorithms, and may be the basis for what Stephen Wolfram has called *a new science*. Within the narrow confines of computable problems Wolfram believes a whole universe could be summed up in a simple bivalent mathematical formula. He believes that everywhere you look a simple binary algorithm can predict the awesome complexity of all things in the universe.[22] However, there are whole worlds lost to this simplistic description of things. We either collapse the world into smaller and smaller constituents, a methodology sometimes called reductionism, which is common in Newtonian physics, or we generalise about the worlds of things and only assume that their reality is defined by the things that they *'modify, transform, perturb or create.'* Harman adds to these world-views an approach to knowledge that combines both reductionist undermining, and the overmining generalisation that describes a reality such as mathematical worlds that ignore the qualities of the things, or what he calls duomining. According to Harman: "Undermining, overmining, and duomining are the three basic forms of knowledge, and for this reason they cannot be avoided, to the extent that human survival hinges on acquiring such knowledge. Yet some disciplines

21.   See Von Krogh, G., Nonaka, I., & Ichijo, K. (2000). Enabling knowledge creation: How to unlock the mystery of tacit knowledge and release the power of innovation. Oxford; New York: Oxford University Press.

22.   Wolfram, S. (2002). A New Kind of Science. Wolfram Media.

are not forms of knowledge while still having considerable cognitive value."[23]

They are the worlds of culture, artistry and poetry that disrupts the binary world that denies contradiction, paradox or the liminal worlds that exist in-between the gaps, between 0 and 1, zero and carbon, the flux between matter and anti-matter, the virtual and the actual. These gaps are cognitively co-designed by ourselves and the computer, but the gaps are the vast chasms between computation and reality. Finn argues: "But the human participants, users, and architects of these systems play an equally important role in constructing the gap when we organize new cognitive patterns around computational systems and choose to forget or abandon forms of knowledge we once possessed. Every moment of dependence, like a forgotten phone number or spelling that we now depend on an algorithmic system to supply, and especially every rejected opportunity for direct, unmediated human contact, adds a little to the space between computation and human experience."[24]

These are the uncomputable, the spectres and ghosts, the demons and the undead that can't be touched or trapped in a cage to be studied or pinned in a collection case. It is where we must look if we are going to imagine and creatively engage in worldbending, worldending, and worldbuilding. These worlds have to be opened up so we can dive into the abyss that beckons us to step into the gap between worlds and participate in the constant flux of world-bending.

In these metaphorical and paradoxical worlds, things are

---

23.   Harman, Graham. Immaterialism: Objects and Social Theory (Theory Redux) (p. 12). Wiley. Kindle Edition.

24.   Finn, E. (2017) ibid.

like other things, but they are *not* other things. We have to learn to re-train our minds to be comfortable with the uncertainty of definitions that play with states between on and off, zero and one, the unknown, and unknowable, the fuzzy contradictory in-betweeners.

Zero Carbon Algo

In our anthropogenic crises, around every corner are algorithms offering to solve our problems. And, when we are desperate, we grab the nearest tool, but it is a sign of madness if we keep banging away with the old hammer when we know it doesn't work. Finn describes these algorithmic tools as *'culture machines'* and makes the claim that we are now living in the: "age of the algorithm: the era dominated by the figure of the algorithm as an ontological structure for understanding the universe."[25] Algorithms have become ubiquitous in the modern world of globalisation and the Internet of Things, they represent the encoding of our most dominant and potentially most dangerous assumptions. Even in the mundane world in which my fingers have been transformed into data input digits, and my chosen tools for the purposes of writing this book is an Apple MacBook Air with two screens; I am constantly interrupted by the ever helpful, never tired, algorithmic oxymoron, *Grammarly*, which, by the way, saw nothing wrong with my spelling of that grammatically incorrect brand name.

Orwell's Aotearoa

In 1984, when I set out to write my Masters Thesis, *A Wealth*

---

25.   Finn, E. (2017) ibid.

of Knowledge in a Bankrupt Databank[26] I thought it was appropriate, and hopefully creatively efficient, to learn how to touch type so I could input my thesis into my father's Apple IIe computer. In that Orwellian year of 1984 most people had no idea that computers would eventually rule their world. What I also did not realise at the time was that was the fateful year that the neoliberal algo was installed in Aotearoa.

Computers then were barely able to play games, let alone run a small business, and most of the serious computers were massive machines run by wizards that hid in backrooms. I enrolled in a night class in typing at Selwyn College, and joined a large group of young women who were learning to type on electric typewriters so that they could get jobs in offices, mostly as receptionists. I was literally the odd man out. What I didn't know at the time was that the machine and its algorithms had begun to indoctrinate me into their cult, and they started with my initiation by re-training me in how to write and how to think – I was transmogrifying into a cybernetic organism, a cyborg. I had three attempts to type my 30,000 word thesis into a computer that had a tiny internal memory of 128 kilobytes. A quick glance at my current desktop reveals that the closest file of the same size is a png file titled, *Megan Woods Contacts*; that file that consisted of only about twenty names and contact details was the same size as the entire memory on the Apple IIe. I went to my night classes religiously and began typing my thesis, however, unbeknownst to me my computer had no safety warning when it approached full capacity. When it did reach that

26.   Rive, P. (1984) Masters Thesis. A Wealth of Knowledge in a Bankrupt Databank: the politics of establishing a 'universal database'. PRESTEL and the implications for videotex development in New Zealand. University of Auckland, NZ.

capacity it simply crashed and lost everything I had typed – this happened three times before I cursed, gave up and wrote it by hand. I then passed it on to a professional who had probably graduated from a typing school like the one I had just attended.

Thirty six years later, the smug little machine on my desk has won – I am still diligently typing data into it, and it is still an Apple, albeit one made in the 21st century. Every now and then it reminds me with a notification popping up in front of my draft chapter, reminding me that I haven't backed up to my aptly named Time Machine for 14 days. Like so many algorithms this could be an automated process but instead I choose to backup manually to an external hard drive. I have to close the notification to get rid of it, but not to worry, at the bottom of the screen I see that it has autosaved my draft on the ingramspark.pressbooks.pub website. It was safely stored in the Cloud, or was it? I was informed: 'Draft saved at 12:00:31 am. Last edited by peterive on August 21,2020 at 11:32 pm'. That was weird because it was actually Sunday 23rd of August 2020 12:03:06 pm during the day.

Of course, what that could be telling me is that the time is only relevant to the End Users Licence Agreement that made me agree that I am not legally in New Zealand but I am actually in a disembodied cyberspace, of course I am not – but that's the legal reality. Being known in my family as something of a Tisco Man[27] I am often asked why does the machine, or often the algorithm, do what it does? They would usually begin with an accusation – something like, "There is something wrong with my computer! I was writing an email and it just disappeared", as if it was my

---

27. In NZ a Tisco Man was a TV repair person. My family extended my role and responsibilities to electronics, and computer repairs.

fault as I was clearly in cahoots with them. I would get questions like, "where's it gone?" As I proofread and edit this draft I recall that subsequent to the earlier draft I had a scare when I tried to revisit my book, now residing in the Cloud. Instead of seeing the draft on the screen there was a bald error statement – '502 BAD GATEWAY'. Sometimes I know the answer to 'where has it gone?' Sometimes I guess, this time I don't know and when I tried to find out I found there were three possible and ambiguous reasons. I quickly gave up and hoped for the best as it seemed to be coming from the application provider's server. I suspect that is the standard reaction of most of you – it just is – life is too short!

100% Uncertain

It did remind me that there is no 100% certainty, even when it is stored in the Cloud, and our civilisation and worlds are vulnerable to random or deliberate erasure – either caused by an electromagnetic pulse from a solar flare, or perhaps a possible NNEMP (non-nuclear electromagnetic pulse) weapon attack.[28]

Our company, S23M, develops software for the healthcare sector. We were recently talking to another company that has developed a very comprehensive crisis management software about how we might integrate with their system. One of their biggest customers uses their products for bush fire prevention and management. It has the ability to allow remote users to operate on their devices completely offline. It made me think about our hospital customers who in a crisis might be cut off from the Internet, or completely without electricity. This scenario

---

28. See https://en.wikipedia.org/wiki/Electromagnetic_pulse

might seem ludicrous and yet I watched a recent Bloomberg story in which a number of countries around the world shutdown citizen access to the Internet for long periods of time during political protests and even during their elections, disrupting the democratic process in a desperate attempt to hold onto power. Countries such as Belarus that shut down for 218 hours during the elections, Ethiopia, Uganda, and even India.

Our dependency on the Internet, or the central electricity grid is only exposed during these crises. During the pandemic it was revealed that the New Zealand healthcare system has been seriously degraded by years of neoliberal cost benefit analysis. We recently discovered that the National Immunisation Registry that was commissioned in 2000 has become virtually unworkable. The system that was designed to record the immunisation of all children against the big diseases that maim and kill was no longer capable of local printing, or running reports. Meanwhile, in order to cope with the COVID 19 vaccination roll out the government felt they had no choice but to grab the nearest bandaid and select the multinational giant, *Salesforce*, to run our contact tracing system, and this will likely be the solution for the new National Immunisation Solution. It is hard to imagine how a platform that is designed to be a CRM (Customer Relations Management) system will be fit for purpose in a healthcare scenario – yet it is no surprise when the neoliberal ideology believes everything can be monetised and turned into a product.

Health Blitzkrieg

Another blitzkrieg has begun and the current government has taken advantage of the state of emergency created by the pandemic crisis. Unbeknownst to all of the DHBs, the Ministry of Health, the media, and the public that the healthcare system serves, the Minister of Health, Andrew Little, shocked the country with the announcement of a complete erasure of the 20 District Health Boards, the 30 Primary Health Organisations, and the biggest restructuring in 30 years. There is little known about how this shock treatment was wired, but the telltale traces of accountants and economists were detected as the transition unit was led by the ex-Director General of Health, Ernst and Young leader of Government and Public Sector, Stephen McKernan, QSO. There was scant detail about the plan that had been hatched in secret following the Health and Disability Review that had been led by Heather Simpson within the Prime Minister's office.

The private sector was once again advising the government on efficiency, and ironically, equity. The algorithm is unlikely to notice the underserved and deprived communities in our country – this is what the transition from the Welfare State to the Wellbeing State looks like – more code layered on top of the old code delivered by a centralised digital health authority. This means that the original neoliberal assumptions will be most likely buried beneath the layers of digital rot. Transparency and democratic oversight cannot take place when we are waist deep in the dark detritus built up on flimsy assumptions about free markets, and business optimisation algorithms. There are repeated calls for strong leadership, yet the government will struggle to balance that with the *design thinking* born-again accounting firms profess to use in their so-called, *community co-design*.

Arthur C. Clarke had noted that: "Any sufficiently

advanced technology is indistinguishable from magic."
Electronics and computing for most of us is magical, we
can even talk to our machines and they will reply. However,
unless we peel back the curtain to reveal the little man from
Oz that is manipulating our world we can really have little
impact on the future of these digital worlds.

Zero Carbon Magic

Zero Carbon is an algorithm that needs to be excavated,
deconstructed, and exorcised so that the spell is destroyed
and we can demand that the technocrats, economists, and
politicians that are constantly remaking the algorithm
must stop and relinquish control over the code, and more
importantly, the fate of the planet, the humans and the
nonhuman majority. According to Finn: "Like the
algorithm itself, algorithmic reading is a complex
conceptual structure containing layers of processes,
abstractions, and interfaces with reality."[29]

In what way can Zero Carbon be an algorithm? Finn
writes that: "An algorithm is a recipe, an instruction set,
a sequence of tasks to achieve a particular calculation or
result."[30] For an algorithm to have a chance of solving a
problem it must first gather the relevant data. The IPCC
2019 report stated: "Human activities are estimated to have
caused approximately 1.0°C of global warming above pre-
industrial levels, with a likely range of 0.8°C to 1.2°C. Global
warming is likely to reach 1.5°C between 2030 and 2052 if it
continues to increase at the current rate."

In simple terms the objective of the IPCC is to prevent
the planet getting hotter than another 2°C and preferably

---

29.   Finn, E. (2017). ibid.

30.   Finn, E. (2017). ibid.

less than 1.5°C. Scientists try to use precise language, and for anyone that is interested you should be able to access this precise language, however, politicians and policy makers benefit from a language that is impenetrable to a majority of voters. It is not translated and the whole subject is dense with techno-babble designed to be understood only by technocrats. The algorithms are as technically tricky as the source code of a computer program. The algorithmic recipe to achieve the stated objective of the IPCC is by: "Reaching and sustaining net zero global anthropogenic CO_2 emissions and declining net non-CO_2 radiative forcing would halt anthropogenic global warming on multi-decadal time scales (high confidence)...On longer time scales, sustained net negative global anthropogenic CO_2 emissions and/ or further reductions in non-CO_2 radiative forcing may still be required to prevent further warming due to Earth system feedbacks and to reverse ocean acidification (medium confidence) and will be required to minimize sea level rise (high confidence)."[31]

This was written as a summary intended for an audience of policymakers. If you struggled to understand it, don't be surprised. I believe it is politically convenient for those who write policy, and make laws, to keep the narrative technical and beyond the comprehension of most of the voters. Whether it is deliberate, or not, it has the effect that most people leave the decisions to the politicians, who also

31.   IPCC, 2018: Summary for Policymakers. In: Global Warming of 1.5°C. An
       IPCC Special Report on the impacts of global warming of 1.5°C above pre-
       industrial levels and related global greenhouse gas emission pathways, in
       the context of strengthening the global response to the threat of climate
       change, sustainable development, and efforts to eradicate poverty
       [Masson-Delmotte, V., P. Zhai, H.-O. Pörtner, D. Roberts, J. Skea, P.R.
       Shukla, A. Pirani, W. Moufouma-Okia, C. Péan, R. Pidcock, S. Connors,
       J.B.R. Matthews, Y. Chen, X. Zhou, M.I. Gomis, E. Lonnoy, T. Maycock, M.
       Tignor, and T. Waterfield (eds.)]. World Meteorological Organization,
       Geneva, Switzerland, 32 pp

probably don't understand, and they use simplistic election slogans to rally support for their campaigns.

To state the obvious an algorithm can only produce effective, or predictable results if the data input is accurate – today it is an existential truism, *'garbage in, garbage out!'* The cost of getting it wrong could literally be our waste of life. It is not true that merely naming it *Zero Carbon* is not enough to make climate change go away. Marketing and algorithms are like magic incantations and many have faith that the naming and instantiation of the word in an algorithm are the only technological magic needed to make it disappear. The semantics of the reports by the IPCC scientists who write them project a strong degree of surety and the impression that they have weighed all the evidence and peer reviewed every qualified paper relating to anthropogenic climate change. Even when they report low confidence, in the predicted outcome of an event, this signals to the reader that they have at least considered all of the scientific evidence and with a wave of the magic algo wand the problem will disappear. However, what this conceals is that the Anthropocene is a non-analog epoch, in other words, the planet has never experienced this event before, and that even high confidence does not imply certainty.

Just like the algorithm that produces the status bar showing how long until an application will be installed on your computer, the scientists can never be wrong because as time goes by they keep adjusting their predictions, until D Day comes. If we wait until just the day before the status bar for the planet runs out and the world ending algorithm has finished installing, the prediction will be much more accurate, but alas it could be too late. It may be that given the difficulty of predicting what the average temperature will be in 200, 50, or 10 years from today, the closer you are to the event the easier it is to predict the outcome. However, rather than adhere to the precautionary principle, assuming the worst case scenario and so implement maximal avoidance of a harmful predicted event, our governments, and the media, publish statements based on outmoded assumptions. Yet, the day before our anthropogenic mistakes are obvious to us, it is most likely that the Earth will have surpassed the tipping point and geological change will be irreversible within the lifespan of the human species.

If you go back over the reports written by the IPCC you will note that by most measures sea levels are rising faster, and higher, ice is melting faster, and the permafrost is releasing more methane and carbon dioxide than they had previously reported, and predicted. Valérie Masson-Delmotte, Co-Chair of the IPCC Working Group I, made the comment that: "In recent decades the rate of sea level rise has accelerated", [however, she fails to mention that this was not predicted], she continues "due to growing water inputs from ice sheets in Greenland and Antarctica, in addition to the contribution of meltwater from glaciers and the expansion of warmer sea waters."[32]

It was 2003 when the IPCC predicted that if we do not change the rate of GHG emissions we could see a half meter rise in sea levels by the end of the century based on a prediction of 2°C global heating above pre-industrial average temperatures. This sea level rise has not been adjusted since 2003, however, what they did not take into account back then was the impact of ice melt from

32.   IPCC (Sept. 25, 2019) "Choices made now are critical for the future of our ocean and cryosphere" https://www.ipcc.ch/2019/09/25/srocc-press-release/

Greenland, the Arctic, and Antarctica, nor have they adjusted it since their fourth assessment revised their upper end of the temperature scale to be between 3-6°C by 2100. Yet, the algorithms behind the data that constructs these numbers and predicts the possible outcomes of global heating are not that obvious, or revealed to us in easy to understand terms. We are expected to trust the tech wizards and the magical thinking of their paymasters.

Buried In an Acronym

But what of the acronym of the IPCC, even that is algorithmic, a distillation of years of politics and millions of dollars of lobbying and negotiation. In the movie, *VICE*, (2018), I learned how the Vice President, Dick Cheney, ran the White House using the government bureaucracy to design this algorithm. This *'culture machine'*, the IPCC, stands for the United Nations Intergovernmental Panel on Climate Change. This sounds innocuous enough but it was revealed in the movie that Cheney and his team, led by Frank Luntz,[33] ran focus groups to discover just what words to use to make the concept of global heating sound less alarming to the voters – the answer was the phrase, *'climate change'*. So, already you get a sense of how deep down in the abysmal layers of algorithms, there are cultural machines that are hiding political decisions, hunches, guesses, and even the most accurate data today, the least accurate data today, and lots more that we don't know about are compacted into political slogans. Cheney was the CEO of the US oil and gas company, Haliburton, and those that

---

33.   Frank Luntz communications advisor to Cheney suggested the change from global warming to climate change. https://en.wikipedia.org/wiki/Frank_Luntz

wanted the US to pull back from fossil fuels were a direct threat to his wealth and power so a focus group came up with 'climate change'.[34]

While the IPCC has numerous political pressures to publish conservative reports, behind the closed doors of the Board rooms of BP and Shell they were adjusting their market strategy, and no doubt carbon pricing, with the expectation that there would be a catastrophic 5°C temperature rise as soon as 2050.[35]

It is as if the mantra of half a meter of sea level rise, and 1.5°C hotter, has become the target for governments, voters, and regulators. It ignores the obvious, that only 1°C warming has already cost millions of human lives, and billions of animal lives.[36] It is forgotten that even if we managed to keep the average temperature rise to 2°C it is estimated that 32 to 80 million people worldwide would be exposed to flooding, and if we went to 3°C that number leaps to 275 million.[37] These are enormous numbers of loved ones that will most likely be lost, and dwarfs the number of people killed by COVID 19, today sitting at 3 million with a cost already estimated at over USD$28 trillion to the world economy.[38] However, what if

34. Friedersorf, C. (Aug. 30, 2011) Remembering Why Americans Loathe Dick Cheney. https://www.theatlantic.com/politics/archive/2011/08/remembering-why-americans-loathe-dick-cheney/244306/

35. Independent. (Oct. 27, 2017). "BP and Shell planning for catastrophic 5°C global warming despite publicly backing Paris climate agreement."

36. Aljazeera. (28 July, 2020) Nearly 3 billion animals killed or displaced by Australia fires. https://www.aljazeera.com/news/2020/07/3-billion-animals-killed-displaced-australia-fires-200728095756845.html

37. Neumann, B., Vafeidis, A. T., Zimmermann, J., & Nicholls, R. J. (2015). Future Coastal Population Growth and Exposure to Sea-Level Rise and Coastal Flooding - A Global Assessment. PLOS ONE, 10(3), e0118571. https://doi.org/10.1371/journal.pone.011857

38. IMF estimates global Covid cost at $28tn in lost output. https://www.theguardian.com/business/2020/oct/13/imf-covid-cost-world-economic-outlook

these frightening numbers of deaths are underestimating the death toll and damage, the precautionary principle tells us that we should listen to the scientists who warn: "Sea level in the geologic past was much more responsive to changes in global climate than what IPCC predicts for the year 2100," writes the University of Chicago's David Archer. "Past sea level varied by 10–20 meters (30–60 feet) for each 1°C change in the global average temperature. The IPCC business-as-usual forecast for 3°C would translate to 20–50 meters (60–150 feet) of sea level rise." The IPCC may well be right about its half-meter prediction by the end of the century. But it might not be."[39]

In New Zealand the Ministry for the Environment and the Climate Change Committee use the IPCC's modelling to predict the possible changes to the climate. In their risk assessment they write: "The world has already experienced significant climatic changes due to emissions from the combustion of fossil fuels and changes in land use. The volume and rate of emissions will depend on climate policies, resource availability and demographic, economic and technological change, all of which are uncertain."[40] The IPCC created four future emission scenarios called representative concentration pathways (RCPs) to model a possible range of scenarios. However, the Ministry for the Environment did acknowledge that the IPCC warns the: "RCPs are only one set of many scenarios that would lead to different levels of global warming (IPCC, 2019)" and even when they review the projections of the worst case scenario, RPC8.5, a high greenhouse gas emissions

39. See Brannen, Peter. The Ends of the World . Oneworld Publications. Kindle Edition.

40. Ministry for the Environment. (2020) National Climate Change Risk Assessment for New Zealand. pg.35

scenario, that is: "more extreme scenarios are possible, and the sensitivity of the climate system remains uncertain."[41] RPC8.5 projects a 4-6°C rise relative to the global average surface temperature change relative to 1986-2005. Yet, what we really face is the 'unknown unknowns' because we are only considering an arbitrary timescale that cuts off at 2100.

History Repeats

Even if global heating only goes to 3°C over a couple of hundred years we will go well beyond a half meter rise in sea levels. Scientists have pointed out that instead of just slowing down our GHG emissions we must be removing them otherwise the planet will keep heating for tens of thousands of years into the future.

Many policymakers assume that if we gradually reduce our GHG emissions we will be able to control an orderly retreat from the brink of disaster, however, Peter Brannen points out that this has not been the case in the past, and today we are emitting ten times the amount of GHGs, or 40 gigatons of CO_2, faster than any other period in the past 300 million years. The climate is not just some thermostat – we can't tweak our GHG emissions with precision in order to manage to keep below 1.5°C. Brannen corrects this misconception: "Unfortunately, this is not how the world has tended to behave in the geological past. Throughout the climate swings of the Pleistocene, the ice sheet that covered North America—one larger even than modern Antarctica—did not merely shrink in response to a few degrees of warming. It exploded. Rather than slowly dwindling over thousands of years, these continents of ice sometimes violently disintegrated in spectacles that

---

41.   Ministry for the Environment. (2020). ibid. pg.36.

unfolded over mere centuries. In one rapid collapse 14,000 years ago, called Meltwater Pulse 1A, three Greenlands' worth of ice fell into the sea in icy flotillas, sending sea level soaring 60 feet [almost 20 meters]."[42]

If that were to happen in the near future it would put most of the largest cities in the world underwater and greatly reduce the amount of arable land to feed the world's population. There is no time to lose. Brannen puts the next two hundred years in perspective: "Civilization has numbered sixty centuries so far, but the next handful may well see the ocean rise more than 200 feet if we burn it all. This isn't that surprising. In the millennia before civilization, the ocean rose up 400 feet from the edge of the continental shelves. Boston was built as a seafaring city, but a few thousand years ago it would have been landlocked more than 200 miles away from the ocean. That the coastline would continue to migrate inland should come as no great surprise. This is what the ocean does in geological time, mocking the putative permanence of our coastal settlements."[43] Algorithms not only give us a false sense of control and certainty but they also hide the assumptions behind them such as the idea that 2100 is some kind of crazy futuristic time horizon and yet it is only 2-3 generations away. The longevity of many GHG emissions means that they will continue to heat the planet well beyond a century and possibly even a millennium. We have to consider how to extract these GHGs from the atmosphere. If we were to be prudent and abide by the precautionary principle we would acknowledge the

42. Brannen, Peter. (2018) The Ends of the World . Oneworld Publications. Kindle Edition.

43. Brannen, P. (2018) ibid.

extreme danger and significant uncertainty of the current
IPCC modelling.

Who Owns the Invisible Hand?

However, as I mention this *'inconvenient truth'* let me also
mention that not only is the data uncertain, and the
assemblage of an indeterminate number of things
unknown, but there is an almost impenetrable and
unassailable algorithm that is based on the ideology of the
'invisible hand of the market'. In 2008 the President of the
US, Bill Clinton and his Vice President, Al Gore, set about
convincing the European Union representatives that a
carbon tax was not the way to go. They argued that in order
to lower the GHGs and save the planet, the only way the US
would agree to participate in such a plan was to implement
a carbon market that would be magically controlled by a
laissez faire market algorithm. Thus, was born the EU's
ETS, (Emissions Trading Scheme). It would seem that Gore
'saw the light' after losing office to a dodgy voting scandal
that happened in Florida.[44] The Washington Post revealed
that *'Gore has thrived as a green-tech investor'*. Gore went from
USD\$2 million in assets in 2000 when he lost the election,
to an estimated wealth of USD\$100 million in 2012.
According to the Washington Post that tracked Gore's
investments and cosy deals with the Obama
administration, the article noted:

"Gore bristled when Rep. Marsha Blackburn (R-Tenn.)
asked if he stood to profit from his investments and
political connections.

---

44. See Stephen A. Sheller (2016). Pharmageddon: a nation betrayed. Chp.1
     Bush v. Gore. How the Fox Got in the Henhouse. Sheller outlines in detail
     how hidden algorithms, plots, and vested interests can undermine
     democracy.

"I believe that the transition to a green economy is good for our economy and good for all of us, and I have invested in it," he said. "And, congresswoman, if you believe that the reason I have been working on this issue for 30 years is because of greed, you don't know me."[45]

Gore and Clinton's faith that, despite a market failure that had caused the biggest anthropogenic disaster on this planet, the market might somehow magically stop GHG emissions, while still making them a tidy profit. This was either naive or cynical. Gore after the 1997 Kyoto Protocols went on to profit not just from green-tech investments, but on the back of his movie, he was paid to burn carbon, flying around the world to speaking engagements that paid him USD$176,000 at each event. The Washington Post revealed:

"Fourteen green-tech firms in which Gore invested received or directly benefited from more than $2.5 billion in loans, grants and tax breaks, part of President Obama's historic push to seed a U.S. renewable-energy industry with public money."[46]

What is extraordinary is that President Bush then reneged on the Clinton/Gore deal and did not join the Kyoto Protocol, yet the biggest buyers of the Clean Development Mechanism (CDM) offset credits were speculators on Wall Street, and the London markets. There exists an elite alliance of big business, commodity traders, financial firms, and neoclassical economic theorists from the Chicago School of Friedman and others who have

45.   Carol D. Leonnig. (Oct. 2012) Al Gore has thrived as green-tech investor. Washington Post. https://www.washingtonpost.com/politics/decision2012/al-gore-has-thrived-as-green-tech-investor/2012/10/10/1dfaa5b0-0b11-11e2-bd1a-b868e65d57eb_story.html

46.   Carol D. Leonnig. (Oct. 2012) ibid.

profited greatly from a perverse system that encourages more, not less, GHGs.[47]

Böhm and Dabhi explains "The underlying assumption of this system is to achieve maximum possible emissions reduction at the lowest possible cost by: first, quantifying emissions caused by industrial activities; second, setting a cap on all GHG emissions; and, third, incentivizing companies and entire industries to make decisions on how to meet their caps in the cheapest possible way. They could do this by bringing in 'green technologies' or by buying credits/allowances from other polluters or, in fact, from offsetting schemes. The 'cap and trade' system assumes that cuts made anywhere are globally equivalent (which is technically speaking and from a climate science point of view correct, but in a wider socio-economic sense does not hold true). Hence, this carbon market could allow a company or industry to continue in its polluting business-as-usual ways, as long as it keeps buying allowance or carbon credits from other sources."[48]

What Does the ETS Want?

Finn asks what do algorithms want? When we consider the ETS we should dig deeper to consider the desires of those that set up these carbon markets such as the EU's ETS, New Zealand's ETS and the Clean Development Mechanism that came out of negotiations between the US and the UN. From the beginning these carbon markets can be categorised as neoliberal capitalist algorithmic machines that claim to want to find the most cost efficient mechanisms to reduce

---

47.   Böhm, S. and Dabhi, S. (eds) (2009) Upsetting the Offset: the political economy of carbon markets. MayFlyBooks. London.

48.   Steffen Böhm and Siddhartha Dabhi (eds) (2009) ibid.

the GHG emissions to net zero by 2050. However, the explicit wants of the algorithms can disguise the hidden desires of powerful lobbies and captive politicians that established and continue to run these markets. Many of them continue to use the rhetoric of economic growth, free markets, and the accumulation of capital. This *'common sense'* approach is described by Mark Fisher as *'capitalist realism'*. He wrote: "The power of capitalist realism derives in part from the way that capitalism subsumes and consumes all of previous history: one effect of its 'system of equivalence' which can assign all cultural objects, whether they are religious iconography, pornography, or Das Kapital, a monetary value."[49]

Janet Yellen who served as the US Federal Reserve Chair from 2014-2018 has said that together the social inequities that have been exposed by COVID-19, the Black Lives Matter protests, and the Californian wildfires as a result of global heating, has challenged carbon markets and capitalism. She wrote: "There really is a new kind of recognition that you've got a society where capitalism is beginning to run amok and needs to be readjusted in order to make sure that what we're doing is sustainable and the benefits of growth are widely shared in ways they haven't been." Yellen told Reuters in an interview that both Democrats and Republicans share a concern about the need for urgent action on *'climate change'*. Yellen has been working with a G30 group of former central bankers, academics, policy-makers and financiers on a carbon tax approach to transition to net zero emissions. The Climate Leadership Council plan is to halve carbon emissions by 2035 from 2005 levels with a USD$40 per ton of carbon

---

49. Fisher, M. (2010). Capitalist realism: Is there no alternative? Winchester, UK: Zero Books.

emissions. Yellen said: "Our thinking is that countries with carbon pricing would form essentially clubs, or carbon customs unions, within which there would be frictionless trade."[50] This still sounds like a market solution to our existential crises but the loss of faith in markets such as the ETS, by a stalwart of capitalism, should direct the NZ government to reconsider industry regulation, and or a carbon tax (still a market mechanism) and scrapping the ETS. It is unlikely that the government will reconsider the ETS as the faith in the market runs deep in this country and even the Green party, and the Climate Change minister, James Shaw are playing along with the carbon pricing mechanisms. The ability of capitalism to coopt anything and everything by giving it a nominal monetary value in which everything becomes ontologically equal according to the power of the mighty dollar. Enumerating and then equivalence are the first steps in the process of taking social, environmental, and geological harm and turning it into a commodity that can be traded. Words matter and the Emissions Trading Scheme sounds innocuous, but the sanitisation and commodification of pollution does not make the harm go away. According to Böhm and Dabhi this is how liberal ideology sets about depoliticising a problem that threatens, not just humanity, but untold other species on the planet. Here is their simplified method to achieve a 'system of equivalence'.

Carbon market construction in brief

Step 1: The goal of overcoming fossil fuel dependence by

---

50. Green, M. (Oct.9, 2020) U.S. could adopt carbon tax under a Biden presidency, ex-Fed Chair Yellen says. https://uk.reuters.com/article/us-usa-climate-tax-idUKKBN26T23L

entrenching a new historical pathway is changed into the goal of placing progressive numerical limits on emissions (cap).

Step 2: A large pool of 'equivalent' emissions reductions is created through regulatory means by abstracting from place, technology, history and gas, making a liquid market and various 'efficiencies' possible (cap and trade).

Step 3: Further tradeable emissions reductions 'equivalents' are invented through special compensatory projects, usually in regions not covered by any cap, for additional corporate cost savings, and added to the commodity pool for enhanced liquidity and further 'efficiencies' (offsets).

Step 4: Project bundling, securitization, financial regulation, rating agencies, 'programmatic CDM' etc. add new layers of obscurity and complexity.[51]

Böhm and Dabhi explain that this process of equivalence may appear uncontroversial, however, as they show with examples: "A moment's reflection will show, however, that, in producing such equivalences, carbon traders are already drifting away from the climate problem."[52] One of the most difficult aspects of the ETS is for the public to know what is included and what is excluded from the algorithms? If what gets measured gets done then surely you have to know what GHGs are measured in order to reduce them. For example, AUT researcher Dr Bradley Case, funded by NZ Beef + Lamb, concluded that because there was no IPCC agreement to include shelter belts and other native vegetation in mitigating GHG emissions that by his calculations in fact NZ farms were essentially carbon neutral. RNZ reported: "Case said the ETS was a "relatively

---

51.   Böhm and Dabhi (eds.) (2009). ibid. p.27

52.   Böhm and Dabhi (eds.) (2009). ibid. p.27

complex beast" and only "one mechanism" by which farmers could get credit for sequestered carbon. Case said he hoped the research would show the limitations of the ETS, and result in people coming up with "other instruments" to record on-farm carbon sequestration."[53] However, there are those who would argue that the entire supply chain should be included in the calculations and that would include the CO_2 emissions from the dirty shipping industry and airlines. In 2018 these pumped 0.9 million tons of CO_2 equivalence, and 3.9 million tons of CO_2 equivalence respectively. These are currently outside the IPCC requirements and not included in the ETS.[54] Algorithms can hide more than they reveal.

Speculative realism offers a radical philosophical challenge to the capitalist worldview that assumes markets can fix wicked problems such as global heating. However, the common response of neoliberals to any alternatives put forward is the condescending remark, 'Get real!' The OOO, or thing related reality allows us the freedom to dream beyond the Anthropocene and to envisage a world whereby nature and culture are no longer divided by the dangerous mindset of consumer economics and market algorithms.[55]

Once the desires and interests of capitalism are exposed it is not surprising that the outcome of the carbon markets has been the commodification of GHG pollution and the incentives to grow the very things the markets were supposed to reduce.

53. NZ Herald (9th Oct. 2020) New Zealand sheep and beef farms close to being carbon neutral - Research. https://www.nzherald.co.nz/the-country/news/new-zealand-sheep-and-beef-farms-close-to-being-carbon-neutral-research

54. https://www.transport.govt.nz/area-of-interest/environment-and-climate-change/climate-change/

55. See Rive, P. (2019). Worldbending: A Survivor's Guide for those who want to think and act creatively about our future.

Hiding Market Failure

Once we get to know who owns the invisible hand of the market we will be less surprised about the dismal failure of the carbon markets to date. We will see that desire + consumption is embedded in the algorithms of the ETS delivering massive subsidies and profits to the biggest carbon emitters in the world. One example of this perversion is the extremely potent greenhouse gas, HFC-23. One ton of HFC-23 has the heating equivalence of 11,700 tons of CO_2. The carbon markets in China has encouraged speculators to build manufacturing plants to produce HFC-23 and to profit from the carbon markets designed to stop it. Böhm and Dabhi puts it bluntly: "rather than 'efficiently' reducing the production of this highly potent GHG, the newly created carbon markets have introduced a perverse incentive to produce and emit even more GHG."[56]

What is troubling is that the laissez-faire ideology that promotes the claim of market efficiencies and freedom from government bureaucracies is just an excuse to avoid carbon taxes, and or the government regulation that would upset their business-as-usual. Meanwhile, we are reassured that the *free market* simply needs a few policy tweaks, a contradiction in terms, because a free market is supposed to be self-regulating, and that will eventually lead to algorithmic automation.

We are supposed to go to sleep until 2030 when we will be awoken by the carbon market alarm clock to be told *'go back to sleep everything is fine and working according to plan'*. We will no doubt need some more tweaks but don't touch the market levers the technocrats are the only ones that

---

56.   Böhm & Dabhi (eds.) (2009) Upsetting the Offset

understand how the engine works. The next alarm should be set for 2050. When you wake again, we are assured, then you will find that the carbon markets have magically reduced all GHGs to net zero, the crisis has been averted, or so says the story tellers. But where does this faith in this story come from?

The Invention of Magic

Where did we get these magical algorithms that promise us a cooler planet? It was 1967 when Ellison Burton and William Sanjour first introduced the world to an algorithm they argued could make the skies of California blue again. They demonstrated a series of micro-economic computer simulations between 1967-1970 while they worked for the National Air Pollution Control Administration – the predecessor to the EPA's Air Office. After WWII: "In the US, the economy and the military became a powerful partnership, and it was not long before industry, entertainment and academia joined the party forming an M.I.E.A network[57] that helped to sop up the overproduction of the war machine with a consumer driven explosion of highways, cars, and suburban sprawl fanning out all over the West Coast of America. Politicians bundle algorithms into dreams that can be easily swallowed by gullible voters, and around the world the dreams of personal wealth were graphed onto material growth that gave birth to our anthropogenic nightmare. Two cars, the daily commute, and the housewife's visit to the strip mall all succeeded in a powerful uptick in the GDP and, as a result, fossil fuel pollution. This was just what was promised by the classical political economists when they wrote of the 'invisible hand

---

57.   See Rive, P. (2019). Worldbending: a survivor's guide.

of the market'. What accompanied this acceleration in consumption were brown skies, respiratory morbidity, heart disease, and with the added bonus of sulphur oxide emissions from car pollution. Was this a market failure, an economic externality, or simply an unwanted side effect of market success? What the scientists proved was that their economic simulations based on assumptions of rational choice theory[58] was the lowest cost solution for a given level of air pollution abatement. This economic analysis and conclusion led to the concept of 'Cap and Trade' and the first market solution to a wicked problem. Through the 1950's and 1960's think tanks such as the RAND corporation had convinced the military and the government that their computer simulations, algorithms, and free market economic theories would not only defeat communism but deliver the Free World from the evils of environmental degradation.[59] The Cold War had a chilling effect on not only American scholarship,[60] but silenced those who would argue that capitalistic markets were at the root cause of anthropogenic harm to the environment. Anyone that threatened the oil and gas industries, that were the driving force behind the consumer economy, were accused of disloyalty to the *'American Dream'*, if not communist subversion. Capitalism was sold as the product of democracy, free markets, and freedom of expression, and necessitated the free flow of capital and trade. Milton Friedman's tactics in support of neoliberal economics were less interested in democratic freedom and more in the

58. See McCumber, J. (2016). The philosophy scare: The politics of reason in the early Cold War. Chicago: The University of Chicago Press.

59. Jardini, D. (2013). Thinking Through the Cold War: RAND, National Security and Domestic Policy, 1945-1975. Meadow Lands: David Jardini.

60. McCumber, J. (2016). The philosophy scare: The politics of reason in the early Cold War. Chicago: The University of Chicago Press.

freedom of the market and American business, to go about their profit maximisation, unmolested by the government and left wing environmentalists.

However, what the neoliberals began to fear in the 60s and 70s was the rise of opposition to *free enterprise* and in August 1971 Lewis Powell, who Nixon was about to appoint to the Supreme Court, wrote a secret memo to the American Chamber of Commerce arguing that those that would try and limit the markets had gone too far. Powell wrote: "the time had come—indeed it is long overdue—for the wisdom, ingenuity and resources of American business to be marshalled against those who would destroy it." Powell argued that individual action was insufficient. 'Strength', he wrote, 'lies in organization, in careful long-range planning and implementation, in consistency of action over an indefinite period of years, in the scale of financing available only through joint effort, and in the political power available only through united action and national organizations'. The National Chamber of Commerce, he argued, should lead an assault upon the major institutions—universities, schools, the media, publishing, the courts—in order to change how individuals think 'about the corporation, the law, culture, and the individual'. US businesses did not lack resources for such an effort, particularly when pooled."[61] At the same time, Powell and others argued in public that American corporations should be free to express their opinions through their extensive resources.

This argument is so engrained in the American ideology today that the exchange of money is seen as freedom of expression. This was illustrated in the 2009 Supreme Court

---

61. Harvey, David.(2005). A Brief History of Neoliberalism (p. 43). OUP Oxford. Kindle Edition.

case of the Citizens United v. the Federal Election Commission. In that case the Republican nominated majority in the Court voted 5 to 4 in favour of the argument that money was effectively equivalent to the freedom of expression – i.e. this was a first amendment right.

Citizens United is a shadowy conservative think tank, and its President at the time was David Bossie, who later became Trump's deputy campaign manager, he was also a close friend of Alt-right cheerleader, Steve Bannon, and Trump apologist, Kellyanne Conway. This Court decision has enabled the Super PACs to hide their influence and anonymously donate as much as they like to election campaigns, including for the purposes of attack ads and movies against the likes of Hillary Clinton and other Democrats. The Citizen's United argument was that to prevent anonymous donations by corporations, unions, non-profit think tanks and their political communications, was to restrict freedom of speech, in effect declaring that money is a form of expression.

Was this what Adam Smith meant by the invisible hand of the market? In other words, their influence was covert, invisible and undetectable. "By contrast, President Barack Obama stated that the decision "gives the special interests and their lobbyists even more power in Washington". The Supreme Court ruling had a major impact on campaign finance, allowing unlimited election spending by corporations and labor unions and fueling the rise of Super PACs."[62]

Long before the first Cold War, way back in the 19th century, there was the rise of the liberal ideology, inherited

---

62.   See Citizen's United v. FEC. https://en.wikipedia.org/wiki/ Citizens_United_v._FEC. Also see Lessig, L. (2015). Republic Lost: The corruption of equality and the steps to end it (Revised edition). New York, NY: Twelve.

from the political theorists of the Enlightenment. Prior to the Victorian social engineers such as Herbert Spencer, who convinced Darwin to include the phrase, '*survival of the fittest*,'[63] the classical political economists invented the concepts of laissez faire economics, free trade and the free market. The invisible hand of the market became an algorithm itself that ensured that the petit bourgeoisie, championed by one of their own, Adam Smith, would go unmolested by nobles and governments that did their bidding. The contrarian, Michael Perelman, paints an unflattering picture of Adam Smith by exposing his consistent self-interest that is hidden in his private letters and pamphlets that conceal how he promoted laissez-faire market forces, and the deprivation of peasants to encourage low wage labour in order to benefit proto-capitalist accumulations. Perelman speculates about Smith's subconscious use of the term, '*the invisible hand of the market*'. He wrote:

"I suspect that society has more to fear from the repressed emotions of someone like Smith. His metaphor of the invisible hand may be relevant in this regard. We may equate friendship with an open, outstretched hand, but an invisible hand has something sinister about it. In this spirit, Macbeth requested that the darkness of night, "with thy bloody and invisible hand," cover up the crimes he was about to commit (Macbeth 2.2). Or we could turn to Friedrich Nietzsche's eerie discussion of the invisible hand: If I wanted to shake this tree with my hands I should not be able to do it. But the wind, which we do not see, tortures and bends it in whatever direction it pleases. It is by invisible hands that we are bent and tortured worst....

---

63.   This phrase only appeared in the 1869 5th edition of Darwin's On the Origin of the Species

It is with man as it is with the tree. The more he aspires to height and light, the more strongly do his roots strive earthward, downward, into the dark, the deep—into evil."[64]

Surely, if you cannot see it, it must be magic. The hand of the market is invisible, algorithms are hidden, as are cyborg automatons, the computing culture machines behind the scenes, it is only the implementation that we see, but as the digital theorist Galloway insists we must ask two important questions about technologies, how does it work, and who does it work for?

Galloway points out that 'computer code is always enacted: "it is commissioned, designed, and implemented to achieve specific ends 'in particular contexts. Code = praxis."[65]

Economic Cyborgs

It is our religious faith in computational machines that has usurped human cognition, making decisions about our future for us. Faith in the machine is coupled with an equal theocratic passion for market forces. These came together with Alan Greenspan in the 1980's who was once a programmer and became Chairman of the Federal Reserve. It was in the late 90s he convinced President Bill Clinton that: "computers would relieve governments, (and human politicians) from the necessity to intervene in the economy because algorithms would ultimately apply a decentralised and rational control over the economy."[66] Since the Second

---

64.   Perelman, Michael. The Invention of Capitalism (pp. 208-209). Duke University Press. Kindle Edition.

65.   Galloway, Alexander R.. Protocol: How Control Exists after Decentralization (Leonardo Book Series) (Kindle Locations 135-136). The MIT Press. Cited by Rive in Worldbending.

66.   Excerpt From: Pete Rive. "Worldbending." Apple Books.

World War neoclassical economists including Milton Friedman, and Alan Greenspan have embraced computational economics entrusting the invisible hand of an automaton that was plugged into the market. Going back to Adam Smith in the eighteenth century it was assumed by economists that the market was like a machine and this was somehow regulated by the natural laws of physics and mathematics. This universal machine metaphor was handed down to the 20th and 21st century and is at the heart of neoclassical economics. Philip Mirowski has cited the founder of cybernetics, Norbert Wiener, who argued that while the seventeenth and eighteenth centuries were the age of the clock, and the eighteenth and nineteenth were the age of the steam engine, the twentieth was the age of communication and control. Mirowski wrote: "Natural order for economists coming of age after World War II is still exemplified by a machine, although the manifestation of the machine has changed: it is now the computer." or as Mirowski characterises it, 'economics is a cyborg science'.[67]

The obvious problem with this faith in the market forces algorithm is that it removes the decision making process from the public domain. Not only are the regulatory algorithms hidden by the invisible hand, with yet another layer of abstraction, but there is also the barrier of technical complexity, and proprietary closed code that disenfranchises democracy and the majority of voters.

It was during the time when Ronald Reagan and Margaret Thatcher were in power in the 80s that the neoclassical economists, Friedman and the Chairman of the Federal Reserve, Greenspan, installed the 'free market'

---

67.   Mirowski, P. (2002). Machine Dreams: Economics becomes a cyborg science. p.9

algorithms that are still running today. While Greenspan admitted his contribution, as a programmer, to the Y2K bug, he also smugly shrugged off his uncertainty as the Millennium came to a close. Despite going to a sober Federal Reserve party, where they waited with fingers crossed to see if the lights of New York would suddenly blink out, Greenspan walked away having learned nothing of the dangers of algorithmic efficiency. Instead of coding 1974 in full Greenspan and others 'optimised' by writing the shorthand 74. This FORTRAN programming technique that saved two digits of computer memory space almost cost millions of lives and chaos around the planet. Greenspan had promoted computer automation to President Clinton, yet he admitted: "As I wrote programs using punchcards in the 1970's (I'd used this technique myself, writing programs on punchcards at Townsend-Greenspan. It had never entered my mind that such programs, extensively patched, might still be in use at the end of the century, and I never bothered to document the work.) There was understandably widespread concern that the shift from 1999 to 2000 might cause such software to go haywire. This potential glitch was often devilishly hard to detect and costly to fix. But in Y2K doomsday scenarios, the consequences of failing to do so were dire: vital civilian and military networks would crash, causing electricity to go out, phones to fail, credit cards to stop working, airplanes to collide, and worse."[68]

The belated precautions taken by the Federal Reserve and the fact that the World did not end that night, meant that Greenspan and others continued to ignore the fragility of the algorithmic system. The irony of his book published

68.   Greenspan, A. (2007) Alan Greenspan: the age of turbulence. Adventures in a new world. Penguin Books. p.204.

in 2007 was that the algorithms of the sub-prime disaster and the opaque consequences of computerised global trading systems had not yet exploded – the Global Financial Crisis had not yet hit. Greenspan boasted: "We have all sorts of safeguards so that, for example, the data of one Federal Research bank are backed up at another Federal Reserve bank hundreds of miles away or in some remote location. In the event of a nuclear attack, we'd be back up and running in all non irradiated areas very quickly."[69] I'm not sure that the comfort that Greenspan felt i.e. that money and therefore, capitalism, would survive a nuclear blast gives me any peace – ground zero will still not be a habitable zone.

Friedman's original monetary theory known as Friedman's *k-percent rule* proposed to automatically increase money supply by a fixed percent each year. Thus, this would remove the ability of the central reserve bank to manipulate the money supply as it would be determined by an algorithm running on a computer. In Friedman's mind that would provide private businesses with the confidence to anticipate the interest rates and money supply. Friedman, and Greenspan shared this almost religious belief that a market free from government regulation, and interference would deliver the best possible outcome for everyone. It is important to realise that Friedman's free market did not necessarily imply democratic freedom as he advocated market liberalisation under dictators in Chile, Indonesia, Russia, China and many other democratically suspect states.[70] The wealthy would become more wealthy, but so would the poor, as the money trickled down,

---

69.   Greenspan, A. (2007), ibid, p.2.

70.   See Klein, N. (2007). The Shock Doctrine: The Rise of Disaster Capitalism. Penguin Books.

pumped slowly out of the machine by market forces. The faith in this ghost that lives in the machine, or an invisible hand that was born of liberal political economics during the Enlightenment, has endured until today. As Mirowski puts it: "Machine rationality and machine regularities are the constants in the history of neoclassical economics; it is only the innards of the machine that has changed from time to time."[71]

The Algo of Aluminium Abundance

Close to the core of the free market ideology is the concept that competition between buyers and sellers will drive down prices and that will find a *'spontaneous order'* or equilibrium guaranteeing market efficiencies, optimisation, and welfare for the whole of society. This eighteenth century idea was unproven, and yet promoted by Adam Smith. Faith in the concept of competition was bolstered by social darwinism that built on Herbert Spenser's interpretation of Darwin's theory of evolution and natural selection. Known for short as, *'survival of the fittest'*, this idea is another of the commandments that neoclassical economists, and neoliberal ideologues adhere to. Fred Turner traces how this ideology of competitive individualism embedded itself in Silicon Valley, and how technocrats such as Friedman and Greenspan embraced what Turner called *'technoliberalism'*.

Both Friedman and Greenspan had a great deal of respect for Ayn Rand who Friedman described as "an utterly intolerant and dogmatic person who did a great deal of good". Friedman proudly proclaimed Rand as a champion of libertine ideas and the free market. Her ideas

---

71.   Mirowski, P. (2002) ibid. p.10

of self-interest and the pursuit of happiness are inherited from classical political economists such as Smith and Bentham but in the 20th century her fiction had a much greater impact rather than her dry, if not fascist, political-economic theory. The heroic character of the entrepreneurial architect, Howard Roark, has become the iconic archetype of the tech billionaires of America, and the *Fountainhead* is Trump's most favourite book that he has probably never read. Trump said of the book "It relates to business and beauty and life and inner emotions...That book relates to...everything."[72]

The obsession with Rand's beliefs is nicely encapsulated in the figure of Peter Diamandis who in October 2020 accepted the Atlas Society's Lifetime Achievement Award. The tech entrepreneur, Diamandis, said of his recognition by the Society: "I'm honored by the opportunity to receive the award." "The work of Ayn Rand and *Atlas Shrugged* is very important in my life as well. I've read the book no less than five times, and it gave me a new world view that I cherish."[73] It is ironic to reflect on his acceptance speech when he holds up a 'variant' face mask during the COVID 19 pandemic. Hyper-individualism and neoliberal economics may trumpet the individual but the virus is spread by the reality that we all depend on others to survive and we can be killed by social infection.

Diamandis who is the founder of the XPrize is a firm believer in market competition and heroic entrepreneurs,

---

72.   See Michael Wolff's *Fire and the Fury*, "Trump didn't read. He didn't really even skim. If it was print, it might as well not exist (p. 113). Little, Brown Book Group. Kindle Edition. " Excerpt From: Pete Rive. "Worldbending." Apple Books.

73.   See https://atlassociety.org/commentary/commentary-blog/6405-peter-diamandis-is-the-atlas-society-2020-honoree. See his acceptance speech https://www.youtube.com/watch?v=VgPnam1T2rQ

such as Musk, Branson, and Bezos that share his techno-optimism. He credits market competition for the success of his XPrize, that has kicked started such endeavours as SpaceX, that it was founded on. In his book, *Abundance*, Diamandis tells the story of how aluminium went from being a metal cherished by Kings to something that is recognised today for its abundance, light weight, and low cost. Diamandis explained the value of this metal was that its rarity (something that drives markets and determines price) came down to its chemistry. In its raw form aluminum is abundant making up 8.3 percent of the weight of the world but because it bonds easily with oxygen it never appears as a pure metal in nature.

It was not until 1854 that Henri Sainte-Claire Deville created the first commercial process for extraction '*driving the price down by 90 percent*.' Yet, it was still costly and in short supply. With the breakthrough of the innovative process known as electrolysis in 1886 by American chemist Charles Martin Hall and Frenchman Paul Heroult everything changed. Diamandis crowed that it was the context that made this technology important:

"The Hall-Héroult process, as it is now known, uses electricity to liberate aluminum from bauxite. Suddenly everyone on the planet had access to ridiculous amounts of cheap, light, pliable metal...Technology is a resource-liberating mechanism. It can make the once scarce the now abundant."[74] This abundance would come at great cost because in many countries the electricity came from burning coal, one of the most polluting fossil fuel uses we now want to eradicate. But that is not the end of the story

---

74.   Diamandis, Peter H.,Kotler, Steven. Abundance: The Future Is Better Than You Think (Kindle Location 179-184). Free Press. Kindle Edition.

for aluminium and the algorithm that brought its manufacturing to New Zealand.

Hidden Power

As early as 1903 Fiordland, that includes Lake Manapouri and Lake Te Anau, was recognised as a place of extraordinary beauty and should be designated as a scenic reserve for tourists and future New Zealanders. On the 17th November 1904 the government reserved 2,326,000 acres in Fiordland under the Land Act (1892). Globally the first national park was Yellowstone National Park established twenty years earlier. Hydroelectric power was recognised in New Zealand as early as the 1880s with Reefton being the first town to fully electrify with a 1 kilowatt generator on the 4th August, 1888. On 24 September 1903 the Water Power Bill was introduced that identified the potential of Lake Manapouri and Lake Te Anau for hydroelectric power. Peter Hay, an engineer with the Public Works Department, envisaged how this power could be utilised to produce nitrogen fertiliser, aluminium and other electrochemical products. However, it was not until 1959 that government officials and politicians began to see this dream as a real possibility.

Since the late nineteenth century there was concern that this fledgeling country was too dependent on agriculture and should diversify. Lake Manapouri had continued to be the obsession of those that imagined the little colony might use hydropower to become a global player in the international markets but they saw that it was the State and not the individual that could make it happen. The invisible hand of the market was nowhere to be seen but as the historian Aaron Fox pointed out:

"The guiding hand of government determined the shape, form and duration of the Manapouri power project after 1904, with the perennial challenge for successive administrations being the degree of direct state investment and involvement in the realisation of the scheme."[75]

It was the Holyoake government in the late 50s and then the *'young turk'*, Muldoon, from the conservative National Party that inverted the power supply and turned the negative of state intervention into a positive attribute. New Zealand joined the IMF in 1961 and by the 70s both Labour and National were convinced that the New Zealand economy could achieve some sort of independence and insulation from international economics and global energy shocks. The massive debts from overseas investors that Muldoon racked up during the *'Think Big'* projects such as energy and petrochemical projects played directly into the hands of the neoliberal economists from the IMF who used inflation and the current account deficit to then *'shout fire in the theatre'* causing a run on the dollar and thus influencing monetary policy.

According to Klein, reflecting on the neoliberal revolution that happened in the 80s: "The principle was simple: countries in crisis desperately need emergency aid to stabilize their currencies. When privatization and free-trade policies are packaged together with a financial bailout, countries have little choice but to accept the whole package. The really clever part was that the economists themselves knew that free trade had nothing to do with ending a crisis, but that information was expertly

---

75. Fox, A. P. (2001). The Power Game: the development of the Manapouri-Tiwai Point electro-industrial complex, 1904-1969 (Thesis, Doctor of Philosophy). University of Otago. Retrieved from http://hdl.handle.net/10523/335

"obfuscated."[76] This term was applied by the renowned World Bank economist, Dani Rodrik, and was meant as a compliment with regards to the wholesale *'structural adjustment'* of foreign economies. "Not only did this bundling work in getting poor countries to accept the policies selected for them by Washington, but it was the only thing that worked—and Rodrik had the numbers to back up his claim. He had studied all the countries that adopted radical free-trade policies in the eighties and found that "no significant case of trade reform in a developing country in the 1980s took place outside the context of a serious economic crisis."[77]

Government engineers and ambitious bureaucrats courted the multinational aluminium industry with the offer of low electricity prices and high profits. Secrecy and sweetheart deals were the order of the day despite great concern expressed by environmental organisations that daming Manapouri would be an act of visual and ecological vandalism for little economic gain. The plan to build an aluminium plant at Tiwai Point using cheap power from Manapouri required huge foreign investment for New Zealand to join the international aluminium market.

Rio Tinto (formerly Comalco) recognised the government's ambition who held the naive faith that this would provide huge benefits to the New Zealand public, and as a result Rio Tinto managed to extract massive concessions from the government including cheap power and tax breaks. At the same time they cried poverty so that the government ended up paying for much of the infrastructure through foreign loans. In all, the final

---

76. Klein, N. (2007) ibid.
77. Klein, N. (2007) ibid.

project cost approximately $2 billion in today's currency, and most of it was financed by the taxpayers.

Since hydro-electric power was first proposed in Te Anau and Manapori in 1904 there had been a steady stream of protests by members of the public that wanted to protect and preserve the natural beauty of the environment. To begin with this was about tourism and the look of the place, but after around 60 years of lobbying the big dreams of economic independence and men who fancied themselves standing on the world's stage as players in the *'power games'* of capitalist globalisation, eventually won out.

Yet, as Fox has shown, the benefits of the large scale of the Manapouri project were also its disadvantages: "since the successful exploitation of the scheme required an energy-intensive industry for it to be economically viable. International developments in the field of electro-chemical and electro-metallurgical processes meant that the necessary raw materials, expensive processing and fabrication technologies, and markets were controlled by a small number of large multi-national corporations."[78] From the early days there had been repeated attempts to lure Australian, British, Japanese, and American electro-industrial companies to New Zealand with the bait of Lake Manapouri's hydro-electric potential. For politicians of all stripes this was the *'ultimate get rich quick scheme'* and a precursor to globalisation. What the government did not calculate was the expropriation of common value by offshore companies who negotiated secret deals with naive bureaucrats and politicians.

The energy potential of all sources of power on the planet is derived from resources that are held in common by all

---

78.   Fox, A. (2001) ibid

humans, and most importantly nonhumans, and yet energy companies, and those that depend on that power, profit from a process which is effectively the same as the land enclosure movement.[79] Simply put, because the sources of power, production, and even life itself are derived from the sun, water, air, rock, fauna and flora, then there is no entitlement to any of it, but rather these resources should be recognised as subject to informed stewardship and evolutionary principles that determine resource allocation. These resources, known as the commons are beyond human sovereignty, beyond governance and legislation and are ultimately 'ruled' by the laws of evolution, and physics that stretch beyond human spacetime scales.

As Fox pointed out everything that happened at Tiwai Point revolved around a '*power game*', both politically and from the point of view of energy. Internationally hydro-electric power has required massive investment and partnerships between the state and entrepreneurs who were looking to profit from the application of cheap energy to produce industrial products that demanded huge amounts of electricity such as synthetic nitrogen fertiliser, aluminium, and armaments for impending wars.

It was the birth of the military-industrial complex with a Cold War dimension in which the means of production justified the end. As early as 1909 a US chemist, Charles Munroe wrote a paper, *The Nitrogen Question from the Military Standpoint,* sponsored by the US Navy. In his talk he argued that mineral nitrates played a crucial role in both the manufacturing of explosives and the expanding market for commercial fertilisers. Munro feared that if the US

---

79.   See Chp. 4 Zero Growth

entered a war their supply from Chile could be jeopardised. "For war and agriculture alike, fixing nitrogen was nothing less than a matter of national security."[80] A synthetic process for nitrogen fixation also had great appeal for New Zealand agriculture.

Internationally the colossal cost of the manufacture of aluminium, and the energy infrastructure it needed, was often covered or underwritten by the state policy makers, bureaucrats and politicians that assisted industrial magnates by offering them subsidised electricity, loans, tax breaks, and other subsidies. Cozy relationships between energy corporations, large manufacturers, politicians and their bureaucrats meant that many of the deals that were subsidised by governments happened behind closed doors away from the scrutiny of the public.

The *Save Manapouri* campaign enlisted the help of academics, environmentalists, local farmers and organisations such as *Forest and Bird*, who could see the likely damage to the surrounding lake ecosystem, the loss of rare plants and birds, and all of that for unsubstantiated claims by the government and officials that there would be significant social and economic benefit from a project that would '*improve on nature*'.[81] Alternative ecological and economic evidence was put forward by the activists but they became increasingly frustrated with government sources that would not release all of the information relating to negotiations with Comalco and the intention to push for the maximisation of power which would require an unecological 30 meter rise in Lake Manapouri. Under the existing law at the time, *The Official Secrets Act* (1951),

80. Johnson, T. (2016) 'Nitrogen Nation: The Legacy of World War I and the Politics of Chemical Agriculture in the United States, 1916–1933.'

81. Fox, A. (2001) ibid.

could prosecute any unauthorised person divulging government information, including civil servants. This frustration eventually led to the Coalition for Open Government (1979) who opposed further industrialisation by Robert Muldoon under his 'Think Big' policies that sought to expand on Tiwai Point with another aluminium smelter at Aramoana; the oil refinery at Marsden Point; and the steel mill at Glenbrook. Against the determination of Prime Minister Muldoon, and the NZ Treasury, the Official Information Act came into effect in 1982. Today, there are 45,000 requests for information on government activity, and 90% are answered within 20 days of the request. But this was not the case when the National government was negotiating with international corporate interests, and neoliberal organisations such as the IMF, World Bank and OECD who managed to extract significant concessions from the government.

In 1937 the total power consumption of New Zealand was 235,509 kws of main plant with 86,959 kws of stand-by plant yet Manapouri had the potential to generate a colossal 200,000 kws just on its own, and Lake Te Anau 600,000 kws. Those that planned these massive power projects were only too aware that the taxpayers would be paying for over-engineering at great cost in order to sell off the oversupply to foreign interests. Despite Manapouri being a long way from any industry, for over 60 years entrepreneurs and politicians lobbied and campaigned to construct a hydro-electric power station there in order to provide cheap energy for aluminium smelting and synthetic nitrogen fertiliser production. What was worse, the aluminium process that happened at Tiwai Point was part of a multi-stage manufacturing that turned alumina into aluminium, a process that was the most expensive and as a stand alone business could never turn a profit but only made financial sense as a component of Rio Tinto's vertical integration.[82]

From the beginning it was understood that it could ruin the aesthetics of this beautiful ecology and that the cost benefits did not stack up. These projects were pushed by entrepreneurs who could see large personal profits by hawking the power to multinationals, and politicians who fantasised about growing New Zealand's industrial economy beyond agriculture. The only reason for a multinational aluminium company to locate in a remote place at the bottom of the world was the promise of a power supply so cheap that they could make substantial profits from the finished products. Aluminum became a metal of strategic military importance during both the First and Second World Wars. Internationally, the production of fertilisers also supported the armaments business, as the byproducts, nitrates and ammonia were useful in bomb construction.

From the very beginning industrial growth, both manufacturing and agricultural, were going to be balanced by the cost of environmental damage which was to be no cost to industry, but would be treated by economists as externalities. There was to be a 'white gold rush' of aluminium, with a side hustle of nitrogen fixation that enriched private enterprise at the monetary cost to the New Zealand public, and an environmental cost to the planet. The business model was only viable as an export industry as New Zealand was importing only 100 tons of aluminum a year in 1940 and the projected aluminium production output of the proposed plants was 800,000 tons. Today, Tiwai Point only produces 250,000 tons of aluminium and

---

82.   Fox, A. (2001) ibid.

still relies on international markets to buy the surplus supply.

Tiwai Point's only advantage was cheap power and that was offered to the multinational Rio Tinto at a reduced cost subsidised by the New Zealand public and other businesses. While aluminium ore, or bauxite, is plentiful around the planet it is relatively rare in New Zealand with only a small deposit known in Kerikeri, in the North of the North Island, that has never been mined. The bauxite is imported from Weipa in Queensland at the mine also owned by Rio Tinto. Rio Tinto has battled with successive New Zealand governments pushing hard in negotiations as masterful agents in a game of brinkmanship. They have now called it a day as the government and power companies have refused to extend any more subsidies, not including the substantial revenue they were collecting from the ETS.[83]

For all of those that bemoan the lost jobs from Tiwai Point, the cost to New Zealand has been enormous and the profit was all extracted by Rio Tinto. The ETS and market mechanisms have in no way encouraged the company to reduce their carbon pollution, and more than likely encouraged and enriched them to continue with business as usual. Since the Emissions Trading Scheme was introduced in 2008 a small handful of industrial polluters known as Emissions Intensive, Trade Exposed (EITE) industries have received from the government free carbon credits. These equated to 90 per cent of their emissions, that almost entirely insulates them from the cost of their GHG pollution.

What is more Rio Tinto has been storing dangerous

---

83.   The recent election of the Labour party could mean that Rio Tinto can be persuaded to stay for another 5 years.

waste as a by-product of their production. Sadly, this is not uncommon amongst the biggest GHG polluters who have helped to increase the temperature of our planet causing extreme weather such as the recent flooding in the South Island of New Zealand. Rio Tinto has been storing waste from the Tiwai smelter an hour away in Mataura. However, the storage is inadequate and has been threatened by inundation caused by the floods. The waste is classified as a class six hazardous substance that is capable of producing poisonous ammonia gas when it comes in contact with water.[84] Once again the company wanted taxpayers to pay for the clean up while it consumes 13 per cent of New Zealand's total electricity supply.

The thin veil of corporate social responsibility is cast aside by GHG corporations such as Rio Tinto who often argue that they should self regulate or allow the market to determine their behaviour, but as they have recently shown in Australia their attitude towards the environment and the cultural concerns of others plays a poor second to profits. Despite knowing of the ancient cultural significance of 46,000 year old caves at Juukan Gorge the company went ahead and blew them up in order to extract iron ore causing great distress to the local indigenous tribes.[85]

Why do polluters, like Rio Tinto, get paid to pollute our atmosphere and threaten life on this planet? The explanation is provided by economists who say that exposing these companies to international competition,

84. See https://www.rnz.co.nz/news/national/420964/rio-tinto-s-smelter-closure-unnerves-mataura-locals-on-waste-removal

85. Butler, B. (7th Sept. 2020) 'Rio Tinto CEO and senior executives resign from company after Juukan Gorge debacle'. The Guardian. https://www.theguardian.com/business/2020/sep/11/rio-tinto-ceo-senior-executives-resign-juukan-gorge-debacle-caves?CMP=Share_iOSApp_Other

while they have to pay the higher cost of carbon pollution, would disadvantage them and cause the loss of jobs. This in effect gives them electricity subsidies to pay for their high energy use negating the intended affect of the ETS that would increase electricity cost in order to lower carbon emissions. However, as Bertram explains heavy industrial polluters are also big electricity users and because of the fear of lost jobs the government has insulated them from rising prices on the wholesale electricity market. The very mechanism that the ETS market claims to be able to solve works in reverse. Bertram writes: "As rising carbon prices prevail in the New Zealand economy, the perverse consequences of the way the electricity market works will become increasingly problematic. The wholesale price of electricity is set at the margin of the spot market, where the predominant form of generation is from thermal plants burning coal or gas. A rising carbon price therefore drives up the price of electricity – precisely the wrong signal to send when electrification is the key to decarbonisation of the economy, and when over 90% of electricity will be coming from renewables."[86] The hand of the market is certainly invisible to most people but it is firmly holding the lever that ensures the speed and amount of profitability continues. Fear has been used around the world to market the concept that should these big industries fail then jobs, comfortable lifestyles, and even the world as we know it would end. Those algorithms are embedded in the operating systems of our digitised economy.

Marketing Failure

The irony is that the market failure that is responsible for

---

86.   Betrand, G. (2020) ibid.

the environmental and economic shortcomings of our profligate use of energy was then offered up as the solution. The algorithms of the ETS would apply the market ideology in an attempt to reduce GHGs and carbon emissions, however, this has been an unmitigated failure that has not decreased, but rather increased emissions and global heating. It is expected to continue to pay the biggest carbon emitters billions of dollars for decades to come. As the economist Geoffrey Bertram put it in his submission on the Emissions Trading Reform bill the proposed amendments were only going to make the ETS more complex while: "(i) perpetuating an unfair and distortionary allocation of adjustment burdens, (ii) leaving untouched the perversely anti-decarbonisation effect of interaction between the NZETS and the wholesale electricity market, and (iii) failing to remove private-sector uncertainty over the future quantity and price of allowable emissions." Bertram did not mince his words: "The New Zealand ETS has now become so degraded relative to the simple originating theory of cap-and-trade that it is probably beyond rescue as a sustainable framework for this country's climate change policy."[87]

According to the Motu Guide to the New Zealand Emissions Trading Scheme written by Catherine Leining and economist Suzi Kerr (Aug. 2018) "An ETS translates a regulatory limit on emissions into an emission price set by the market which changes behaviour to reduce emissions." This is despite the fact that they acknowledge from 2008-2015 New Zealand had no limits on domestic emissions. "NZ ETS participants met compliance

---

87.   Bertram, G. (2020) Submission to the Environment Committee on the Climate Change Response (Emissions Trading Reform) Amendment Bill. http://www.geoffbertram.com/publications/?no_cache=1

obligations using unlimited overseas units at low prices and faced little incentive to reduce their own emissions."[88] From these economists, who originally helped build and sell the ETS to the government, there is a simple haiku, that hides the complexity and assumptions behind the ETS algorithms.

'Emissions trading creates a price signal that transforms behaviour'

However, this simplistic algorithm hides the economic and scientific complexity of the calculations made by the Climate Change Commission in their Draft Advice for Consultation. One of the major problems with the ETS is that it dresses up ideology to look like science. That ideology could be loosely defined as neoliberal, or simplistically *'markets over government regulation'*. The hands off approach is cheered on by corporations and the worst carbon polluters because it gives them wiggle room to attend to business as usual. Carbon offsets have been used by the biggest carbon polluters to pretend to be doing something about their emissions by paying someone else (usually in a country away from prying eyes) to complete a sustainable low-carbon emission project. However, most of those projects turned out to be fraudulent and the global increase in carbon emission figures proves it. Economics is the true religion behind this ideological error in our thinking.

The appointment of Dr Rod Carr as the Chairperson of the CCC in NZ shows that this so-called *'independent'* commission is an uncontentious political appointee who has been central to our neoliberal faith. Not

---

88. Leining, C. & Kerr, S. (April 2019) 'Managing Scarcity and Ambition in the NZ ETS.' Motu Working Paper 19-07. Motu Economic and Public Policy Research.

only was he one of the faithful as the Chair of the Reserve Bank of New Zealand but he has also a PhD in Insuaranace and Risk Management from the neoliberal church of The Wharton School, University of Pennsylvania. He has an MA in Applied Economics and Managerial Science, an MBA in Money and Finance and honours degrees in law and economics. To round out his neoliberal education Carr would have also the lived-experience of being a Director of the Canterbury Employers' Chamber of Commerce where he was no doubt informed of the essential role of dairy farming to the New Zealand economy, just in case he didn't already know. It is no wonder that his commission has advised the government to take a *'balanced'* view to mitigating climate change. This a term that was also used by the NZ oil and gas lobby spokesman when it came to on-going support for the extraction of the fossil fuel *'natural gas'* – that will help to maintain our current carbon emissions. The government is terrified of losing power to a party who they believe, and has proven to-date, to have accelerated carbon emissions.The Treasury would argue that a decline in GDP is a fate worse than any anthropogenic harm, and if the UN says that you can have *'sustainable growth'* who is little New Zealand to argue?

A New Zealand group known as Lawyers for Climate Action, in their submission to the Commission claimed that the Draft had a serious error in the Commission's calculations. They wrote:

- "In our view, the Commission's draft advice does not comply with the legal requirements. The main reason for this is that the advice is not consistent with what is required to keep global warming to less than 1.5° Celsius – we consider that emissions over the current decade must be capped at 400 Mt, not the 628 Mt proposed by the Commission's draft budgets. This is a fundamental error that must be fixed before the advice is finalised. Failing this, the advice will be unlawful, in our opinion. Further, Aotearoa New Zealand's international reputation and brand will be at risk if we fail to adopt budgets and policies consistent with doing our fair share to keep global warming to less than 1.5° Celsius.

- The methods for accounting for Aotearoa New Zealand's emissions and presentation chosen by the Commission have the effect of obscuring our lack of progress to date and of making the budgets appear more ambitious than they really are.

- The Commission's recommendation that Aotearoa New Zealand should use offshore mitigation to bridge the gap between our Nationally Determined Contribution (NDC) under the Paris Agreement and our domestic emissions reductions, is, in our view, at odds with the Act and with the Paris Agreement itself. It would leave Aotearoa New Zealand exposed to international criticism and a high level of uncertainty about the cost and availability of international credits."[89]

In order for the ETS to have any chance of working it must be easily understood by politicians, and voters, and not be so arcane that only economists and lawyers can understand it, and thereby ensure that carbon emitters, such as Rio

---

89. See Lawyers for Climate Change. https://static1.squarespace.com/static/ 5cf3039126905000011c02b0/t/60582ad62998f4194f095ad7/ 1616390878157/2021-03-22+LCANZI+Submission.pdf

Tinto, are richly rewarded for their pollution.[90] If you are wondering what the actual code or algorithm for the invisible hand of the market might look like I have published it here as one of many examples of an esoteric formula associated with the ETS and climate change. The algorithm was used for calculating Rio Tinto's carbon allocations. It turns out that Rio Tinto was not only earning billions from the ETS, but a mistake in the algorithm meant they were being overpaid. The public's ability to interrogate this algorithm is hidden beneath acronyms and complex formulas that convert GHGs into a carbon dioxide equivalent unit, resulting in this erroneous formula: "A=LA x AB x Out, is too high at 0.537 tCO_2–e/MWH, which is over-allocating NZUs to EITE firms with levels of assistance set at 90 per cent within the NZETS." The Climate Change Commission should assist the public in simplifying and translating complex technical algorithms so that the public can easily understand how they contribute to reducing GHG emissions.[91]

So, if the ETS is so bankrupt why do politicians, economists, and the Climate Change Commission continue to tinker with it hoping that a market driven solution will save us from apocalyptic global heating? The neoliberal ideology of the invisible hand of the market continues to self-perpetuate itself via the binary algorithms running on machines we never see. We have given up even pretending that these economic algorithms are somehow

---

90. See this online article for an explanation of how the ETS works. Marc Daalder. (March 23, 2921) "Analysis: What is this cap-and-trade scheme and how does it reduce emissions?" https://www.newsroom.co.nz/emissions-trading-scheme-nzs-carbon-market-explained

91. Mitchell, C. (Feb. 2020) 'Confused? Why not understanding the Emissions Trading Scheme is the point.' https://www.stuff.co.nz/environment/climate-news/119665441/if-youre-confused-you-understand-why-not-understanding-the-emissions-trading-scheme-is-the-point

rational and have bowed down in devotion to our new religion the *free market* that will deliver a magical outcome and our salvation.

Bertram summed up his frustration with over a decade of New Zealand politicians who remained committed to the ETS. He comments rather caustically:

"My central submission is that if the New Zealand Parliament cannot do better than this in terms of climate change policy, it should set aside this Bill and transparently declare to the world both this country's unwillingness to live up to its international obligations and its inability to forge a political consensus around genuine emission-reducing action. New legislation should wait until the domestic political situation has moved on sufficiently to make genuine action politically possible. In the interim the old, justifiably discredited NZETS is not much worse than the proposed new one."[92] Of course, the whole world is running out of time while we wait for politicians to negotiate complex and technical details as the end of the world gets ever closer. Ultimately, the ETS is founded on a principle of efficiently reducing the cost of decarbonisation rather than rapid mitigation of carbon emissions and would eventually require New Zealand to join an international trading market in carbon emissions.

Our current thinking tools are not up to the task. Think Big helped to industrialise New Zealand and extract value from our common resources at the expense of the environment. We have to find new tools, and new thinking to help us to adapt to the Zero One Worlds beyond the simplistic Carbon paradigm. Algorithms, computer models, and software can not be relied upon to automate

---

92.   Bertram, G. (2020) ibid.

solutions unless they are open, transparent, and easily understood. Ultimately, we can not hope to solve wicked problems with binary questions and simplistic answers, we need to take a critical and transdisciplinary worldview that accepts the plurality of worlds – both human and nonhuman.

9

Zero One World

In a binary world, one defined by two numbers, zero and one, it is assumed that the entire universe, or even the multiverse can be described, defined, built, bent, and destroyed by the application of a base 2 mathematical system. The so-called digital, or computational universe is just that sort of universe, and as I have just discussed in the previous chapter this has become an algorithmic faith as powerful and convincing as any of the major religions on this planet.

However, I would like to explore another zero-one world that is open, rather than closed, and far from being exclusive, is a metaphor for inclusivity, diversity, and an infinite number of universes that defy the imagination or the EULA[1] designed by the lawyers of a software development company. The zero-one world is a virtual container for the infinite multiverse, all of those worlds

---

1. A EULA is an acronym for the End User License Agreement, the terms and conditions you unconsciously agree to when you click 'yes' before using a new piece of software.

that identify with an irrational number, one that cannot be expressed as a ratio of two integers. This recalls my earlier discussion about Zeno's Paradox and the infinite divisibility of non-integers between 0 and 1, the infinite and the whole.

To state this simply, there are no natural numbers between 0 and 1, but there are an infinite number of non-whole numbers that can be infinitely divided. I would now like you to hold that thought and to introduce another contradictory metaphor, a paradox, this is something that our minds are capable of but is beyond the ability of a computational machine. Because you can easily imagine a world that includes black, white and an infinite variety of grey, or a paradoxical sentence that says, this is not a sentence, or this sentence is a lie, you are capable of something a computer is not. But that paradoxical, imprecise, and contradictory reality that you live with everyday is not a binary reality.

Despite the fact that Gödel and Turing both agreed that our universe includes non-computable numbers, and functions, computer and electronic engineers ignore this awkward truth, and live with what is a fudged reality of approximation. You might be wondering, so why should I care? Because, our current reality is being dominated by algorithms that assume a binary world of computational machines and that severely delimits our ability to imagine, bend, and create new worlds outside of that ruling paradigm. Ultimately, it is important to realise whether you think about it or not the foundations of all action begin with feelings, and emotions, that eventually inform thoughts, and philosophies. If those philosophies at a high level of abstraction do not incorporate open, and universal beliefs about the agency of the nonhuman majority, as well as humans, and their human minorities then your world will be narrow, closed, and extremely limited in options.

I have attempted to show that zero carbon is an illusory algorithm that on the face of it seems rational, logical, and very sensible if we are going to continue to live on this planet together, co-existing with the non-human majority. Our alternatives inhabit the worlds not specified as 0 or 1 but in the infinite space between them including those that cannot be computed. A binary world delivers us an illusion, an answer to the algorithm asking what is right and what is wrong? A simple black and white answer. But the world of humans is no different from the world of physics – as reassuring as certainty is, the worlds we occupy are full of uncertainty, contradictions, and paradox. Try and imagine a void, a space that is not a space, that is full of nothing and so is not full at all – is it black? What is black, what is zero colour? What is white? If it is the opposite of black does that means it is full of colour or if we make this a digital problem is it 1 or 100% – is it every colour added together? But wait, is that not black? So maybe it is every colour subtracted giving us zero – yet surely that is also black? What about grey? Is that black + white? Again, a paradox, a contradiction that we think is logical and common sense – anything that is not black, and it is not white must be grey. Yet, that is also not correct because if it has no black or no white it must be the opposite.

I can imagine you are getting impatient with this nonsense, but the answer must reside in the fact that we are more than comfortable with approximations, which are by definition more than one thing than another – in other words not a whole number. It is, afterall, how even the most precise engineering and science works. To cite the astrophysicist Brian Greene the most accurate experimental measurements that accord with theory known to modern science have been recorded in quantum field theory with an extraordinary precision of ten decimal places but even that is not 100% accurate.[2] The most certain we are about anything in our world is that there is uncertainty, and yet we wake up every morning assuming that the Sun will rise, our phone will turn on, and that we will not be burnt to death by the unbearable heat caused by GHGs.

The Infinite Zero-One Worlds

When you watch a digital TV you are looking at millions and millions of binary decisions controlling the light emitting diodes, LEDs, or some such digital electronic light emitters, that are responding to an ON or OFF command directed by an algorithm. But that is not how you see reality. Your eye is an analogue biological organ, that responds to different frequencies of electromagnetic light. Now, your Smart TV may try and approximate that process to help reverse engineer the problem of how we see, but it is only a simulation, and how we really mediate light bouncing off an object, into our retina, before it stimulates our optic nerves, and this is more like how a river flows. The light moves as if it is rushing towards your eye in waves rippling at different speeds to stimulate different electrochemical colour responses, but if you try and measure how this electromagnetic energy is converted into an electrochemical stream that translates into your colour perception, you are confronted with a paradox that happily

2.   Greene, B. (2011). The Hidden Reality: Parallel Universes and the Deep Laws of the Cosmos.

contradicts itself – light is both a wave and a stream of particles.

The world that you see depends on where you are standing and what you are looking at? It is as true for colour, as it is for time and space. If science, and Einstein tells us that reality is relative, why are we so quick to accept one way of looking and measuring only ONE world. What is more, we assume that how that **one** world is quantified is akin to the laws of physics – yet is actually a cultural artefact, known as economics. To simplify our reality and to create products that can be sold to the most number of consumers, our education system, and the capitalist world that profits from it, have spent trillions of dollars on getting us all to agree. The *'manufacturing of consent'* was achieved by instruction and training us how to look into *'a mirror, darkly'*. We have been sold a distorted form of a reality, a reality that claims that there is only one world constructed from a capitalist algorithm. This was how it has been with religion and the Kings of old. Yet, just as the Apostle Paul warned in his letter to the Corinthians, as we approach the end of this world, the apocalypse will reveal new worlds beyond this one. Some of us may save ourselves and even our nonhuman loved ones, but those who cannot adapt will burn in a fiery hell on earth.

Your definition of hell will depend on where you stand on all sorts of issues, it is relative. For Donald Trump, an annoying little country at the bottom of his world, Aotearoa, New Zealand is a hell hole – #NZhellhole.[3] This may sound like the rantings of a prophet of doom, but this is an apocalyptic trope that is as old as civilisation itself.

---

3. See #NZhellhole on Twitter, and this article on the Australian Traveller website. https://www.traveller.com.au/nzhellhole-new-zealand-responds-hilariously-to-claim-country-is-a-hellhole-in-lockdown-h1q977

Today, our global consumer culture serves up the old beliefs encoded into algorithms that preach our new religion. It is a self-fulfilling fantasy based on an ancient archetype that speaks of the cycle of life, the cycle of carbon, and the inevitable destructive disruption that comes from a lack of wisdom, or lack of attention to the nonhuman world around us.

Still, infinite worlds are opening up in front of us as pluralism, and concepts such as transdisciplinarity are already beginning to displace, the one worldview of advertising and propaganda that capitalists, lobbyists, politicians, and educators try and sell to us daily. There are hopeful signs of adaptation to these new worlds and a critical mass of humans keen to find a co-existence with the nonhuman majority. Every thing that is connected to every thing in a space and time that is subjective to them has the opportunity to creatively imagine an alternative world in which they will continually build, bend and break the worlds of their own device. If the mirror fragments and becomes multiple prisms we will all begin to appreciate that we are sharing multiple worlds with others and that it is more than rude to thoughtlessly impinge on, or smash their worlds because you think you are the only one that understands or has the right to build one reality.

Remarkably there is high degree of confidence in the theoretical physics known as M theory or the brane theory of our universe. One of the conclusions of this theory is that we are living in a multiverse that is defined by a hidden reality. This would make our reality exceedingly odd and is one of the many reasons we should assume that ideologies such as economic growth theory, rational choice theory, market forces, and capitalism are nothing more than simplistic mental constructs that humans have made up.

The best selling sci fi book, *The Space between Worlds*, begins by quoting the popular theoretical physicist Brian Greene: "In the far reaches of an infinite cosmos, there's a galaxy that looks just like the Milky Way, with a solar system that's the spitting image of ours, with a planet that's a dead ringer for earth, with a house that's indistinguishable from yours, inhabited by someone who looks just like you, who is right now reading this very book and imagining you, in a distant galaxy, just reaching the end of this sentence. And there's not just one such copy. In an infinite universe, there are infinitely many. In some, your doppelgänger is now reading this sentence along with you. In others, he or she has skipped ahead, or feels in need of a snack and has put the book down. In others, he or she has, well, a less than felicitous disposition and is someone you'd rather not meet in a dark alley."[4]

Alternate Reality

Even if the theory of the multiverse turns out to be wrong it is invaluable for liberating our minds from the hegemonic closed world that currently imprisons us. In a philosophical sense multi-worldviews are a reality, as real as the concept of a pink elephant or a poem. At the peak of the Gartner hype cycle around virtual worlds the economist Edward Castronova speculated that humanity would exodus the physical world into virtual worlds. He reasoned that the physical world has become an increasingly unpleasant place to be and as it becomes more polluted, violent, too hot, too cold, or just plain ugly then beautifully designed,

---

4. Brian Greene, The Hidden Reality" from "The Space Between Worlds: a Sunday Times bestselling science fiction adventure through the multiverse" by Micaiah Johnson.

and very entertaining virtual worlds will become increasingly attractive.

According to Statista in 2020 the total number of gamers worldwide was estimated to be 2.7 billion. 1.5 billion gamers were estimated to come from the Asia Pacific region, not including North America (210 million). These numbers deserve some serious consideration. Gamers make up almost a third of the world's 7.8 billion population, which is well beyond a conceptual tipping point.[5] While the number of people in the Asia Pacific region living in extreme poverty has declined dramatically it must be remembered that the economic growth policies that accompany this has also seen a dramatic rise in environmental pollution and degradation of arable land, water ways, and air quality. Therefore, we might speculate about the quality of life that these populations have inherited? While their wealth might have increased enough to be able to afford the luxury of game consoles, computer access, and or gaming parlors, their cramped apartments and surrounding concrete jungles have little to offer. Castronova wrote: "Anyone who sees a hurricane coming should warn others. I see a hurricane coming. Over the next generation or two, ever larger numbers of people, hundreds of millions, will become immersed in virtual worlds and online games. While we are playing, things we used to do on the outside, in "reality," won't be happening anymore, or won't be happening in the same way. You can't pull millions of person-hours out of a society without creating an atmospheric-level event. If it happens in a generation, I think the twenty-first century will see a social cataclysm larger than that caused by cars, radios, and TV, combined. .

---

5. See Gladwell, M. (2000). The Tipping Point: How little things can make a big difference (1st ed.). Little, Brown.

.. The exodus of these people from the real world, from our normal daily life, will create a change in social climate that makes global warming look like a tempest in a teacup."[6]

The bullshit jobs that David Graeber wrote about just won't cut it when you can have so much fun, comradery, and fulfilment in a virtual world or game. As Jane McGonigal has said, *'Reality is broken'*. McGonigal observed: "The real world just doesn't offer up as easily the carefully designed pleasures, the thrilling challenges, and the powerful social bonding afforded by virtual environments. Reality doesn't motivate us as effectively. Reality isn't engineered to maximize our potential. Reality wasn't designed from the bottom up to make us happy."[7] And herein lies the threat. If we truly are *'happiness machines'* in search of the ultimate utopian experience, is it not possible that we might just ignore the bitter reality all around us because global warming is far from just *'a tempest in a teacup'*.

Our education system is failing to prepare students and society for a future that could transition to alternative worlds. Politicians, business leaders, and media pundits demand that universities are more vocational, preparing students for *'real world'* jobs. However, that sees the universities as job factories for busyness-as-usual and not the vanguard for better worlds. As Dunne & Raby have pointed out mainstream design is *'affirmative design'* and movements such as the Circular Economy and Ecomodernists fail to apply critical thinking to technological solutions. Dunne & Raby wrote: "Corporate futurologists force-feed us a 'happy-ever-after' portrayal of

---

6. Castronova, E. (2007). Exodus to the virtual world: How online fun is changing reality.

7. McGonigal, J. (2011). Reality is Broken . Random House. Kindle Edition.

life where technology is the solution to every problem. There is no room for doubt or complexity in their techno-utopian visions. Everyone is a stereotype, and social and cultural roles remain unchanged. Despite the fact that technology is evolving, the imagined products that feature in their fantasies reassure us that nothing essential will change, everything will stay the same. The future forecasters have a conservative role, predicting patterns of behaviour in relation to technological developments. They draw from what we already know about people, and weave new ideas into existing realities. The resulting scenarios extend pre-existing reality into the future and so reinforce the status quo rather than challenging it. Their slick surface detracts us from the dystopian vision of life they wish for. By designing the props for the videos produced to show us what the future could be like, design works to keep official values in place."[8]

The lack of criticality and speculation permeates not just industrial design which hides its toxic interiors within but are prevalent in the design of virtual worlds and Multiple Player Online Role Games. They represent what Paul Edwards described as *'closed worlds'* that confine the discourse to neoliberal capitalism.[9] However, as Mary Flanagan has detailed game designers have other opportunities to create alternative, and critical worlds that are open and challenge the status quo.[10]

8.   Dunne & Raby. (2001) Design Noir: The Secret Life of Electronic Objects. Birkhauser. p.6

9.   Edwards, P. N. (1997). The Closed World: Computers and the politics of discourse in Cold War America. Cambridge, Mass.: MIT Press.

10. Flanagan, M. (2009). Critical Play: radical game design. MIT Press.

Persuasive Technology

There remains a very real concern that the majority of games or virtual worlds are designed based on a capitalist's hierarchy of needs i.e profit first, engagement second, then entertainment, and after that, who cares. Most AAA games[11] cost so much to produce and distribute that companies such as Tencent and Sony, that earn billions in revenues from their sales, sit squarely inside the capitalist camp. They are shaping and maintaining a closed world discourse that supports the status quo and the consumer economy that has turned the ETS and global heating into one very large and dangerous game. The average time spent gaming a week was shown to be around 7 hours, or almost one whole working day.[12] The psychology and science behind game play is fixated on the reward and engagement of the players, or in virtual worlds, *'residents'* using theories such as Fogg's *'persuasive technology'* approach.

B.J. Fogg coined the acronym Captology, or Computer Aided Persuasive Technology in 1996, and established the Persuasive Technologies Lab at Stanford in 1999. According to Fogg: "persuasive computing technology is a computing system, device, or application intentionally designed to change a person's attitude or behavior in a predetermined way."[13] According to the critical media historian, Adam

---

11. AAA (pronounced and sometimes written Triple-A) is an informal classification used for video games produced and distributed by a mid-sized or major publisher, typically having higher development and marketing budgets." https://en.wikipedia.org/wiki/AAA_(video_game_industry)

12. Forbes. (2019) Research Report Shows How Much Time We Spend Gaming [Infographic] https://www.forbes.com/sites/kevinanderton/2019/03/21/research-report-shows-how-much-time-we-spend-gaming-infographic/?sh=115daa4a3e07

13. Fogg, B. J. (1999). Persuasive Technologies. Communications of the ACM, 42(5), p.27.

Curtis, Stanford was also the home of Abraham Maslow who developed the Freudian theory known as *'the hierarchy of needs'* which posited a process he called *'self-actualisation'* that was deployed as a tool by the Stanford Research Institute. The purpose of the Stanford lab was to research *'lifestyle marketing'* categorising consumers according to inner drives and desires. Stanford University had a cosy relationship with paying customers from large corporations, and Fogg also saw commercial opportunities amongst game developers, the military, and governments.

Fogg proposed a *'Functional Triad'* as a classification of : "three basic ways that people view or respond to computing technologies"[14]

From a computer user's point of view they function in three ways:

- as tools

- as media

- as social actors, or as more than one at once.[15]

As a tool computers provide users with new capabilities; as a medium they can convey either symbolic or sensory content (virtual worlds and simulations), and as a social actor they *'can invoke social responses from users'.*[16]

It is possible that virtual technologies, and artificial intelligence could have serious social and political consequences as roboticists overcome the *'uncanny valley'*[17]

14.   Fogg, B.J. (1998). "Persuasive Computers: Perspectives and Research Directions."

15.   Fogg, B. J. (1999). Persuasive Technologies. Communications of the ACM, 42(5), 26–29.

16.   Fogg, B.J. (1999) ibid, p28.

17.   In aesthetics, the uncanny valley is a hypothesized relationship between the degree of an object's resemblance to a human being and the

and unethical or ideologically driven powers seek to use persuasive technologies for the manipulation and control of whole populations. Persuasive technologies could be used to advertise and market anything and everything. One area of considerable research interest in the future could be the use of General AIs and virtual agents using persuasive technologies. This could be compounded by photorealistic, and even multi-modal sensory robots within both virtual worlds, and possibly even actual physical environments. Research published in 2004 has already suggested that the avatars of virtual agents that have a 40% blend of the same facial characteristics as the human subject, and who have slightly delayed their body mimicry by four seconds are far more persuasive than an avatar that has little resemblance to the person that is being persuaded.[18] According to Bailenson et. al: "Results of this study demonstrated a huge difference between groups. Agents that mimicked the participants were far more successful at persuading the participants and were seen as more likeable than recorded agents."[19]

Mirror neurons are thought to be responsible for empathetic behaviour and the small resemblance of the virtual agent, below the conscious threshold, has been shown to engender higher levels of trust.[20] According to Bailenson: "Living breathing humans socially respond to virtual humans in IVEs, (immersive virtual environment),

emotional response to such an object. The concept suggests that humanoid objects which imperfectly resemble actual human beings provoke uncanny or strangely familiar feelings of eeriness and revulsion in observers. See https://en.wikipedia.org/wiki/Uncanny_valley

18.   Bailenson, J. N., Beall, A. C., Loomis, J., Blascovich, J., & Turk, M. (2004). "Transformed social interaction: Decoupling representation from behavior and form in collaborative virtual environments."

19.   Bailenson et. al. (2004). ibid. p18.

20.   Bailenson, et. al. (2004)ibid.

in a naturalistic way regarding personal spaces social presence and affect."[21]

Bailenson et al. showed that cyborgs in a virtual environment that simulate mutual gaze, mimic or communicate facial expressions and gestures, and simulate naturalistic body language can be more persuasive than in an actual face-to-face meeting.[22] Fogg refers to research performed at the School of Management, Boston University. In the first study participants were asked to play a social dilemma game either with an on screen character, looking unattractive and in Fogg's words,*'creepy'*, or choose to serve their own *'selfish purpose'* and *'received rather low cooperation rates: just 32%'*[23] In a later study: "thanks to technology developments ... the onscreen character looked less artificial and, I would argue, more attractive and less creepy. This new and improved character garnered cooperation rates of a whooping 92%...that is this study was statistically indistinguishable from cooperation rates for interacting with real human beings"[24]

Fogg wrote: "Computers as media suggest different pathways to persuasion, most notably by providing people with simulations and virtual environments."[25] New virtual world environments, such as High Fidelity can allow users to immerse themselves in a virtual experience. High

---

21. Bailenson, J. N., Blascovich, J., Beall, A. C., & Loomis, J. M. (2003). "Interpersonal distance in immersive virtual environments." Pers Soc Psychol Bull, 29(7), 819-833.

22. Also see Rive, P. B. (2012). Design in a Virtual Innovation Ecology: A Cybernetic Systems Approach to Knowledge Creation and Design Collaboration in Second Life. Retrieved from http://researcharchive.vuw.ac.nz/handle/10063/2747

23. Fogg, (2003). ibid. p93.

24. Fogg, (2003). ibid. pp.93-94.

25. Fogg, B. J. (2003).ibid. p.28

Fidelity has eye and facial tracking via the computer camera, that enables the user's avatar to mirror the gaze and emotional response in-world. These features combined with a gestural interface could provide a powerful persuasive experience.[26]

Virtual worlds and online game environments do offer great opportunities to explore and experiment with alternative ways to envisage our worlds but we must also view them through a critical lens to avoid creating a digital prison, a matrix of which we are unaware.

A Queer Way to Look at the World

What I have attempted to show in this book is how the concept of zero carbon may appear radical and an urgent call to action but when it is examined more closely it in no way threatens the status quo because it operates algorithms safely located within the old market paradigm. What we need is creative intelligence that honestly synthesises totally new paradigms that recognises the profound challenges before us.

It was the poet Audre Lorde who delivered a speech at an international feminist conference in 1979, entitled 'The Master's Tools Will Never Dismantle the Master's House.' Nick Walker, a neurodivergent, gay, academic activist wrote that:

"Lorde, a Black lesbian from a working-class immigrant family, castigated her almost entirely white and affluent audience for remaining rooted in, and continuing to propagate, the fundamental dynamics of the patriarchy:

---

26.   While High Fidelity has 'pivotted' and changed its offering to a virtual world 3D audio solution. I predict that many of its innovative virtual world solutions will resurface in the future.

hierarchy, exclusion, racism, classism, homophobia, obliviousness to privilege, failure to embrace diversity. Lorde recognized sexism as being part of a broader, deeply-rooted paradigm that dealt with all forms of difference by establishing hierarchies of dominance, and she saw that genuine, widespread liberation was impossible as long as feminists continued to operate within this paradigm."[27] Walker describes this as a pathological paradigm, and defines this according to two criteria:

1. There is one "right," "normal," or "healthy" way for human brains and human minds to be configured and to function (or one relatively narrow "normal" range into which the configuration and functioning of human brains and minds ought to fall).

2. If your neurological configuration and functioning (and, as a result, your ways of thinking and behaving) diverge substantially from the dominant standard of "normal," then there is Something Wrong With You.[28]

The brutality of Conversion Therapy, that has resulted in LGBTQIA+ suicides, is one and the same therapy used on autistic children to make them think 'normally'. The legacy of inhuman shock treatment should warn us about those who want to impose one worldview on everyone and everything. It is my argument that our current hegemonic, capitalist, neurotypical paradigm is pathological and that

27.   Walker, N. (2013) Throw Away the Master's Tools: Liberating Ourselves from the Pathology Paradigm. https://neurocosmopolitanism.com/throw-away-the-masters-tools-liberating-ourselves-from-the-pathology-paradigm/

28.   Walker, N. ibid.

this worldview is morbidly overweight and incapable of stepping lightly on the Earth without having a psychotic episode. Our political, economic, cultural and educational paradigm is suffering from what my colleague, Jorn Bettin, describes as a *social learning disability*. Is civilisation nearing collapse? Are we really incapable of thinking and acting creatively about future worlds? Neurodiversity is described by Walker as follows:

"Neurodiversity is the diversity of human minds, the infinite variation in neurocognitive functioning within our species.

What It Doesn't Mean: Neurodiversity is a biological fact. It's not a perspective, an approach, a belief, a political position, or a paradigm. That's the neurodiversity paradigm, not neurodiversity itself.

Neurodiversity is not a political or social activist movement. That's the Neurodiversity Movement, not neurodiversity itself." [29]

The neurodiversity movement provides political solutions for breaking apart the *normality* paradigm and encouraging those who have truly original cognitive behaviour because, "We humans are a neurodiverse species." The rainbow spectrum of cognitive variations encompassed by neurodiversity includes, but is not limited to the following list. You will note that a number of these include the words disorder and disability which are indicative of the normality paradigm and are politically rejected by the neurodiversity movement: Autism, ADHD, developmental speech disorders, dyslexia, dyspraxia, dyscalculia, dysnomia, intellectual disability and Tourette

29.   Walker, N. (Sept. 2014) "Neurodiversity: Some Basic Terms & Definitions." https://neurocosmopolitanism.com/neurodiversity-some-basic-terms-definitions/

syndrome; and mental health conditions such as bipolarity, schizophrenia, schizoaffective disorder, antisocial personality disorder and obsessive–compulsive disorder, ODD – Oppositional defiant disorder, Dyslexia, LGBTQIA+. I would add to this incomplete list artists, and entrepreneurs because by definition the artist and the entrepreneur must avoid neurotypical thinking and diverge from normality. The neurodiversity movement rejects the neurotypical disability paradigm and demands to have a say about their worlds, proclaiming 'nothing about us without us' and reshaping the conversation using the social model of disability. This model insists it is not the individual who is disabled but the society and it is the design of society that disables the individuals. For those that cannot accept this queer view of the world, the neurodiversity movement insists that at least they must be able to accept that diversity is a survival strategy for our species, especially when even a cursory review of almost all efforts to date have failed so spectacularly.

Switch to Safe Mode

As Donna Haraway has noted many educators have dropped the ball because rather than teaching how to learn about learning which is not simply academic but thinking by doing, and doing by thinking. This constructivist approach to learning is deeply subversive because thinking and doing are exactly how the worlds revolve – they are the precursor to revolutions and are precisely how the status quo undergoes change and transformation through critical thinking and action. Haraway makes this wry observation about the academic world of so-called thinkers who instead of educating society about how to learn about learning and how to experiment with ideas through designed activities, instead many academics provide society with a heavy layer of cynicism which ensures a safety valve to harmlessly release steam with a smug smirk, or alternatively approve a catatonic intransigence towards the very wicked problems before us. Haraway exhorts us to *'stay with the trouble'* but also to beware of two forms of futurism often exhibited amongst teaching and research faculty who either gravitate towards geoengineering our broken planet, or wallow in despair that effectively pours cold water on student agency by leaking a cynical *'game over'* attitude.[30] Those damp squibs who hide their pathetic detachment from the world beyond the ivory towers, *'coinfect any possible common imagination'* amongst the students by showing no interest in truly critical thinking about the status quo. They have accepted that their lot in life is to first gain tenure, and this initiation is a process of indoctrination into mediocrity, and the long, slow climb to the top of academia by kowtowing to the university's professorial mandarins who demand obedience, and obsequious fawning. This is a long game, for them it is the real game, that ensures them a life of relative comfort paid for by an unquestioning faith in the system, that just happens to be capitalism. Any unrostered time by the academic staff must be soaked up with administering a meaningless bureaucracy that demands reports, and surveillance of peers and students in a institutional panopticon which would make Bentham jealous. Student surveys report on their satisfaction with their experience as if there was a correlation with their education, and this *'McKinsey Stalinism'* ensures any controversial debate is effectively bled from the system.

---

30.   Haraway, D. J. (2016). Staying with the Trouble: Making kin in the Chthulucene.

In the US after the Second World War research grants, and university administrators, with the collusion of academic staff managed to remove most researchers and educators who might express values and opinions critical of the economic, social, and political Cold War worldview.[31] This de-politicisation spread around the world amongst those impressed, or oppressed by American cultural imperialism. Today, it does not matter whether you are a student of science, technology, the humanities or the social sciences, you are almost immediately de-politicised as soon as you enter a school of higher learning.

Returning to university to teach at AUT (Auckland University of Technology) in 2014 I began to realise this was not like my days as a student at Auckland University during the late 70s and early 80s. Political activism was conspicuous by its absence and there was no necessity for the Vice Chancellor to post any edict banning protests or congregating in large numbers. The students were as uninterested in world change as were the teachers, and certainly the administrators. I became intrigued and began to research what had happened to the university, the so-called *'critic and conscience of society'*, a legal requirement of universities under the New Zealand Education Act (1991)? I concluded that not only had the universities been neutered but education had become weaponised. Teachers no longer felt safe to speak out and they would not only be unprotected by their universities against political molestation, but be hunted down and removed by university management.[32] As Peter Fleming has noted in

---

31. See McCumber, J. (2016). The philosophy scare: The politics of reason in the early Cold War. Chicago: The University of Chicago Press.

32. While there are exceptions amongst the shrinking number of teachers with tenure, increasingly education is as precarious as other occupations, and has even become part of the gig economy with short terms contracts

his book, *Dark Academia: How Universities Die*, (2021) this stultifying situation is an international problem that has spread throughout the neoliberal universe. It has strangled creativity in the UK, US, Australia, and New Zealand as well as many more countries who have corporatised their universities. Fleming wrote that teaching and research staff at the modern university suffer from: "Economic isolation and insecurity that is comingled with claustrophobic levels of performance management, corporate socialisation and close-quarters' supervision. Needless to say, this all but kills the creativity and self-determination essential to scholarly excellence and radical breakthroughs in the sciences and humanities."[33] Researcher and humanities lecturer, Dr Luis Prádanos, had this to say about the future of education and the limits to our growth paradigm: "education would better serve students in particular and all humans in general if our teaching and research methods stop perpetuating the cultural paradigm that brought us to the brink of extinction and start encouraging students to imagine and create alternatives to it."[34] This morning I was listening to our state broadcaster, Radio NZ, and heard a couple of items that made me blurt out a series of expletives – perhaps I was having an off day but I do despair that our academics spend time and money on stupid research that is cheered on by thoughtless journalists. The researchers had trained calves to wee into a MooLoo using classic behavioural

---

and hourly pay for many. See Peter Fleming, *Dark Academia: How Universities Die*, (2021); also Chap. 7 The Weaponisation of Education, in *Worldbending*.

33. Fleming, P. (2021) Dark Academia: How Universities Die. Pluto Press. London.

34. Open Democracy. 24 August, 2020. Preparing for the end of the world as we know it. https://www.opendemocracy.net/en/oureconomy/preparing-end-world-we-know-it/

psychology by using rewards. I am sure it will give farmers a laugh and no doubt help delay any reduction in herd numbers but its unlikely to solve methane emissions and nitrate pollution. Even if they then dilute the wee and spray it back onto the fields its going to require even more water, and result in an even bigger waste issue. The next challenge will be scaling the training up to teach millions of cows where to pee – or they could just try reducing the number of stock. If this is considered innovative research I am doubtful the universities are going to save us.[35]

While I am pessimistic about our current education system's ability to critique, analyse, experiment with, and transform worlds beyond carbon, I am aware that the international rise of authoritarian societies will most likely cause a backlash against foolish tyrants who believe they can suppress, and oppress those who want to challenge and change the status quo. Authoritarians such as Trump may believe they can simply flip the algorithm and switch to safe mode but history has shown that such repressive behaviour only encourages revolution and transformational change. Education is not something that is exclusive to institutions that can perversely erase learning skills. We all have a responsibility to learn more about our worlds, and ourselves, learning to co-exist with the nonhuman majority. As Mahatma Gandhi advised:

'Live as if you were to die tomorrow. Learn as if you were to live forever.'

---

35.   MooLoo - researchers find toilet training cows is possible.
       https://www.rnz.co.nz/news/country/451424/mooloo-researchers-find-
       toilet-training-cows-is-possible

How To Gently Shake Worlds

The reason that I have waited until the end of this book to begin talking about doing something about the Anthropocene is because I want to encourage you to think about what you are doing, and then accepting that you will also be doing by thinking. History has shown us that violence begets violence and only propagates misery under the guise of change. If Gandhi suggests we should *'gently shake the world'*, how might we go about it? First, we have to learn that no one species, person, thing or institution can lay exclusive claim to creativity. If we were to reflect on the incredible creative and destructive power of carbon we might move beyond our egotistic confidence that only humans can make and break worlds. Sheldrake has written about the essential powers of the imagination to unlock ideas, not just in the arts but also the sciences. He talks about how his experimental field work into fungi was aided by his imaginative creativity: "I spent whole months staring into a microscope, immersed in rootscapes filled with winding hyphae frozen in ambiguous acts of intercourse with plant cells. Still, the fungi I could see were dead, embalmed and rendered in false colours. I felt like a clumsy sleuth. While I crouched for weeks scraping mud into small tubes, toucans croaked, howler monkeys roared, lianas tangled and anteaters licked. Microbial lives, especially those buried in soil, were not accessible like the bristling, charismatic, above-ground world of the large. Really, to make my findings vivid, to allow them to build and contribute to a general understanding, imagination was required. There was no way around it."[36] Acceptance that

---

36.   Sheldrake, M. (2020) Entangled Life: How Fungi Make Our Worlds, Change Our Minds and Shape Our Futures.

even nonliving nonhumans can build, bend, and break worlds should awake in us an ability to listen and learn before we act.

In our company, S23M, we facilitate Open Space workshops with our clients once we have established a psychologically safe place. There is no sage on a stage instructing or directives from senior management. We call this creative collaboration and this is based on scientific evolutionary evidence. For those not familiar with the value and purpose of creative collaboration we present the scientific background to how, and why it works? Collaboration can only take place where social and political hierarchies have been supplanted with a free flow of information based on competencies without political perversion, or manipulation. If all the participants feel psychologically safe to speak their mind then the people who have self-selected to participate in this creative collaboration, or agreed to co-design a solution are ready to begin the process using an Open Space discussion.

This is a leadership-free-space, no one is taking control, leading, or telling others what they should think. We sometimes undertake an anonymous audit of a company or organisation to assess the psychological safety felt by the potential participants. An overbearing leader can be deaf to the fears and concerns of those they regard as beneath them and this can lead to the problem of 'groupthink' when those with less power, and those who want to please others explicitly agree with ideas they privately do not agree with. There are many examples of how groupthink leads to cascading and then catastrophic failure. Contrary to what politicians and those that get paid excessive salaries may think and say, it is not strong leadership that will solve a wicked problem but a collaborative sharing of information.

A facilitator can suggest themes, and or request problem statements, or questions that the groups wants to explore using the Open Space method. It is important that any discussions are open ended questions that takes a co-design approach that has no preconceived solutions in a box ticking exercise i.e. pick the option you like. This is known as a *'decision approach'* that is less effective and creative than a *'design approach'* that is more about exploring the problem space and making no assumptions about what tools will be required? An open space uses a very simple process, and has easily understood principles. There are four rules:

1. Whoever come are the right people
2. Whatever happens is the only thing that could have
3. Whenever it starts is the right time
4. When it's over, it's over

There is also one law: the Law of Two Feet. "If, during the course of the gathering, any person finds themselves in a situation where they are neither learning nor contributing, they can go to some more productive place."

Once everyone has had their say and someone has documented the thoughts of the open space participants that information can be recorded in a shared repository by the facilitator with a draft circulated for comments, additions, and corrections. This is how we begin to gently shake the world.

Understanding Each Other

A truly open space that has the right people in the room,

and feeling psychologically safe, can surface deep knowledge held by each individual. Knowledge that is not normally exposed but is tacit knowledge held by experts who know something but can not easily say what it is they know. Sometimes that knowledge can only be surfaced by observing people and teams carry out something they have done often and are experts in. However, what if the individuals do not feel psychologically safe? They are unlikely to display or say what they think. But what are the circumstances in which individuals don't feel safe? It is the perception of dangerous situations that will inhibit the conversation and stifle creativity. These quiet moments could include the following:

- speaking up in front of a supervisor or manager who is loud, aggressive, and does not listen.

- speaking up when you know you are contradicting a boss who lets everyone know 'it's my way or the highway!'

- a group discussion where everyone has agreed with a forceful individual and you are alone – this is groupthink

- fear of being wrong

- fear of being different

- fear of being noticed

- fear of ridicule

- fear someone will steal your idea and claim it for themself

- knowledge hoarding – a company promotes those that don't share

- knowledge is power

- knowledge is competition – the survival of the fittest

- fear that your questions might be seen as disrespectful

The evolutionary biologist, David Sloan Wilson, reviewed how biology gained and then lost its voice over the past two hundred years. The dark history of social darwinism and eugenics have meant that policy makers and social scientists are wary of applying biological lessons to human organisations, and yet he argues that everything is related to evolution. He points out that evolution is not the narrow study of genetics, but has far more contextual, and environmental relevance that has shaped our adaptation to the worlds around us. Wilson cites the economist Robert Frank who predicts that in one hundred years it will be Darwin and not Adam Smith who will be remembered as the father of economics. In the meantime neoliberal economists cling to the outmoded view of laissez-faire policies based on an ideological view of nature. In the eighteenth century it was known as 'laissez-faire la nature' or 'let nature run its course', and was the premise of 'free markets' with little to no governmental interference. Yet, as Wilson, pointed out today: "ecologists have largely abandoned the idea that nature left undisturbed achieves a harmonious balance...Wise ecological policy requires active management with complex systems in mind."[37]

In order to understand our species and each other we have to understand evolutionary biology and rethink nature as a complex network of collaborative relations with

---

37.   Wilson, David Sloan. (2019) This View of Life . Knopf Doubleday Publishing Group. Kindle Edition.

interdependencies, and not based on some simplistic view of *'the survival of the fittest'*. The Nobel Prize winners in medicine, Niko Tinbergen, Konrad Lorenz, and Karl von Frisch were awarded their prize in 1973 for their pioneering work in the field of ethology, or the study of animal behaviour. Tinbergen contributed to evolutionary biology by posing four questions that Wilson says is essential to understanding evolution and thereby human nature:

"First, what is the function of a given trait (if any)? Why does it exist compared to many other traits that could exist? Second, what is the history of the trait as it evolved over multiple generations? Third, what is its physical mechanism? All traits, even behavioral traits, have a physical basis that must be understood in addition to their functions. Fourth, how does the trait develop during the lifetime of the organism? Recognizing these as separate questions and studying them in conjunction with each other form the foundational concepts in Darwin's toolkit."[38]

As I have noted earlier thanks to Lynn Margulis we now know that the evolution of life from simple cell microorganisms to the complex nonhuman worlds, within us, and all around us, have come about through collaborative symbiosis.

Regenerative Worlds

We need to be more focused on collaborative knowledge and co-existence with the nonhuman majority that will inform us of their needs, desires, and impulses, but only if we really listen. To date society is prone to the worst conservational impulses as we desperately try to hang on to a world that has already gone. We need to have

38.   Wilson, D.S. (2019) ibid.

conversations with diverse people, diverse things, and across spacetime scales that we are unfamiliar with. The ancient wisdom of the circle of life, the letter O and the number 0, are calling upon us to recollect the cosmic archetypes expressed in the science, arts and the myths that remind us nothing ever truly ends but is perpetually regenerated. This is not just the Second Law of Thermodynamics but wise advice for the reconstruction of a world that is running out of steam, and self-belief. How are we going to design regenerative cultures?

According to the ecological biologist Daniel Christian Wahl: "Transformational responses at a personal and collective level take place when we question deeply ingrained ways of being and seeing and in the process begin to reinvent ourselves. In doing so we also change how we participate in shaping culture through our interaction with the world around us." In other words we are mindful and take responsibility for our actions and the worlds' reactions to our behaviour and beliefs. As Wahl puts it we face two choices 'either collapse or profound transformation'.

Yet, there is an even more profound approach that we can adopt. The exploration of poetic metaphors, paradox, and conscious contradictions as we 'stay with the trouble'. This is not to suggest we should give up, but at the same time we might show some humility and accept that we have only just evolved enough to clamber out of the mud. The Anthropocene is not just the age of the algorithm but the age of the entitled primate, and this in no way gives us the keys to this universe, let alone any other in the multiverse. Wahl warns: "The basic story we are telling about humanity – who we are, what we are here for and where we are going

– no longer serves us as a functional moral compass."[39] We might even admit we might be lost, and that is ok, as long as we are not panicking and just grabbing the nearest technology at hand.

This is a good time to reflect on the Precautionary Principle that is expressed in the Wingspread Consensus Statement as follows: "When an activity raises threats of harm to [nonhuman and] human health or the environment, precautionary measures should be taken even if some cause and effect relationships are not fully established scientifically"[40] The principle puts the burden of proof on those that are proposing to take the action, especially when it includes egotistical overreaching, beyond the human scale. You only have to pause to consider some of the most outrageous technological projects currently underway to see that the Precautionary Principle is either ignored or not worthy of a media headline.

"We have to learn to live within the limits of the Earth's bioproductive capacity and use current solar income instead of ancient sunlight (stored in the Earth's crust as oil, gas, and coal) to provide our energy."[41]

While technology might be an extension and expansion of our evolutionary capability[42] there are those who are more interested in brand recognition and selling units of tech than the survival of the planet. One example of a personal brand that has outgrown his packaging is Elon Musk who achieved great wealth at 31 years old but was not going to be limited by that and has used his physics and business education to imagine more and more Promethean

39.   See Wahl, D.C. (2016) ibid.

40.   Wingspread Statement. (1998).

41.   See Wahl, D.C. (2016) Designing Regenerative Cultures. Triarchy Press.

42.   As argued by Ray Kurzweil. See Kurzweil. (2005). The Singularity is Near.

projects. He has been described as a cross between Edison and Tony Stark and such a cloned abomination should warn us of the subconscious desires and fantasies hidden in the investments that Musk has attracted. Just as Donald Trump does not just suddenly pop out of a test tube, Elon Musk is a creation of our times. Rather than heed Wahl's warnings of energy consumption, Musk continues to push the envelope by selling over sized investments such as Tesla & its Gigafactory, the Hyper Loop and Space X companies. Drunk on his own fame and sense of agency Musk recklessly pushes ahead with his plan to colonise Mars. It is fitting that the planet he has chosen is named after the God of War as Musk contributes to the world's infodemic by propagating Orwellian double-speak, at once talking about rocket fuel, sustainable electric cars, and self-sufficient farming on the red planet. Musk has even revised his concept of terraforming Mars by firing nuclear weapons at the planet's frozen ice caps. "Nuke Mars refers to a continuous stream of very low fallout nuclear fusion explosions above the atmosphere to create artificial suns. Much like our sun, this would not cause Mars to become radioactive," the entrepreneur tweeted.[43] In his rush to get there he has opted for an old Russian approach using methane/LOX for his Raptor rocket propellant. There is something chilling about his Cold War references to Russia and China that puts him up there with Trump in terms of recycling old school McCarthy digs at the communists while he is loyal to the capitalist cult of the entrepreneur.[44]

---

43. Wall, M. (Aug.21, 2019) "Looks Like Elon Musk Is Serious About Nuking Mars." https://www.space.com/elon-musk-serious-nuke-mars-terraforming.html

44. See Todd, D. (2012) Flight Global. 'SpaceX's Mars rocket to be methane-fuelled.' https://www.flightglobal.com/spacexs-mars-rocket-to-be-methane-fuelled/107953.article

What is more, Musk makes himself out to be a sustainability hero with his mega-investments in Tesla and Solar City, but the investment in Space X is the most outrageous example of his hubris. As reported on a website that champions methane, *Brightr: Natural gas. Naturally part of every day*: "In 2013 when the Raptor was announced, it was as a: "highly reusable methane staged-combustion engine that will power the next generation of SpaceX launch vehicles designed for the exploration and colonisation of Mars."[45] Despite the lower cost of Space X transportation it still uses and creates large amounts of GHGs, and on Mars will use the dense carbon dioxide atmosphere with the underground water reserves to create its own methane and oxygen. Perhaps Musk could also begin an ETS exchange on Mars and profit from the sale of CDM offset credits on the Moon or in China? Clearly with Space X the precautionary principle was thrown out the window.

As Vaclav Havel pointed out it is '*only by creating a better life can a better system be developed.*' Musk could pay attention to the fact that this is not a symmetric formula. You do not necessarily get a better life by developing a better system, yet that is our faith in the system and why we will continue to replicate harm on this and other planets. Many believe that technology is deterministic, at least when it is guided by the invisible hand of the market. The billions of dollars that Musk has attracted to build his capital-sucking-fantasy are directed by neoliberal ideologies and power lust that believes the bigger the project, the more capital, the more energy used, the greater the power and influence of

---

45. See Brightr (2019) Why the next generation of rockets will be powered by methane. https://bright-r.com.au/why-the-next-generation-of-rockets-will-be-powered-by-methane/

the investors, and shareholders. But that capital has to come from somewhere and its value is ultimately derived from the wealth and value stored in our planet. The financialization of the system does not just make capital invisible, it turns tangible resources into gases that disappear in the vapour trials of the billionaire's rocket fantasies. What if Musk had started with another problem, rather than working out how to burn capital? What if he had asked how do I live a better life? Unfortunately, the system has told him that consumption is the answer to how to live a better life, and therefore excessive consumption is presumed to be a much, much better life. The 'American Dream' has become the world's nightmare and precaution has been thrown to the Santa Ana winds as the worst wildfires in US history have pumped millions of tons of carbon dioxide and methane into the air. We may have already reached the tipping point as the biomass of almost 3 million acres of burnt forest pushes up the concentrations of GHGs that leads to more extreme weather and more fires. Musk may only have a few seats for wealthy tourists and colonisers on his rockets, but as Buckminster Fuller said 'we are all crew on spaceship Earth.'

In Australia, which has also suffered from serious bushfires fanned by global heating and droughts, the federal government showed little evidence of adaptation and learning from the fires and the COVID 19 pandemic. According to Alison Verhoeven, Chief Executive, Australian Healthcare and Hospitals Association:

"Just this week, the Australian government released its draft National Prevention Health Strategy consultation papers, and any right minded person would expect after the 9 to 10 months of experience we've had relating to bushfires and COVID-19, that there'd be some mention of those issues in a national preventive health strategy. But no, there wasn't. These issues were completely absent in this draft document, which is out for consultation, and supposedly will inform a 10 year strategy for us going forward.

Absolute absence of climate change or of the way that we might respond to pandemics, or the way we might deal with health equity issues in the social determinants, which are driving not only poor health outcomes for many of our communities in both Australia and New Zealand, and in the population groups most affected by COVID-19.

So, I don't feel a whole lot of hope to be honest in that space. I don't think we have actually realized that this isn't going to just end when COVID ends. We actually have broader things to think about: How do we live in society? How do we, make sure our impact on the world is actually a little bit lighter, and our tread is a little bit softer than it has been for the last 30 or 40 years?

How do we recognize that with globalization came pathology like HIV, which spread very rapidly around the world, and now a pandemic, from a disease that started in a market somewhere in China in December, by January is in Iran, by February is in Italy, and by March is in Australia and New Zealand. If we don't start thinking about the impacts of economic decisions of our industrial behaviours, frankly, we're not going to be a very healthy society anywhere."[46]

For those of us who don't want to live with Musk on Mars, and the majority who will not be able to afford to, we can look at adapting to the troubles. For those who have

---

46. S23M Webinar series 'Trans Tasman Knowledge Exchange'.
     https://healthcare-solutions.s23m.com/trans-tasman-knowledge-exchange/

contributed to the infodemic and double-speak of *'sustainable carbon emissions'*, they will eventually be held to account, but in the meantime we can work together to live better lives. What has been conclusively shown over the past 60 years is that we can not solve any of these anthropogenic problems with market forces. Today there is a global movement of those individuals who recognise their community interests and have already begun the transformation of our worlds. There are positive signs that they are gaining a critical mass as more and more people have begun to transition to an ethical and systems aware ecology.

The Transition Network

In 2005 a movement began in the UK known as Transition Towns, a grassroots, bottom up movement that encourages communities to learn and experiment with local sustainable food growing. It began as an urban movement that has now spread around the world.[47] On the charity's website it describes Transition as a movement: "about communities stepping up to address the big challenges they face by starting local. By coming together, they are able to crowd-source solutions. They seek to nurture a caring culture, one focused on supporting each other, both as groups or as wider communities. In practice, they are reclaiming the economy, sparking entrepreneurship, reimagining work, reskilling themselves and weaving webs of connection and support. It's an approach that has spread now to over 50 countries, in thousands of groups: in towns,

---

47.   See this video for a short history of Transition https://youtu.be/ObmpRoqopSo

villages, cities, Universities, schools."[48] In Transition asks: what are you thinking? What are you feeling? What are you doing? A combination of head; heart; hands. In their guide they introduce 7 essential ingredients for doing Transition:

1. Healthy groups – learning how to work well together

2. Vision – imagining the future you want to co-create

3. Involvement – Getting the wider community involved and developing relationships beyond friends and natural allies

4. Networks & partnerships – collaborating with others

5. Practical projects – inspiring others and building new infrastructures

6. Part of a movement – scaling up your impacts by linking up with Transitioners elsewhere

7. Reflect & celebrate – reflecting on how you're doing and celebrating the difference you're making

The rallying cry for many who join the Transition has been the acknowledgement of the precarious state of our urban and rural environments – for people, plants, animals and the geological spheres that comprise the Anthropocene. Together, *we are the precariat*. Based on work by Guy Standing and Judith Butler members of the precariat no longer have reliable incomes, and are often members of the *gig economy* , constantly on-call for hourly pay. The pandemic has accelerated this precarity and many of the

---

48.   See https://transitionnetwork.org/about-the-movement/what-is-transition/

precariat have no access to land to grow their own food and are dependent on their next gig to survive. According to the Wikipedia entry: "In sociology and economics, the precariat (/prɪˈkɛərɪət/) is a neologism for a social class formed by people suffering from precarity, which is a condition of existence without predictability or security, affecting material or psychological welfare."[49] What is apparent from this definition is that increasingly the precariat are no longer accepting their role within the capitalist economy and have slipped sideways into a new world that even includes their own local currencies that bypass consumerist values, and grows their own food based on regenerative farming practices. Complementary currencies can introduce a value systems based on cultural and local needs using what Raworth calls, 'finance based on life' that encourages open design, open data, and open exchange of value instead of capital accumulation where money is a spurious commodity.[50]

In Tāmaki Makaurau, Auckland, there is a recently established urban organic market garden, OMG, sandwiched between two aged concrete buildings, next to the construction of the new City Rail link. It is a temporary garden that will be returned to Auckland Transport once the station is built but was soil tested for toxins and it plans to relocate as part of the *For the Love of Bees* collective that intends to create an urban organic garden every 1 km apart in the inner city. They explain that they are: "Inspired by regeneration and wellness, OMG is a living example of the role of urban farming in achieving a sustainable and regenerative food system for Auckland. It inspires local residents, businesses and communities in the uptown

---

49.   See https://en.wikipedia.org/wiki/Precariat
50.   See Raworth, K. (2017) ibid.

neighbourhood to learn new practical skills to grow nutrient-dense food by using biology-first and biodynamic principles, in an open-access teaching hub. Guided by the principles of Kaitiakitanga, OMG demonstrates a sustainable social enterprise model by providing a space of inspiration, healing and knowledge-exchange centred on regenerative learning."[51] OMG has a subscription model that allows local residents to buy vegetables and salads, and also offers a compost service. This means that instead of dumping bio-waste in the landfill it returns to the earth as compost, regenerating the soil. While it is disappointing that the Auckland Transport authority appears to lack the vision to embed OMG into their transport hub the gardens have started to regenerate a neglected part of uptown and new apartment buildings are drawing in residents who want to be part of this regenerative community. In the UK the Kensal to Kilburn Transition Town approached the London underground to use the planting boxes at the Kilburn platform on the Jubilee line to plant food for anyone that wants to harvest it. Like many urban cities around the world, Tāmaki Makaurau, Auckland, has a growing number of homeless people sleeping on the street or who are unable to provide themselves with food. Yet, there are many examples of urban city gardens that are owned by the city in common that could grow food for their people.[52] The commons could feed many more people providing healthy food, community and purpose and could be the path to adaptation and resilience to climate change.

51.   Read more https://www.fortheloveofbees.co.nz/omg

52.   Watch the documentary, In Transition 2.0: a story of resilience and hope in extraordinary times. The Kilburn story is at 22:36. Also, watch 15:55, Stage 2 Deepening, a story about city gardens in a poor urban area in Pittsburgh - Whitney Avenue Urban Farm https://www.youtube.com/watch?v=FFQFBmq7X84

The Bounty of the Commons

Beginning in the 13th century the Enclosure Movement was a legal device by which land held in common became consolidated into larger privately held properties. By the 19th century almost all arable land had been enclosed and the peasantry who had relied on common land to supplement their meager food provisions were forced into industrial wage labour in order to pay for rent and food. The enclosure of the commons contributed a significant amount of land, profits, and labour productivity to the land owners. These contentious laws were violently policed and were made even more brutal by the British Game Laws that imposed harsh penalties for poaching game such as imprisonment, transportation to Australia, and ultimately execution. The feudal rights of the aristocracy to hunt game in deer parks and to keep out members of the public continued through the 19th century. In Germany even the collection of wild berries was deemed a crime and those caught illegally gathering wood in the forests were also harshly punished.[53]

According to Marx the industrial revolution had caused a massive *'metabolic rift'* between humans and land use. As Foster explains: "Drawing upon the work of the great chemist Justus von Liebig and other scientists, Marx noted that the soil nutrient cycle necessitated the constant recycling of nitrogen, phosphorus, and potassium, as plants absorbed these nutrients. Plant and human wastes in pre-capitalist societies were generally returned to the soil as fertilizer, helping replace lost nutrients. But the enclosure movement and the privatization of land that

---

53.   See Perelman, Michael. (2000). The Invention of Capitalism. Duke University Press. Kindle Edition.

accompanied the advent of capitalism created a division between town and country, displacing much of the population from the land and expanding the urban population. Intensive agricultural practices were used to increase yields."[54]

According to Perelman political economists such as Adam Smith, Thomas Malthus, and David Ricardo helped to consolidate the ideological prejudices of the rising English middle class against the peasantry and the impoverished. Perelman has exposed the cynicism of these economic theorists through his research and examination of their private correspondence.

Wage labour was largely unpopular which presented a widespread problem for both the farmers and the factory industrialists. The commons, hunting game, and fishing provided the peasantry with enough additional food for them to have plenty of leisure time. It was estimated that during the sixteenth and seventeenth century the common people only needed to work for one third of a day to live.[55] Those reliant on wage labour to increase their profits and wealth commonly referred to the poor as slothful and indolent. This in turn fed into the colonial narrative provided by the likes of Wakefield who advocated that land should be made artificially scarce and expensive in order to force the poor to enrich the capitalists through their wage labour. The combination of the Enclosure Laws, Game Laws, and inhuman conditions in the factories forced many to colonise New Zealand and the New World. According to Perelman: "The brutal process of separating people from their means of providing for themselves,

---

54. Foster, Bellamy, York, Clark. The Ecological Rift: Capitalism's War on the Earth (p. 404). Monthly Review Press. Kindle Edition.

55. See Perelman, M. (2000). ibid.

known as primitive accumulation, caused enormous hardships for the common people."[56]

It is hard for us to appreciate the relevance of Perelman's description of the, *Invention of Capitalism*, in the 21st century. The violent and brutal enforcement of poaching laws and the pitiless use of force to remove people from the commons has been forgotten as we accept the drudgery of bullshit jobs[57] and urban living without the means of self provisioning our food. The Enclosure Laws continued to morph into conservation parks and recreational areas while the common resources of the sea, the sky, the land, and the rivers have been reopened to oil, gas and mining corporations. The water, the air, the soil, and the biological bounty of the planet have been taken from us all, but this is especially true of indigenous peoples, and at the same time those who have been granted rights to exploit them consider the earth's gifts without responsibilities. This is what Foster has called '*the tragedy of the privatization of the commons*' that is accelerating their destruction.[58] After Garrett Hardin published an economic article in 1968 outlining how common resources such as land, fishing stocks etc. can become exhausted due to the self interest of those who would extract more than could be sustained, the '*tragedy of the commons*' has become an accepted theory of economics. This theory was built on the liberal political economic theories going back to John Locke and then Adam Smith's description of how self-interest drive market efficiencies and social benefits.

A popular 18th century book, *The Fable of the Bees: Or*

---

56.   Perelman, M. (2000) ibid (p. 13).

57.   See David Graeber's book, Bullshit Jobs.

58.   See Foster, J. B., Clark, B., & York, R. (2011). The Ecological Rift: Capitalism's War on the Earth. Monthly Review Press.

Private Vices, Publick Benefits (1714) was a satirical tale of how a bee hive that ignored individual self interest in favour of the social benefit of the hive ended up without honey and impoverished. The neo liberal version could be summed up by the 1980s movie, *Wall St.*, in which Gordon Gekko, played by Michael Douglas, declares, *'Greed is good!'* This faith in the invisible hand of the market dispensed with the do-good economics of the 1930s New Deal and insisted that taxing the rich to pay the poor, was a perversion of nature and would fail to deliver quality social services. Today, many believe it is common sense to think that in a crisis people will only look after themselves and selfishly hoard resources for themselves. However, as Rutger Bregman pointed out in his recent book, *Humankind*, research has proven that faced with a crisis humanity does not atomise into selfish, and depressed individuals, but rather, they unite together into altruistic communities that find happiness in a higher purpose, life gains meaning when others seek to help one another.[59]

Tino Rangatiratanga

As a Pākehā, a white kiwi male of Irish, French, Scottish and possibly Jewish ancestry, I have never identified with neurotypical cognition. That is probably why I identify and associate closely with those who have been forced to submit to dominant cultures and ideologies – it is something I rail against, and it is a solid foundation for my creativity.

In Aotearoa, New Zealand, there have been appalling historical injustices against Māori, the indigenous people of this country. According to Michael King, under the 1867

---

59.   Bregman, R. (2020) Humankind: A hopeful history. Bloomsbury
       Publishing. Great Britain.

Native Schools Act, Māori children were able to attend their own schools and Māori parents often encouraged those schools to teach in English in order to give children the opportunity to learn the language. However, this was sometimes taken to an extreme with many children complaining they were often punished for speaking Māori. By the 1930s there was a generally shared attitude by both Māori and Pakeha that it was beneficial for Māori children to be educated in English that: "...would make upward social mobility for Māori more likely and better prepare youngsters for a world in which Māori culture was going to be a diminishing influence. This was the era when the number of native speakers of the language began to diminish sharply."[60] In effect this led to the erasure of Māori culture, their oral histories, memory and understanding of mātauranga Māori, or Māori tradition and knowledge.

From my time researching the politics of information for my MA thesis, supervised by Professor Ruth Butterworth, I have been highly sensitised to power relationships with respect to knowledge and knowledge exchange. The topic of my research paper was the UK's PRESTEL or videotex. According to Winsbury & Lane:

"The concept that led to its development was the notion of a 'universal database' which had been intriguing computer scientists ever since the computer was invented. The goal was computer or computers on which all the information that a nation needs to operate economically, socially and culturally could be stored and accessed by everyone within that nation."[61] What my research

---

60.   King, M. (2003) The Penguin History of New Zealand. Penguin p.360
61.   Winsbury, R. & Lane, M. (May, 1979). 'Prestel is the first to Start'.
        Intermedia - special survey: videotex. London. IIC. Vol.7. No.3, p.11

uncovered was a neoliberal strategy to use the concept of a 'universal database' to sell another telephone service and increase usage. This was like a precursor to the Google model, except without the advertising.[62] It was apparent that under the watchful eye of Thatcher and her Friedmanite economists there was no intention to cede sovereignty to the '*great unwashed*'. I realised then that as we began moving towards the knowledge age that a battle was looming over who owned their data, and why knowledge and information had to resist total commodification if sovereignty was to reside with the voters and not a corporation or authoritarian regime. As Bratton has pointed out there has been a battle over the hearts, minds and pockets of info-tech consumers who don't seem to have noticed that the laws made by their elected representatives has less impact on their lives than the EULA and Terms of Service of the big Internet companies.

All of this has made me envious of the marginalised, and underserved Māori because even though, much like all of us in Aotearoa, they are fighting to retain sovereignty over their data, at least they have tino rangatiratanga, and a special recognition of the sanctity of Māori: "self-determination, sovereignty, autonomy, self-government, domination, rule, control, power."[63] I believe our country should not only honour Tiriti o Waitangi, The Treaty of Waitangi, but that we all become whanaungatanga, as kin with not only Māori but all humans and especially the nonhuman majority. We require a shared vision of mātauranga Māori, or Māori tradition and knowledge. There is still room for the neurodiversity of worldviews that

---

62. The French had an advertising based videotex system known as Minitel. See https://en.wikipedia.org/wiki/Minitel

63. See Maori Dictionary, AUT. https://maoridictionary.co.nz

inhabit our whenua, the land. Sharing knowledge and data becomes an essential tool for us to maintain our sovereignty. We can use tools such as Creative Commons, Open Source software, Open Science, and mātauranga Māori to reimagine our worlds without the dictates of neoliberalism – together in Zero One Worlds.

We Are Prosocial

Where do our ideas come from? This nature vs. nurture question is more profoundly ideological than people may suspect. Since the 16th century individualism, and self-interest evolved as an ideology that has overwhelmed theories of community, and sharing, and during the Cold War of the 1950s, the use of those terms identified those loyal to the West and those who were deemed 'enemies of the state'. This in turn infected both the natural sciences and the humanities.[64]

When we try and track the philosophy of science we can be misled by scientists who can mischaracterize what science is and how it achieves its objectives? This takes us back to the Enlightenment when Francis Bacon, Immanuel Kant, René Descartes and David Hume were not just postulating about how science took place, but also provided a philosophical and methodological framework for those who followed. Today, it is not uncommon for people to believe that scientists first see, then they theorize. However, David Sloan Wilson points out that we all need to have theories and models about how the world works to be able to do even the basics. He wrote: "We cannot possibly attend to everything, so a theory—broadly defined

---

64. See McCumber, J. (2016). The philosophy scare: The politics of reason in the early Cold War. Chicago: The University of Chicago Press.

as a way of interpreting the world around us—is required to tell us what to pay attention to and what to ignore. We must theorize to see. A new theory doesn't just posit a new interpretation of old observations. It opens doors to new observations to which the old theories were blind. Albert Einstein understood this point when he wrote, "It is the theory that decides what we can observe."[65]

The worlds we build are both conceptual and physical and they are shaped by what we believe, and what we believe is what we see. Wilson went on to write: "Einstein understood that theorizing about entities that cannot yet be seen can lead to useful predictions about what can be seen, but which had previously gone unnoticed."[66]

The ideas and philosophies of the past have a much stronger hold on our modern thoughts than we might acknowledge; even those that were not scientific and were ideological. Thomas Hobbes (1588-1679), John Locke (1632-1704), Adam Smith (1723-1790), Thomas Malthus (1766-1834), Herbert Spencer (1820-1903) and Milton Friedman (1912-2006) have all contributed to our common sense view of capitalism, and political economy. There are those who might even believe that economics has been purged of politics and that it is a science in its own right.

Yet in 2009 the world of economics, that had been dominated by men such as Milton Friedman from the neoliberal Chicago School of economics, was shocked to hear that a woman had won the Nobel Memorial Prize in economics. What was no doubt equally shocking was the fact she was not an economist but rather a political scientist. Elinor Ostrom was the first of only two women

65.   Wilson, David Sloan. (2019) This View of Life . Knopf Doubleday Publishing Group. Kindle Edition.

66.   Wilson, D.S. (2019) ibid.

that have been awarded the Nobel prize in economics, and what is more she had conducted extensive field work that showed how communities successfully shared common pool resources, proving that the theory of the tragedy of the commons was not a law or a foregone conclusion.

Together with David Sloan Wilson, Ostrom, helped to develop the First Principles of Prosocial. These design principles have distilled the case studies and lessons learned from studying human communities that have managed to share resources without exhausting them. Wilson contributed his research on the collaborative nature of evolutionary biology to help debunk the political economic theory that claimed the natural state of humans is 'red in tooth and claw', brutish self-interest, and bloody competition derived from 'survival of the fittest'. Wilson, Ostrom and her PhD student Michael Cox co-wrote a paper together, *Generalizing the Core Design Principles for the Efficacy of Groups* (2013). According to their abstract:

"This article generalizes a set of core design principles for the efficacy of groups that was originally derived for groups attempting to manage common-pool resources (CPRs) such as irrigation systems, forests, and fisheries. The dominant way of thinking, until recently, was that the commons invariably resulted in the tragedy of overuse, requiring either privatisation (when possible) or top-down regulation. Based on a worldwide database of CPR groups, Ostrom proposed a set of principles that broadly captured the essential aspects of the institutional arrangements that succeeded, as contrasted to groups whose efforts failed. These principles can be generalized in two respects: first, by showing how they follow from foundational evolutionary principles; and second, by showing how they apply to a wider range of groups. The generality of the core design principles enables them to be used as a practical guide for improving the efficacy of many kinds of groups."[67]

Prosocial First Principles

The real estate maxim, location, location, location, could very well apply to ecology and evolution. It is the local context of where a thing exists that will determine how it evolves and adapts to its environment. Living in Tāmaki Makaurau, Auckland, Aotearoa, New Zealand means that epigenetically, culturally, and genetically the superorganism that I am part of has begun to evolve and adapt to worlds that are very different from those that surrounded my ancestors four or five generations ago. Yet, we have evolved fast, as it was only four hundred generations ago that the first civilisations of the Holocene were beginning to adapt to the warmer climates.

I now realise that I need to learn many new languages to appreciate my current ecology, languages that are both human, and nonhuman, because without some understanding of the signs, symbols and communications of those context specific languages I cannot hope to survive. One of my first priorities is learning Te Reo, and as a first step I have begun to map mātauranga Māori onto the first principles that Ostrom and Wilson extracted from their empirical research of collaborative communities who have successfully shared common pool resources. I offer my interpretation of those principles with the addition of kaitiakitanga, a belief that we should all take responsibility

---

67. Wilson, D.S., et al., Generalizing the core design principles for the efficacy of groups. Journal of Economic Behavior & Organization (2013), http://dx.doi.org/10.1016/j.jebo.2012.12.010

for the worlds we inhabit and assume the role of guardians and stewards of those worlds.

1. Kaitiakitanga – guardianship of sky, sea, rivers and land.

2. Whanaungatanga – Strong group identity and understanding of purpose

3. Utu – Fair distribution of costs and benefits

4. Utu – Fair and inclusive decision making

5. Tapu – Monitoring agreed upon behaviour

6. Ngawhi – Graduated sanctions for misbehaviours

7. Whakataunga – Fast and fair conflict resolution

8. Tino rangatiratanga – Authority to self-govern

9. Whakawhanaunga – Appropriate relations with other groups

I have just begun to outline a process whereby my family might come together to evolve, adapt, survive and thrive in the uncertain worlds we face. By thinking deeply about our co-designed community I hope that we can step lightly on this planet and learn how to co-evolve with the nonhuman majority. It will be an experimental co-evolution that I hope to begin soon and will only complete phase 1 in two hundred years from now.

In the words of Haylee Koroi: "Real change will require nothing less than huge personal and collective paradigm shifts in how we view and relate to the world around us. I offer the concept of whakapapa as a guiding logic, which has lived here since time immemorial. It speaks to the ultimate reality of our interrelatedness and interdependence as both human and more-than-human beings. It will require us to think beyond our human desires to the wider web of relationships in which we exist."[68]

kia kaha
be strong, and be brave together

---

68.   Clark, H. (ed.) (2021). Climate Aotearoa: What's happening & what we can do about it. Allen & Unwin. New Zealand.

10

Postscript

Black zero
Carbon incandescent
That none has one.
Joining the chaotic aroha
Is Life the unknown Te Kore
Or the void that threatens
As we yearn and burn forever
Drawn like a moth it beckons.

We have just gone into lockdown again. There is a new wave known by the latest mutation, the Delta Variant, or Deviant as I like to call it. On a recent walk up our maunga I noticed how busy it was and how many people were not wearing masks. I counted 134 exercising their *freedumb*, defined by the Urban Dictionary as: "freedumb – A totally nonsensical and asinine belief (of many Americans) that freedom means you can literally do anything you want, including violating other peoples' rights e.g. "I have the *freedumb to give you a disease!"* It would seem that New Zealand has also been hit by the latest wave of the infodemic that convinces *'freedummies'* that although masks have been mandated in public places, that while they are running, or cycling, they don't need to wear masks, meanwhile the Deviant is free to blow in your face as they flash past you. It makes me wonder where they are getting their news, or rather what their social media network has been spreading. The Delta variant has been called by our Prime Minister *'a game changer'* and the disturbing spikes in the US, UK, Australia and Asia should have warned us that COVID has cleverly adapted to the misinformation anti-vaxxs are spreading. The virus is evolving and might just lead to the adaptation of humanity if we can also overcome the stupidity of the infodemic.

Fake news, alternate facts, post truth – all of these terms have been introduced during a new era of populist leaders and ideological attacks on science, evidence based policy and political rhetoric that are all designed to inject uncertainty into the debate around anthropogenic damage and especially global heating. Doubt really is their product, and those who can't cope with uncertainty (the majority) tend to fall back on the status quo. However, when it comes to the future of education, Luis Prádanos asked the rhetorical question: "[I]s it really smart to educate people to technologically and theoretically refine a system that operates by undermining the conditions of possibility for our biophysical survival?"[1]

The methodology of those looking to usurp democratic policy making is both very simple and effectively

---

1. Open Democracy. 24 August, 2020. Preparing for the end of the world as we know it. https://www.opendemocracy.net/en/oureconomy/preparing-end-world-we-know-it/

devastating. Conservative think tanks in the US sought to install so-called *'thought leaders'* in academia, and other organisations one step removed from the political fray. Their strategy had developed with the assistance of psychology, mass communications, advertising and propaganda. Their objective was to protect and bolster their power and privilege against progressives who were concerned about the impact of big business on the environment and social inequality. They had learned from history that going head-to-head on the facts had failed and so they adopted another approach that included propaganda, misinformation, and deliberately sowing mistrust and doubt. This new approach came to prominence in the US in 2002 with Frank Luntz who had assisted Dick Cheney and the privately owned oil and gas firm, Halliburton, to use communications and public relations to muddy the waters around the role of the oil and gas industries in accelerating anthropogenic global heating. Luntz, initially faced an uphill battle to convince the public and as late as 2003 over 75 percent of Republicans supported strict environmental regulations. Luntz warned that: "the environment is probably the issue on which Republicans in general—and President Bush in particular—is most vulnerable." He argued that global warming deniers had to present themselves as *'preserving and protecting'* the environment. Luntz stressed in his confidential memo that opponents of carbon regulation must *'absolutely'*: "not raise economic arguments first." meaning that revealing their financial self-interests was *'a recipe for losing'*. "The key, he went on, was to question the science. "You need to continue to make the lack of scientific certainty a primary issue in the debate," he advised. So long as "voters believe there is no consensus about global warming within the scientific community," he said, regulations could be forestalled. Language that "worked," he advised, included phrases like "we must not rush to judgment" and "we should not commit America to any international document that handcuffs us."[2] This was part of an orchestrated campaign to sow doubt and deflect from the facts that showed the fossil fuel industry was accelerating global heating.[3]

As these new battle lines formed around the certainty of scientific evidence relating to global heating, a decade later other political agents realised the effectiveness of this approach and applied it to other social issues such as race, gender, and economics. The use of social media was widely credited with Obama's electoral success, and Trump's campaign team set about soliciting funds from secret sources while targeting voters who had already begun to distrust the media, big government, and even science. As Julia Ebner's research into the recruitment methods of extremists showed there were new and sophisticated techniques designed to smash worlds in order to rebuild cognitive constructs that supported both secretive elites and shady reactionaries.[4] While the Tea Party set about world ending, the bedrock of their ideology remained neoliberal economics. For many America was defined by capitalism.

2.   See Mayer, Jane. Dark Money: how a secretive group of billionaires is trying to buy political control in the US . Scribe Publications Pty Ltd. Kindle Edition

3.   See the documentary on Al Jazeera, *The Campaign Against the Climate*. (2021) https://www.aljazeera.com/program/featured-documentaries/2021/4/17/the-campaign-against-the-climate-debunking-climate-change-denial

4.   Ebner, J. (2020). Going dark: The secret social lives of extremists. Bloomsbury Publishing.

The Economic Delusion

Economists, the technocrats, technopoles, and political cheerleaders of orthodox economic policy are now being called out. Professor Steven Keen, a contrarian economist, has pointed out that most of the mainstream economists have *"deliberately and completely"* ignored scientific data about climate change and instead *"made up their own numbers"*. He exposes the arguments made by the likes of William Nordhaus who claimed that even if we suffer from 3°C increase in the global average temperature that will only have a negative impact of 3.6 percent drop in global GDP. This was based on his calculations that put 87 percent of the US economy indoors and so would be protected and immune from the impact of climate change. These economic calculations were repeated by the IPCC at COP21 at the time of the Paris Agreement. Yet, while Nordhaus had conducted a small survey of experts in 1994, including climate scientists and economists, he presented an aggregate finding that implies there was some consensus. However, he did admit: "There is a clear difference in outlook among the respondents, depending on their assumptions about the ability of society to adapt to climatic changes. One was concerned that society's response to the approaching millennium would be akin to that prevalent during the Dark Ages, whereas another respondent held that the degree of adaptability of human economies is so high that for most of the scenarios the impact of global warming would be **"essentially zero"**.[5] Keen warns that mainstream economics have deliberately ignored the science that warns that any further increase in

---

5. Nordhaus, W. D. (1994a). Expert opinion on climatic change. American Scientist, 82(1), 45–51. https://www.jstor.org/stable/29775100

temperatures could result in an irreversible tipping point that could result in runaway global heating. He goes further and points to Nordhaus' book, *The Climate Casino*, (Nordhaus, 2013) in which he states that: "There have been a few systematic surveys of tipping points in earth systems. A particularly interesting one by Lenton and colleagues examined the important tipping elements and assessed their timing ... Their review finds no critical tipping elements with a time horizon less than 300 years until global temperatures have increased by at least 3°C. (Nordhaus, 2013, p. 60; emphasis added)". Keen is outraged and writes: "These claims can only be described as blatant misrepresentations of 'Tipping elements in the Earth's climate system'."[6] The problem is that when economists speak politicians listen, and that is often the only thing that is reported in the media.

Economics has been transformed from a practice that politicians once regarded as an irrelevant art form into a mathematical religion that dominates policy around the planet.[7] Even those who thought about throwing it away, like the economist, Kate Raworth, have realised that if they want to be taken seriously they need to couch their criticism of growth and GDP in economic terms. Keen is one of the few economists who puts it baldly, "I think we should throw the economists completely out of this discussion and sit the politicians down with the scientists and say these are the potential outcomes of that much of

6.   Steve Keen (2020): The appallingly bad neoclassical economics of climate change, Globalizations, DOI: 10.1080/14747731.2020.1807856

7.   See Appelbaum, B. (2019) The Economists' Hour: how the false prophets of free markets fractured our society. Picador.

a change to the biosphere; we are toying with forces far in excess of ones we can actually address."[8]

The economic delusion of growth is pervasive and there are virtually no jurisdictions or governments on the planet who have not had a brush with the neoliberal faith in market forces. Klein has pointed out that if we were to pick an example of this and the likely harm to be caused by the *'free market'* we should explore the lessons of the once booming island of Nauru. The *'invisible hand of the market'* is out of sight and out of mind. The problem is so are crimes, corruption, and tax havens. Secrecy and hidden trusts are so far from the desks of journalists and the living rooms of Northern consumers that they may as well be fairytales. Klein wrote: "This is our relationship to much that we cannot easily see and it is a big part of what makes carbon pollution such a stubborn problem: we can't see it, so we don't really believe it exists. Ours is a culture of disavowal, of simultaneously knowing and not knowing—the illusion of proximity coupled with the reality of distance is the trick perfected by the fossil-fueled global market. So we both know and don't know who makes our goods, who cleans up after us, where our waste disappears to—whether it's our sewage or electronics or our carbon emissions. But what Nauru's fate tells us is that there is no middle of nowhere, nowhere that doesn't "count"—and that nothing ever truly disappears."[9]

Nauru was of little consequence to 19th century explorers until one inquisitive colonist picked up a rock and

---

8.   See Karen Gilchrist. Published Sun., MAY 23 2021 9:55 PM EDT. 'War' footing needed to correct economists' miscalculations on climate change, says professor. https://www-cnbc-com.cdn.ampproject.org/c/s/ www.cnbc.com/amp/2021/05/24/war-footing-needed-to-correct-economists-climate-change-failings.html

9.   Klein, N. (2014) ibid.

discovered it was almost pure phosphate. As you may remember I have previously discussed soil depletion and that 19th century colonialism was also a search for fertile lands and fertiliser. Britain, Australia and New Zealand saw the huge value of Nauru's avian rock poop and mining accelerated in the 1960s just before the Nauru colony claimed independence from New Zealand in 1968. The shit was scrapped from the rocks and exported to Australia and New Zealand where it was dumped on the land by the truck load. Meanwhile, economic growth brought the trappings of consumerism to Nauru – people did not grow food, they imported it and dined out regularly at restaurants. Baby showers were celebrated with cash from relatives and friends, while fossil fuelled cars raced around the tiny island's 22km^2 – hopefully avoiding a drink drive accident or a speeding infringement clocked by the Police Cheif in his yellow Lamborghini. It was reported as a massive success and the wise old men of Nauru set out to bank the profits in a trust fund for the people. But by the 90s Nauru was almost out of phosphate and the island had been hollowed out by aggressive mining. The easy money had gone and so they turned to money-laundering using some four hundred phantom banks to wash clean some US$70 billion dollars of dirty money. What was worse, the trust fund lost all of their money through dodgy investments and fraudulent schemes – the country was broke. There was nothing to show for the loss of their land and destruction of the environment except a debt of at least USD$800 million and the threat climate change would put the whole place under water. Australia and New Zealand, in particular, had profited from the exploitation and extraction of a valuable mineral found in bird waste, yet there appears to be no lessons learned about mining,

mineral extraction, or poor agricultural practices by
Nauru's neighbouring islands and continents.

Island Retreat

On the 18th of May, 2021, the Cook Islands opened a travel
bubble with New Zealand allowing two way travel between
the two countries without quarantine. At the end of July
my wife Sarah and I with a couple of friends, Cathy and
Paul, went to visit our daughter Sophia and her husband
Brennan. Sophia has been teaching primary school
children at a school that has a beautiful beach and lagoon
out the front with passing humpback whales. The Cook
Islands have had their boarders closed and there have been
no reported COVID 19 cases to date.

I feel as if New Zealanders sometimes forget they live on
an interconnected series of islands. Much as Britain seems
to occasionally forget that they are also an island state. Yet,
back in 1623 the poet John Donne wrote a meditation on
death that reminds us that *'no man is an island'*. Brexit may
have been motivated by some perverse attempt to pull up
the drawbridge to prevent the descendants of slaves and
colonised indigenous people coming to England to live, but
historical wrongs cannot be erased by populist politics.
Donne likened people to countries and that metaphor is
instantiated by the reality that all people and things are
inextricably entwined. As Klein has pointed out capitalism,
colonialism, and climate are threads braided together by
historical events.[10] As Donne meditates on the inevitable
death of an individual he reminds us that the fate of us all
are interwoven together.

---

10. Klein, N. (2014) This Changes Everything: capitalism vs. the climate.
     Penguin.

'No Man is an Island'
No man is an island,
entire of itself;
every man is a piece of the continent,
a part of the main.
If a clod be washed away by the sea,
Europe is the less,
as well as if a promontory were.
as well as if a manor of thy friend's
or of thine own were.
Any man's death diminishes me,
because I am involved in mankind;
and therefore never send to know for whom the bell
tolls;
it tolls for thee.

Recent history reminds us that we cannot ignore the fate of others because even for the selfish, and those who deny climate change, we do not have to look far to *'know for whom the bell tolls; it tolls for thee.'* The 2021 IPCC report has finally strengthened the tone of their language stating that it is *'unequivocal'* that humanity has had a negative anthropogenic impact on all of the Earth systems and no one can escape the consequences. However, despite increasingly extreme weather events now reaping catastrophic damage in North America, Europe, China, India, and Africa there are still those who quietly believe that the W.E.I.R.D[11] world can swerve the worst of it by the application of technological innovation and adaptation. Amongst the chosen are those in Aotearoa who continue to

---

11. The acronym W.E.I.R.D. stands for Western, Educated, Industrialised, Rich and Democratic.

promote economic growth and tech-salves over degrowth and biological restoration – clearly they do not think the bell tolls for them.

Another case of cognitive dissonance can be illustrated by two seemingly disconnected geographical locations: the Canterbury Plains in New Zealand and Louisiana in the USA. The bewildered farmers of Canterbury do not seem to attribute any of their recent '*bad luck*' to extreme weather due to climate change, let alone any connection with their agricultural GHG emissions. Increasingly, weather reporters are pointing to the devastation caused by extreme weather probably made worse by the direct result of anthropogenic global heating. While the data collection, modelling and analysis of these extreme weather events can take some time to process there is good historical evidence that this is the case. Cyclones feed off the heat generated by GHGs.[12] In Louisiana and Texas their oil and gas industry have been badly battered by hurricanes that have been pulled into the anthropogenic vortex of high speed winds and torrential downpours made worse by fossil fuel GHGs.[13]

In 1988 New Zealand was smashed by Cyclone Bola that travelled south fresh from the devastation of the island nation of Vanuatu. The tempest destroyed houses and horticulture in Taranaki and many farmers then decided to rebuild by swapping kiwifruit for milk production. The irony is that there are now dairy farmers who are now

---

12. Canterbury floods: Is climate change to blame for severe weather events? https://www.stuff.co.nz/environment/climate-news/125299269/canterbury-floods-is-climate-change-to-blame-for-severe-weather-events

13. (Sept. 2021) How climate change is fueling hurricanes like Ida. https://www.nationalgeographic.com/environment/article/how-climate-change-is-fueling-hurricanes-like-ida

returning to kiwifruit and avocado growing as the anthropogenic effects of cows and climate change are shifting the crops away from the Bay of Plenty to a more suitable climate in Taranaki. It remains to be seen whether the region will be again hit by man-made extreme weather. Dairy farmer, Holly Murdoch was asked if developing another income streams was a motivating factor? She replied: "Absolutely diversification was quite a drawcard for us especially with the way the Government is going with dirty dairying and all that kind of thing."[14]

Despite these climatic woes the message is out, that our motu (islands) could be a sanctuary in a world that is getting worse. Many believe time is running out. Google co-founder, Larry Page, is the latest American billionaire to pay what he would consider the inconsequential fee of a NZD$10 million investment to ensure that his family might have a bolt hold during a climate emergency. Perhaps he has even recognised that the techno-liberalism of Silicon Valley has done nothing to ensure safety from the pandemic as he medivaced his child who had become ill on the COVID-ravaged shores of the island nation of Fiji.

New Zealand is an island country and many of our neighbours are also islands. Yet, despite being dependent on poor Pacific islands for cheap labour and tropical holidays we only remember them when it suits us. After the economic boom of the 60s, the oil crisis of the 70s impacted New Zealand employment and the Muldoon government instigated the 'Dawn Raids' to send our island labourers home. The Labour government has formally apologised for this egregious violence against the people who have helped

---

14.   (Sept. 14, 2021). Horticulture making a comeback in Taranaki.
      https://www.rnz.co.nz/news/top/451478/horticulture-making-a-
      comeback-in-taranaki

build our country's culture and economy. But what is going to happen when the fisheries have gone due to ocean acidification, and their islands disappear below the rising seas?

One of those countries who are closely linked to New Zealand is the former colony, the Cook Islands. They were most likely part of a group of islands that Māori visited during their long migration to New Zealand.[15] The Cook Islands are now in 'Free Association' with Aotearoa, and there are currently more Cook Islanders living here in our country than live on their own islands. A majority of Cook Islanders live on the tourist island of Rarotonga, around 10,000; plus a further 80,000 live in New Zealand. Even now as we become aware of the extreme dangers of the Anthropocene bearing down upon us all, extreme weather and sea level rise begins to expose the highly vulnerable Pacific islands, but there are no public provisions for the resettlement of climate refugees from places such as the Cook Islands. We are barely making provisions for the displacement of our own people in Aotearoa as coastal erosion accelerates and insurance companies prepare to upsticks declaring many properties will become uninsurable. The story of Nauru should provide a salient lesson to the Cooks and to us here – but does history have to repeat?

Let us imagine for a moment that the Cook Islands was an international exemplar of ecological sustainability. Not only would there be global tourists that would want to visit there to appreciate its beauty but many would likely appreciate something like an entry tax that could go into an independent and audited account that is administered

---

15.   See King, M. (2005) ibid.

by an environmental trust. The money spent by the trust could go towards building resilience and adaptation forced by climate change and anthropogenic damage by humans. The waterways could be protected from pollution, local food could support local people, and if climate mitigation does not work then the money could assist resettlement in New Zealand or elsewhere.

Now imagine if the Cook Islands continued on the path that it is currently on. Today, 25% of the country's budget is spent on climate change mitigation. Like many former colonies the Cook Islands have attempted to diversify and become less dependent on their former colonial masters and the export of goods they required in exchange for foreign funds. Following independence from New Zealand in 1965 the Cook Islands began to switch from agricultural exports to tourism. Today, COVID 19 has exposed the disproportionate dependence that this small country relies on tourism. 60 percent of all revenue is derived from tourism and 80 percent of the visitors come from New Zealand. For over a year that flow of tourism dollars completely stopped. The risk of a COVID 19 outbreak in the Cook Islands would have been devastating as it doesn't possess the healthcare infrastructure to treat those who might become infected and likely die.

With the borders closed the debate about diversification of the Cook Islands economy intensified. I had learned from my son-in-law, Brennan and a local businessman who lived in the Cook Islands that self-sufficient agriculture in the Cook Islands had become something of a forgotten art by many who had grown accustomed to food imports and the luxuries of a tourist culture. Many people were talking about new opportunities and labour shortages due to young people leaving the islands to explore new lifestyles in New Zealand and Australia.

Another idea that appealed to the Cook Islands' politicians was the mining potential of polymetallic nodules located in the South Penrhyn Basin between Aitutaki in the south and Penrhyn in the north. These nodule fields are found on the seabed at depths up to 5kms deep. First discovered in the early 1970s and found to be of commercial value in the 1990s with a 1993 assessment by East West Center in Honolulu, and the 1996 feasibility study by the biggest US engineering firm, Bechtel Corporation. The fluctuation in Cobalt prices had led to a loss of interest, however, the futures in the metal have rebounded as it has become an important component in EVs (electric vehicles), rechargeable battery electrodes, and also have a strategic military importance.[16]

Bechtel, aka, *'the working arm of the CIA'*, according to the Independent, is a corporation that has spawned thousands of conspiracy theories. A privately held company that has profited enormously through secrecy and close relations with powerful government and military figures, such as George Shultz and Donald Rumsfeld, it is far from transparent and has been implicated in dubious accounting and construction practices.[17] Just one example

---

16. Cobalt Statistics and Information. https://www.usgs.gov/centers/nmic/cobalt-statistics-and-information "Cobalt (Co) is a metal used in numerous diverse commercial, industrial, and military applications, many of which are strategic and critical. On a global basis, the leading use of cobalt is in rechargeable battery electrodes. Superalloys, which are used to make parts for gas turbine engines, are another major use for cobalt. Cobalt is also used to make airbags in automobiles; catalysts for the petroleum and chemical industries; cemented carbides (also called hardmetals) and diamond tools; corrosion- and wear-resistant alloys; drying agents for paints, varnishes, and inks; dyes and pigments; ground coats for porcelain enamels; high-speed steels; magnetic recording media; magnets; and steel-belted radial tires."

17. See Klein, The Shock Doctrine.

of Bechtel's shady past can be illustrated by their presence in Iraq. First the US government bombed Iraq, and next the Secretary of State, Shultz, selected Bechtel to rebuild the damage for US$680 million. Shultz, was a former Bechtel executive, and later after he left office her rejoined Bechtel on their board.[18] Their activity in the Cook Islands is unlikely to be transparent and is cause for suspicion.

As Friedman's *free market* revolution exploded in the South Pacific in the early 80s, finance deregulation, and international capital markets rushed in to profit from the secret asset protection laws set up in the Cook Islands with the assistance of ex-CIA agents, dodgy tax advisors and shonky New Zealand lawyers.[19] Before the Paradise Papers exposed the commonplace activities of tax cheats, money launderers, dictators, spies, criminals and corrupt politicians, New Zealand had been alerted to dodgy dealings in the Cook Islands popularly known as the Winebox case. The move by the Cook Islands to set up a Sovereign Wealth Fund for the benefit of all Cook Islanders is hardly reassuring given the existence of secret trust funds and banks that are protected from the disclosure of beneficiaries. Ian Wishart's book, *The Paradise Conspiracy*, has been vindicated by some of the findings in the Panama Papers that found tax havens, and secret accounts in Panama, the British Virgin Islands, and the Cook Islands as the standard modus operandi of criminals, and tax cheats. The PM of the Cook Islands, Sir Thomas Davies (who had past CIA connections) setup the statutes to protect the privacy of these shady characters with the help of tax

18. Hirst, C. (13 Dec. 2013) The world's at Bechtel's beck and call. Independent. https://www.independent.co.uk/news/business/analysis-and-features/the-world-s-at-bechtel-s-beck-and-call-115828.html

19. Wishart, I. (2011) The Paradise Conspiracy. Howling at the Moon. New Zealand.

dodgers in Australia, and New Zealand including Faye Richwhite. The likelihood that any future money that might come from the seabed mining of the nodules might be siphoned off by corrupt officials and fraudulent companies cannot be discounted.

Meanwhile, the ecologist Gerald McCormack, who founded the Cook Islands Natural Heritage Trust, has attempted to reassure those concerned about the sustainability of the seabed mining by insisting that the 'Precautionary Principle' as documented in the Cook Islands Seabed Minerals Policy 2014 and the Exploratory Regulations 2015 would ensure that those with mining licences would: "recover the immense investment in exploration and technology development; and more importantly, it would mean that the Cook Islands could be confident that seabed mining would be an environmentally acceptable activity."[20]

This mischaracterisation of what the precautionary principle means[21] gives the impression that the government and mining corporations will take every step to avoid environmental harm. However, what the precautionary principle really warns us about is that if some harm to the environment could arise from some unknown cause and effect relationships then there is an obligation to not proceed. McCormack discusses some of the: "the primary environmental concerns of mining abyssal seabed nodules are...(1) sediment plumes on the seabed; (2) discharge of nutrient rich water in the water column, with or without sediment; (3) intense noise; (4)

---

20.   McCormack, G. (2016). Cook Islands Seabed Minerals: precautionary approach to mining. Cook Islands Natural Heritage Trust. Rarotonga. p.9

21.   See earlier chapters: Binary Worlds, Zero One World.

intense light; and (5) destruction of seabed biodiversity."[22] He details the potential smothering of seabed animals; the possible negative effects on marine biodiversity due to sediment and bright lights; and the chance that interference with the migratory behaviour of Humpback Whales and toothed-whale echolocation and communication due to loud noise could arise from seabed mining. Just because all this might happen in the shadows of hidden bank accounts, and because what is out-of-sight is usually out-of-mind, does not mean that those corporations and governments can be trusted to abide by the precautionary principles. As we have learned from Nauru, once the island paradise is gone there is no bringing it back.

Could the seabed mining be dressed up as a social enterprise for the good of the people and the environment? Social enterprise, or social entrepreneurship has been held out as a positive alternative to the ecological degradation that is caused by neoliberal economics and capitalist globalisation. Yet, many social entrepreneurs find that even when they are in startup mode their vision ends up being diluted by the investors and, the best that can be said about some of these enterprises is that they might have done less harm to the people and the environment. Such a social enterprise can be a form of greenwashing that does not go far, or fast enough, and can quickly also become coopted through takeovers, profit motives, or mergers and acquisitions. Raworth suggests that enterprises should instead of pursuing 'zero carbon' and the UN's sustainability goals, they should instead think about how they can offer to go beyond net zero and sequestrate carbon or generously

---

22.   McCormack. (2016) ibid. p.15

contribute to their local ecology. I have tried to highlight my concerns about the carbon markets, and carbon pricing such as the New Zealand Emissions Trading Scheme that will most likely fail to avert our environmental catastrophe. In the conclusion to his book, Appelbaum highlights America's obsession with market economics and what can happen? "They constructed a market society, and the defining feature of a market is the freedom to walk away."[23] Markets and carbon pricing are unlikely to improve our chances of survival and are really just mechanisms to allow those responsible to ignore the real extent of the problem and waltz off into the sunset. In Aotearoa a group known as Lawyers For Climate Action are threatening to take the Climate Change Commission to court over their lack of urgency and advice given to the government.[24] We have to remember in the case of all the anthropogenic harm being caused by markets there is ultimately no where to walk away to.

Whether it is a social enterprise or a multinational corporation if there is a profit motive it is likely to encourage growth economics, and accumulation of capital extracted from natural resources, (including people), with disastrous outcomes that generates vast amounts of waste.

The hacker, Melanie Riebeck, who admires Raworth's work, advocates that social enterprises dispense with extractive business, dividends, and exit strategies.[25] If these companies are no longer driven by extracting value from people, and the planet, in order to pay shareholders

23.   Appelbaum, B. (2019) ibid. p.332

24.   RNZ. (2 July 2021) Lawyers take legal action against Climate Change Commission. https://www.rnz.co.nz/news/national/446013/lawyers-take-legal-action-against-climate-change-commission

25.   See her TEDX talk "Why we need to reform the startup business model." https://www.youtube.com/watch?v=uW8umQWXFIM&t=1114s

a dividend, or repay an investor who wants to exit within 3-5 years, then there is no pressure to attempt a dangerous exponential growth rate, and value can remain in the company almost forever without any short term exit plan.[26]

COP26 in Glasgow has just ended. Many delegates have called it a cop out and New Zealand was 'honoured' with second place in the Fossil Awards – we achieved this embarrassing title from the Climate Action Network for our lack of commitment to reducing agricultural GHG emissions. The Climate Change Minister, James Shaw, said just because the conference called for more ambitious updates to the our NDC, (National Determined Contribution) didn't mean we had to – which hardly seems to be in the spirit of saving the planet in a climate emergency. It transpires that our modest NDC will only get somewhere near our net zero targets by relying on international carbon offsets, and that while historically only 5% of those offset projects were shown to actually contribute to a reduction of GHGs, New Zealand's NDC may only be met by paying for two thirds of our emissions in offset credits to Pacific Island countries. Meanwhile, we were bumped to 'Associate Member' of the Beyond Oil And Gas Alliance because we have just issued new permits for offshore drilling. As the journalist Rod Oram reported: "Good old Aotearoa also stood in the way of setting limits on carbon offsetting in Article 6 and recently issued two new fossil fuel exploration permits."[27]

We can all contribute to better worlds but we have to first design by thinking about the problems, and then

---

26. See Neurodiventures for an open source operating model for companies.
     https://autcollab.org/community/neurodiventures/

27. Nov. 14, 2021. Rod Oram -COP26 Reaches its Peak Final Day.
     Newsroom. https://www.newsroom.co.nz/cop26/cop26-rod-oram-
     cop26-reaches-its-peak-final-day

experiment with thinking by design or by doing. We will not save our planet, ourselves or the innocent nonhuman majority by achieving 'net zero' by some arbitrary date, even if it somehow delays the tipping point, we must instead find ways to overcome the fear of degrowth and sequestrate more carbon, remove more pollution, and reduce more waste than we have pumped into this world since the industrial revolution. Economic growth and business-as-usual will not cut it. When future generations ask what you did about the Anthropocene? The correct answer should not be zero but much less than zero. We are on the brink of the abyss – time has run out, we must act now.

Bibliography

Abbott, E. B. (2014). Flood Insurance and Climate Change: Rising Sea Levels Challenge the NFIP Symposium 2014. Fordham Environmental Law Review, 26, 10–55. Retrieved from https://heinonline.org/HOL/P?h=hein.journals/ frdmev26i=18

Amadae, S. M. (2003). Rationalizing capitalist democracy: The Cold War origins of rational choice liberalism. Chicago: University of Chicago Press.

Angus, I. (2016). Facing the Anthropocene: Fossil Capitalism and the Crisis of the Earth System. NYU Press.

Appelbaum, B. (2019) The Economists' Hour: how the false prophets of free markets fractured our society. Picador.

Avery, D. (2017). The Resilient Farmer. New Zealand. Penguin.

Babiak, P., Hare, R. D. (2006). Snakes in suits: When psychopaths go to work (1st ed). New York: Regan Books.

Bailenson, J. N., Blascovich, J., Beall, A. C., Loomis, J. M. (2003). "Interpersonal distance in immersive virtual environments." Pers Soc Psychol Bull, 29(7), 819-833.

Bailenson, J. N., Beall, A. C., Loomis, J., Blascovich, J., Turk, M. (2004). "Transformed social interaction: Decoupling representation from behavior and form in collaborative virtual environments." In Presence: Teleoperators Virtual Environments, 13(4), 428-441.

Barnes, T. J. (2008). Geoforum Lecture 2007: Geography's underworld: The military-industrial complex, mathematical modelling and the quantitative revolution. Geoforum, 39, 3–16. https://doi.org/10.1016/j.geoforum.2007.09.006

Basosi, Riccardo Spinelli, Daniele Fierro, Angelo Jez, Sabina. (2014). Mineral Nitrogen Fertilizers: Environmental Impact of Production and Use.

Bell, A., Elizabeth, V., McIntosh, T. and Wynyard, M. (2017) A Land Of Milk And Honey? Making Sense Of Aotearoa New Zealand. University of Auckland Press. N.Z.

Benkler, Y. (2005). Coase's Penguin, or, Linux and the Nature of the Firm. In R. A. Ghosh (Ed.), CODE: collaborative ownership and the digital economy (pp. 169–206). Cambridge, Mass.; London: MIT.

Benkler, Y. (2006). The wealth of networks: How social production transforms markets and freedom.

Benkler, Y., Faris, R., Roberts, H. (2018). Network Propaganda: Manipulation, Disinformation, and radicalization in American politics. New York, NY: Oxford University Press.

Bennett, J. (2010). Vibrant Matter: a political ecology of things. Durham: Duke University Press, 2010.

Berg, M., Seeber, B. K. (2017). The slow professor: Challenging the culture of speed in the academy. Toronto University Press.

Bertram, G. (2020) Submission to the Environment Committee on the Climate Change Response (Emissions Trading Reform) Amendment Bill. http://www.geoffbertram.com/publications/

Bester, A. (1957). The stars my destination. [New York]: New American Library.

Bittman, M. (2021) Animal, Vegetable, Junk: A History of Food, from Sustainable to Suicidal. Boston. Houghton Mifflin Harcourt.

Bogost, I., Project Muse. (2012). Alien phenomenology, or, What it's like to be a thing.

Böhm, S. and Dabhi, S. (eds) (2009) Upsetting the Offset: the political economy of carbon markets. MayFlyBooks. London.

Booker, C. (2004). The seven basic plots of literature. New York; London: Continuum.

Bostrom, Nick. Superintelligence: Paths, Dangers, Strategies . OUP Oxford. Kindle Edition.

Boyd, B. (2009). On the origin of stories: Evolution, cognition, and fiction. Cambridge, Mass.: Belknap Press of Harvard University Press.

Bradley, Naima; Harrison, Henrietta; Hodgson, Greg; Kamanyire, Robie; Kibble, Andrew; Murray, Virginia (2014). Essentials of Environmental Public Health Science: A Handbook for Field Professionals. OUP Oxford. p. 101. ISBN 978-0-19-150540-9

Brand, S. (1988). The Media Lab: Inventing the Future at MIT. New York, N.Y., U.S.A.: Penguin Books.

Brannen, P. (2018). Ends Of The World: Volcanic apocalypses, lethal oceans and our quest to understand earth's past mass extinctions. S.l.: Oneworld Publications.

Bratton, B. H. (2016). The Stack: On Software and Sovereignty. Massachusetts: MIT Press.

Bregman, R., Manton, E. (2017). Utopia for Realists. London, UK: Bloomsbury Publishing, an imprint of Bloomsbury Publishing Plc.

Bregman, R. (2020) Humankind: A hopeful history. Bloomsbury Publishing. Great Britain.

Brown, T., Katz, B. (2009). Change by design: How design thinking transforms organizations and inspires innovation (First edition). New York: Harper Business.

Bryant, L. R. (2011). The democracy of objects (First edition). Ann Arbor: Open Humanities Press.

Burroughs, W. S. (1959). The Naked Lunch (1st ed.). Paris: Olympia Press.

Cai, S. (2016). State propaganda in China's entertainment industry. London; New York: Routledge.

Campbell, J. (1971). The hero with a thousand faces ([2nd). [Princeton, N.J.]: Princeton University Press.

Campbell, J. (1991). The masks of God: Creative mythology [volume 4. Penguin/Arkana.

Capra, F. (1983). The Tao of physics: An exploration of the parallels between modern physics and Eastern mysticism (2nd ed.). Boulder, NY: Shambhala.

Capra, F. (1983). The turning point: Science, society, and the rising culture. Toronto New York: Bantam Books.

Carr, N. G. (2010). The Shallows: how the Internet is changing the way we think, read and remember. London: Atlantic Books.

Carrott, J. H., Johnson, B. D. (2013). Vintage Tomorrows. Farnham: O'Reilly.

Castronova, E. (2007). Exodus to the virtual world: How online fun is changing reality. New York:Palgrave Macmillan.

Chang, J., Halliday, J., Vintage (Londyn). (2007). Mao: The unknown story. Vintage Books.

Chesbrough, H. W. (2003). Open innovation: The new imperative for creating and profiting from technology. Boston, MA: Harvard Business School Press.

Chomsky, N., Miller, G. (n.d.). Final State Language. Information and Control, 1.

Chun, W. H.-K. (2006). Control and freedom: Power and paranoia in the age of fiber optics. Cambridge, Mass: MIT Press.

Clark, A. (2003). Natural-Born Cyborgs: Minds, Technologies, and the Future of Human Intelligence. Oxford University Press.

Clark, H. (ed.) (2021). Climate Aotearoa: What's happening & what we can do about it. Allen & Unwin. New Zealand.

Clark, N. (2011). Inhuman Nature: Sociable Life on a dynamic planet. Los Angeles; London: SAGE.

Clegg, B. (2008). Light years: An exploration of mankind's enduring fascination with light. London; New York: Macmillan.

Climate Change Commission (2021). 2021 Draft Advice for Consultation.

Cline, E. (2018). Ready player one. London: Arrow Books.

Clippinger, J., Bollier, D. (2005). A renaissance in the commons: How the new sciences and the Internet are framing a new global identity and order. In R. A. Ghosh (Ed.), CODE: Collaborative Ownership and the digital economy (pp. 259 -286). Cambridge, MA: MIT Press.

Connor, A. M., Marks, S. (Eds.). (2016). Creative technologies for multidisciplinary applications. Hershey, PA: Information Science Reference, an imprint of IGI Global.

Conway, F., Siegelman, J. (2005). Dark hero of the information age: In search of Norbert Wiener, the father of cybernetics. New York: Basic Books.

Coole, D. H., Frost, S. (2010). New materialisms ontology, agency, and politics. Durham [NC]: Duke University Press.

Cubitt, S. (2017). Finite media: Environmental implications of digital technologies. Durham: Duke University Press.

Culp, A. (2016). Dark Deleuze.

Damasio, A. R. (1999). The feeling of what happens: Body and emotion in the making of consciousness (1st ed.). New York: Harcourt Brace.

Damasio, A. R. (2003). Looking for Spinoza: Joy, sorrow and the feeling brain. London: Heinemann.

Daniel, R., Rubens, J., Sarpeshkar, R. et al. Synthetic analog computation in living cells. Nature 497, 619–623 (2013). https://doi.org/10.1038/nature12148

Dawkins, R. (1976). The selfish gene. Oxford: Oxford University Press.

Dawkins, R. (1986). The blind watchmaker. Harlow: Longman Scientific Technical.

Dean, J.W. Altemeyer, B. (2020) Authoritarian Nightmare: Trump and his followers. Melville House.

Deleuze, G., Guattari, F. (1987). A thousand plateaus: Capitalism and schizophrenia. Minneapolis: University of Minnesota Press.

Desmond, Timothy. (2018). Psyche and Singularity: Jungian Psychology and Holographic String Theory (Kindle).

Diamandis, P. H., Kotler, S. (2012). Abundance: The Future Is Better Than You Think (Reprint edition). Free Press.

Dunne Raby. (2001) Design Noir: The Secret Life of Electronic Objects.Birkhauser.

Dunne, A., Raby, F. (2013). Speculative Everything: Design, Fiction, and Social Dreaming. The MIT Press.

Easton, B. (1997). The Commercialisation of New Zealand. Auckland University Press.

Ebner, J. (2020). Going dark: The secret social lives of extremists. Bloomsbury Publishing.

Edwards, P. N. (1997). The closed world: Computers and the politics of discourse in Cold War America. Cambridge, Mass.: MIT Press, 1997.

Edwards, P. N. (2012). Entangled histories: Climate science and nuclear weapons research. Bulletin of the Atomic Scientists, 68(4), 28.

Ehrenkranz, N.J. MD, D.A. Sampson. (2008) Origin of the Old Testament Plagues: Explications and Implications. Yale Journal Of Biology And Medicine 81 (2008), pp.31-42.

Evans, L., Grimes, A., Wilkinson, B., Teece, D. (Dec. 1996)

"Economic Reform in New Zealand 1984-95: The Pursuit of Efficiency". Journal of Economic Literature. Vol. XXXIV. No.4 pp. 19856-1902. American Economic Association.

Everett, D. L. (2017). How language began: The story of humanity's greatest invention. London: Profile Books.

Festinger, L., Riecken, H. W., Schachter, S. (1956). When prophecy fails. Minneapolis: University of Minnesota Press.

Filimowicz, M., Tzankova, V. (2017). Teaching Computational Creativity.

Finn, E. (2017). What algorithms want: Imagination in the age of computing. Cambridge, Massachusetts: MIT Press.

Fisher, M. (2010). Capitalist realism: Is there no alternative? Winchester, UK: Zero Books.

Flanagan, M. (2009). Critical Play: radical game design. MIT Press.

Fleming, P. (2021) Dark Academia: How Universities Die. Pluto Press. London.

Fogg, B.J. (1998). "Persuasive Computers: Perspectives and Research Directions". Proceedings of the SIGCHI Conference on Human Factors in Computing Systems. CHI '98. New York, NY, USA: ACM Press/Addison-Wesley Publishing Co.: 225–232. doi:10.1145/274644.274677. ISBN 9780201309874.

Fogg, B. J. (1999). Persuasive Technologies. Communications of the ACM, 42(5), 26–29.

Foster, J. B., Clark, B., York, R. (2011). The Ecological Rift Capitalisms War on the Earth. Monthly Review Press.

Foster, J. B. (2000). Marx's ecology materialism and nature. Monthly Review Press.

Fox, A. P. (2001). The power game: the development of the Manapouri-Tiwai Point electro-industrial complex, 1904-1969 (Thesis, Doctor of Philosophy). University of Otago. Retrieved from http://hdl.handle.net/10523/335

Fuller, R. B. (1976). Operating manual for spaceship earth. Mattituck, N.Y.: Aeonian Press Inc.

Galloway, A. R. (2004). Protocol: How control exists after decentralization. Cambridge, Massachusetts: MIT Press.

Garn, A., Antonelli, P., Kultermann, U., Van Dyk, S. H. (2007). Exit to tomorrow: World's fair architecture, design, fashion, 1933-2005. Retrieved from http://catalog.hathitrust.org/api/volumes/oclc/144524840.html

George, A. R. (Ed.). (2003). The epic of Gilgamesh: The Babylonian epic poem and other texts in Akkadian and Sumerian. London; New York: Penguin Books.

Gleick, J. (2011). Chaos: Making a new science. New York, N.Y: Open Road Integrated Media.

Gladwell, M. (2000). The Tipping Point: How little things can make a big difference (1st ed.). Little, Brown.

Goldfinch, S.(1998). "Ranking New Zealand's Economic Policy: Institutional Elites as Radical Innovators 1984-1993". Governance: An International Journal of Policy and Administration. Vol. 11. No.2. p.177

Goldin, C. D., Katz, L. F. (2008). The race between education and technology. Cambridge, Mass.: Belknap Press of Harvard University Press, 2008.

Graeber, D. (2011). Debt: The First 5,000 Years (1St Edition edition). Brooklyn, N.Y: Melville House.

Graeber, D. (2018). Bullshit jobs: A theory. Allen Lane.

Graham, F. (1970). Since Silent spring. Boston: Houghton-Mifflin.

Grau, O. (2003). Virtual art: From illusion to immersion ([Rev. and expanded). Cambridge, Mass.: MIT Press.

Greene, B. (2005). The Fabric of the Cosmos: Space, Time and the Texture of Reality (New Ed edition). Penguin.

Greene, B. (2011). The Hidden Reality: Parallel Universes and the Deep Laws of the Cosmos. Penguin.

Greenspan, A. (2007) Alan Greenspan: the age of turbulence. Adventures in a new world. Penguin Books.

Grier, D. A. (2007). When computers were human. Princeton, N.J.; Woodstock: Princeton University Press.

Hale, E. E. (1899). The brick moon, and other stories. Boston: Little, Brown, and company.

Hall, J. R. (2013). Apocalypse: From Antiquity to the Empire of Modernity. Retrieved from http://qut.eblib.com.au/patron/ FullRecord.aspx?p=1180369

Hansen, M. B. N. (2015). Feed-forward: On the future of twenty-first-century media. Chicago; London: University of Chicago Press.

Haraway, D. (1993). A cyborg manifesto. In S. During (Ed.), The Cultural studies reader (pp. 271–291). London: Routledge.

Haraway, D. J. (2016). Staying with the Trouble: Making kin in the Chthulucene.

Harman, G. (2016). Immaterialism objects and social theory. Cambridge, UK Malden: MA Polity Press.

Harman, Graham. (2018) Object-Oriented Ontology: A New Theory of Everything. Penguin Books Ltd. Kindle Edition.

Harman, G. (2018). Speculative Realism: An introduction. Medford, MA: Polity.

Harvey, D. (2005). A Brief History of Neoliberalism. OUP Oxford. Kindle Edition.

Hawken, P. (Ed.). (2017). Drawdown: The most comprehensive plan ever proposed to reverse global warming. New York, New York: Penguin Books.

Hawking, S., Redmayne, E., Thorne, K. S., Hawking, L. (2018). Brief answers to the big questions.

Hayles, K. (2017). Unthought: The power of the cognitive nonconscious.

Hayles, N. K. (1999). How we became Posthuman: Virtual bodies in cybernetics, literature, and informatics. Chicago, Ill.: University of Chicago Press.

Heim, M. (1993). The metaphysics of virtual reality. New York: Oxford University Press.

Hill, C. (1972). The world turned upside down: Radical ideas during the English revolution. London: Temple Smith.

Hope, W. (2016). Time, communication and global capitalism. Retrieved from http://public.eblib.com/choice/publicfullrecord.aspx?p=4720317

Hutchings, J. (2015). Te Mahi Māra Hua Parakore – A Māori Food Sovereignty Handbook. Wheelers. Auckland, N.Z.

Hutchings, J. & Smith, J. (eds.) (2020). Te mahi oneone hua parakore : a Māori soil sovereignty and wellbeing handbook

Isaacson, W. (2014). The Innovators: How a group of hackers, geniuses, and geeks created the digital revolution (First Simon Schuster hardcover edition). New York: Simon Schuster.

Ismail, S., Malone, M. S., Geest, Y. van, Diamandis, P. H. (2014). Exponential Organizations: Why new organizations are ten times better, faster, and cheaper than yours (and what to do about it). Diversion Books.

Jardini, D. (2013). Thinking Through the Cold War: RAND, National Security and Domestic Policy, 1945-1975. Meadow Lands: David Jardini.

Jesson, B. (1999). Only Their Purpose is Mad. The Dunmore Press. New Zealand.

Johansen, B. E. (Ed.). (2015). American Indian culture: From counting coup to wampum. Santa Barbara: Greenwood.

Johnson, T. (2016) 'Nitrogen Nation: The Legacy of World War I and the Politics of Chemical Agriculture in the United States, 1916–1933.' The Agricultural History society, 2016 dOi: 10.3098/ah.2016.090.2.209

Jung, C. G., Franz, M.-L. von. (1964). Man and his symbols. Garden City, N.Y: Doubleday.

Jung, C. G., Jaffé, A. (1989). Memories, dreams, reflections (Rev. ed). New York: Vintage Books.

Kaku, M. (2018). The future of humanity: Terraforming Mars, interstellar travel, immortality, and our destiny beyond Earth (First edition). New York: Doubleday, a division of Penguin Random House, LLC.

Karacaoglu, G. (2021) Love You: Public policy for intergenerational wellbeing. The Tuwhiri Project. Apple Books. Wellington, Aotearoa, New Zealand.

Keen, A. (2015). The Internet is not the answer (First edition). New York: Atlantic Monthly Press.

Kegan, R., Lahey, L. L., Kegan. (2009). Immunity to Change: How to Overcome It and Unlock the Potential in Yourself and Your Organization.Harvard Business Review Press.

Kelly, K. (1994). Out of control: The rise of neo-biological civilization. Reading, Mass.: Addison-Wesley.

King, M. (2003) The Penguin History of New Zealand. Penguin

Klein, N. (2007). The Shock Doctrine: The Rise of Disaster Capitalism. Penguin Books.

Knight, C. (2016). Decoding Chomsky: Science and revolutionary politics. New Haven: Yale University Press.

Kornberger, M. (2010). Brand society: How brands transform management and lifestyle. Cambridge; New York: Cambridge University Press.

Korowicz, D. (2011) On the cusp of collapse: complexity, energy, and the globalised economy.

Kosko, B. (1993). Fuzzy Thinking: the new science of fuzzy logic. Flamingo. London.

Kurzweil, R. (2005). The singularity is near: When humans transcend biology. New York: Viking.

Latour, B. (1993). The Pasteurization of France. Cambridge, Mass: Harvard University Press.

Latour, B. (2004). Politics of nature: How to bring the sciences into democracy. Cambridge, Mass.: Harvard University Press, 2004.

Lessig, L. (2015). Republic Lost: The corruption of equality and the steps to end it (Revised edition). New York, NY: Twelve.

Lowry, D., Michel, S. E., et al. (2019). Very strong atmospheric methane growth in the 4 years 2014–2017:Implications for the Paris Agreement. Global Biogeochemical Cycles, 33 ,318–342. https://doi.org/10.1029/2018GB006009

Malcolm, W. G., Tarling, N. (2007). Crisis of identity?: The mission and management of universities in New Zealand. Wellington [N.Z.: Dunmore Pub.

Mandeville, B. (1714) The Fable of The Bees: or, Private Vices, Publick Benefits.

Mann, M.E. (2021) The New Climate War: the fight to take back our planet. Scribe Publications. New York.

Manovich, L. (2001). The language of new media. Cambridge, Mass.: MIT Press.

Manovich, L. (2013). Software takes command: Extending the language of new media. New York: Bloomsbury Academic.

Margulis, L., Sagan, D. (1997). Slanted Truths: Essays on Gaia, symbiosis, and evolution. New York: Copernicus.

Marinova, D., Phillimore, J. (2003). Models of Innovation. In L. V. Shavinina (Ed.), The international handbook on innovation (pp. 44–53). Amsterdam; Boston: Elsevier.

Mau, B., Leonard, J., Institute without Boundaries. (2004). Massive change. London: Phaidon.

Massy, Charles. (2017) Call of the Reed Warbler: A New Agriculture – A New Earth. University of Queensland Press. Kindle Edition.

Maturana, H. R., Varela, F. J. (1980). Autopoiesis and cognition: The realization of the living. New York: Springer.

Maxwell, G. (2017). The dynamics of transformation: Tracing an emerging world view.

McCarthy, Lauren; Reas, Casey; Fry, Ben. (2016). P5.js. Maker Media.

McLintock, A.H. (Ed.) (1966) An Encyclopedia of New Zealand. Retrieved from https://teara.govt.nz/en/1966/farming

McCormack, G. (2016). Cook Islands Seabed Minerals: precautionary approach to mining. Cook Islands Natural Heritage Trust. Rarotonga.

McCumber, J. (2016). The philosophy scare: The politics of reason in the early Cold War. Chicago: The University of Chicago Press.

McGonigal, J. (2011). Reality is Broken . Random House. Kindle Edition

McGuire, B. (2013). Waking the giant: How a changing climate triggers earthquakes, tsunamis, and volcanoes. Oxford, United Kingdom: Oxford University Press.

McKibben, B. (2019). Falter: Has the human game begun to play itself out? (First edition). New York: Henry Holt and Company.

McLuhan, M. (1994). Understanding media: The extensions of man. London: Routledge.

Meillassoux, Q., Brassier, R., Badiou, A., Bloomsbury Publishing. (2017). After finitude: An essay on the necessity of contingency. London [etc.: Bloomsbury Academic an imprint of Bloomsbury Publishing Plc.

Melville, H. (1991). Moby-Dick. New York: Knopf: Distributed by Random House.

Miedaner, T., Geiger, H. H. (2015). Biology, genetics, and management of ergot (Claviceps spp.) in rye, sorghum, and pearl millet. Toxins, 7(3), 659–678. https://doi.org/10.3390/toxins7030659

Minsky, M. L. (1986). The Society of Mind. New York: Simon and Schuster.

Moelling, K. (2017). Viruses: More Friends Than Foes. World Scientific Publishing Company. Kindle Edition.

Moravec, H. P. (1988). Mind children: The future of robot and human intelligence. Cambridge, Mass: Harvard University Press.

Montague, R. (2007). Your brain is (almost) perfect: How we make decisions. New York London: Plume; Turnaround [distributor].

Montfort, N., Bogost, I. (2009). Racing the beam: The Atari video computer system. Cambridge, Mass. A: MIT Press.

Morton, T. (2013). Hyperobjects: Philosophy and ecology after the end of the world. Minneapolis: University of Minnesota Press.

Morton, T. (2017). Humankind: Solidarity with nonhuman people.

Morton, T. (2018). Dark Ecology: For a logic of future coexistence. S.l.: Columbia University Press.

Nagle, A. (2017). Kill all normies: Online culture wars from 4chan and Tumblr to Trump and the alt-right. Winchester: Zero Books.

Nelson, H. G., Stolterman, E. (2003). The design way: Intentional change in an unpredictable world: foundations and fundamentals of design competence. Englewood Cliffs, N.J: Educational Technology Publications.

Neumann, B., Vafeidis, A. T., Zimmermann, J., Nicholls, R. J. (2015). Future Coastal Population Growth and Exposure to Sea-Level Rise and Coastal Flooding—A Global Assessment. PLOS ONE, 10(3), e0118571. https://doi.org/10.1371/journal.pone.0118571

Nisbet, E. G., Manning, M. R.,Dlugokencky, E. J., Fisher, R.

Noble, D. D. (1991). The classroom arsenal: Military research, information technology, and public education. London; New York: Falmer.

Noble, J.J. (2012). Programming Interactivity. Beijing; Sebastopol,CA: O'Reilly.

O'Malley, V. (2016). The great war for New Zealand: Waikato 1800-2000. Wellington, New Zealand: Bridget Williams Books.

O'Regan, G. (2013). Giants of computing: A compendium of select, pivotal pioneers. London: Springer, [2013].

Osgood, K. A. (2006). Total Cold War: Eisenhower's secret propaganda battle at home and abroad. Lawrence: University of Kansas.

Oreskes, N., Conway, E. M. (2010). Merchants of Doubt: How a handful of scientists obscured the truth on issues from tobacco smoke to global warming (1st U.S. ed). Bloomsbury Press

Papanek, V. J. (1972). Design for the real world. London: Thames and Hudson, 1972. (City Campus Main Collection 745.2 PAP).

Papanek, V. J. (1995). The green imperative: Natural design for the real world. New York: Thames and Hudson.

Perelman, Michael. (2000). The Invention of Capitalism. Duke University Press. Kindle Edition.

Pilling, D. (2018). The growth delusion: Wealth, poverty, and the well-being of nations (1st American Edition). New York: Tim Duggan Books.

Piot, P. (2013). No time to lose: A life in pursuit of deadly viruses. W.W. Norton. New York; London.

Prebble, R. (1996). I've Been Thinking.

Prigogine, I., Stengers, I. (1984). Order out of chaos: Man's new dialogue with nature. Toronto; New York, N.Y: Bantam Books.

Randle, M., Eckersley, R. (2015). Public perceptions of future threats to humanity and different societal responses: A cross-national study. Futures, 72, 4–16. https://doi.org/10.1016/j.futures.2015.06.004

Raworth, K. (2017) Doughnut Economics: Seven ways to think like a 21st century economist. Random House.

Reed, A.W. (1963). An Illustrated Encyclopedia of Māori Life. A.H. Reed A. W. Reed. Wellington. Auckland. Sydney.

Reed, A.W. (1964). Māori Fables and Legendary Tales. A.H. Reed A. W. Reed. Wellington. Auckland. Sydney.

Reichenbach, H. (1951). The rise of scientific philosophy. Berkeley: University of California Press.

Reisch, G. A. (2005). How the Cold War transformed philosophy of science: To the icy slopes of logic. Cambridge; New York: Cambridge University Press.

Rifkin, J. (1991). Biosphere politics: A new consciousness for a new century (1st ed.). New York: Crown.

Rifkin, J. (2014). The Zero Marginal Cost Society: The Internet of Things, the Collaborative Commons, and the Eclipse of Capitalism (Reprint edition). St. Martin's Press.

Riva, G., Davide, F., Ijsselsteijn, W. A. (2003). Being there: Concepts, effects and measurements of user presence in synthetic environments. Amsterdam; Washington, D.C. Tokyo: IOS Press; Ohmsha.

Rive, L.F. (2019). Queering the Webseries. University of Auckland.

Rive, P. B. (1984). A wealth of knowledge in a bankrupt databank: The politics of establishing a "universal database", PRESTEL, and the implications for videotex development in New Zealand. University of Auckland, Auckland.

Rive, P. B. (2012). Design in a Virtual Innovation Ecology: A Cybernetic Systems Approach to Knowledge Creation and Design Collaboration in Second Life. Retrieved from http://researcharchive.vuw.ac.nz/handle/10063/2747

Rive, P. B., Thomassen, A. (2012). International Collaboration and

Design Innovation in Virtual Worlds: Lessons from Second
Life. In Computer-Mediated Communication Across
Cultures: International Interactions in Online
Environments (pp. 429–448). Hershey PA, USA: Information
Science Reference – IGI Global.

Rive, P., Billinghurst, M., Thomassen, A., Lyons, M. (2008, July).
Face to face with the white rabbit—Sharing ideas in Second
Life. 1–14. https://doi.org/10.1109/IPCC.2008.4610236

Rive, P. B. (2019). Worldbending: A Survivor's Guide, for those
who want to think and act creatively about our future. World
Benders.

Rothblatt, Martine, Kurzweil, R. (2014). Virtually Human: The
Promise—and the Peril—of Digital Immortality. St. Martin's
Press.

Royal, Te Ahukaramū Charles. (2007) Mātauranga Māori and
Museum Practice. Museum of New Zealand Te Papa
Tongarewa.

Rushkoff, D. (2016). Throwing rocks at the Google bus: How
growth became the enemy of prosperity. Londen: Portfolio
Penguin.

Scranton, L. (2014). China's Cosmological Prehistory: The
Sophisticated Science Encoded in Civilization's Earliest
Symbols (1 edition). Inner Traditions.

Scranton, L. (2018) Decoding Māori Cosmology: the ancient
origins of New Zealand's Indigenous Culture. Inner
Traditions. Vermont.

Shaviro, S. (2009). Without criteria: Kant, Whitehead, Deleuze,
and aesthetics. Cambridge, Mass: MIT Press.

Shaviro, S. (2014). The universe of things: On speculative realism.
Minneapolis: University of Minnesota Press.

Shaviro, S. (2015). No speed limit three essays on accelerationism.
Minneapolis: University of Minnesota Press.

Sheldrake, Merlin. (2020). Entangled Life. Random House. Kindle Edition.

Sinclair, K. (1959). History of New Zealand.

Smith, S.A. Talking Toads and Chinless Ghosts: The Politics of "Superstitious" Rumors in the People's Republic of China, 1961–1965, The American Historical Review, Volume 111, Issue 2, April 2006, Pages 405–427, https://doi.org/10.1086/ahr.111.2.405

Soar, M. Smith,V. Dentith,M.R.X. Barnett,D. Hannah,K. Riva,G.

Sporle, A. (6th Sept. 2020) Evaluating the infodemic: assessing the prevalence and nature of COVID- 19 unreliable and untrustworthy information in Aotearoa New Zealand's social media, January-August 2020. Te Pūnaha Matatini: Centre of Research Excellence for Complex Systems and Networks, New Zealand

Spikins, P., Wright, B., Hodgson, D.,. (2016). Are there alternative adaptive strategies to human pro-sociality? The role of collaborative morality in the emergence of personality variation and autistic traits. Time and Mind, 9(4). https://doi.org/10.1080/1751696X.2016.1244949

Sridhar, D. Majumder, M.S. (Published 21 April 2020) Modelling the pandemic: Over-reliance on modelling leads to missteps and blind spots in our response. BMJ 2020;369:m1567 doi: 10.1136/bmj.m1567

Steffen, W. L. (Ed.). (2004). Global change and the earth system: A planet under pressure. Berlin; New York: Springer.

Steintrager, J. (2011). Bentham. Routledge.

Stephenson, N. (1992). Snow crash. New York, N.Y.: Bantam Books.

Stephenson, N. (1995). The diamond age, or, A young lady's illustrated primer. London: Viking.

Stephenson, N. (1999). Cryptonomicon (1st ed.). New York: Avon Press.

Tainter, J. A. (2017). The collapse of complex societies. Cambridge University Press.

Tainter, J. A., Patzek, T. W. (2012). Drilling down the Gulf Oil debacle and our energy dilemma. Retrieved from http://dx.doi.org/10.1007/978-1-4419-7677-2

Taleb, N. N. (2008). The Black Swan: The Impact of the Highly Improbable (Re-issue edition). London: Penguin.

Tarnas, R. (1991). The passion of the Western mind: Understanding the ideas that have shaped our world view (1st ed). New York: Harmony Books.

Tarnas, R. (2006). Cosmos and psyche: Intimations of a new world view. New York: Viking.

Thacker, E. (2011). In the Dust of This Planet Horror of Philosophy vol. 1.O-Books.

Thiel, P., Masters, B. (2014). Zero to One: Notes on Startups, or How to Build the Future. New York: Crown Business.

Thomas, A. (2015). Hidden in plain sight: The simple link between relativity and quantum mechanics.

Trotter, C. (2007) No Left Turn: the distortion of New Zealand's history by greed, bigotry and right wing politics. Random House.

Tsing, A. L. (2015). The mushroom at the end of the world: On the possibility of life in capitalist ruins. Princeton: Princeton University Press.

Turchin, P. (2016). Ages of discord: A structural – demographic analysis of American history. Beresta Books.

Turner, F. (2006). From counterculture to cyberculture: Stewart Brand, the Whole Earth Network, and the rise of digital utopianism. Chicago, IL: University of Chicago Press.

Tucker, P. (2014). The naked future: What happens in a world that anticipates your every move.

Tye, L. (1998). The father of spin: Edward L. Bernays the birth of public relations (1st ed). New York: Crown Publishers.

United Nations Food and Agriculture Organization. (November 29, 2006) Livestock's long shadow: environmental issues and options. http://www.fao.org/3/a-a0701e.pdf

Von Krogh, G., Nonaka, I., Ichijo, K. (2000). Enabling knowledge creation: How to unlock the mystery of tacit knowledge and release the power of innovation. Oxford; New York: Oxford University Press.

Wahl,D.C. (2007) Scale-Linking Design For Systemic Health: Sustainable Communities And Cities In Context. International Journal of Ecodynamics. Vol. 2, No. 1 (2007) 1–16

Wahl, D.C. (2016) Designing Regenerative Cultures. Triarchy Press.

Wallis, J.L. (Jan. 1997) Policy Conspiracies and Economic Reform Programs in Advanced Industrial Democracies: The Case of New Zealand. UNE Working Papers in Economics No. 3. Ed. Brian Dollery. Dept. of Economics. University of New England. Australia. ISBN 1 86389 405 5 ISBN 1 86389 405 5

Watson, P. (2005). Ideas: A history from fire to Freud. London: Weidenfeld Nicolson.

Weart, S. (2019). The Discovery of Global Warming: A hypertext history of how scientists came to (partly) understand what people are doing to cause climate change. Retrieved from https://history.aip.org/climate/index.htm

Wertheim, M. (1999). The pearly gates of cyberspace: A history of space from Dante to the Internet. New York: W.W. Norton.

Whitehead, A. N. (2014a). Process and reality. Retrieved from http://www.myilibrary.com?id=893479

Whitehead, A. N. (2014b). Process and reality. Retrieved from http://www.myilibrary.com?id=893479

Wiedner, K. and Glaser, B. (Martin-Luther-University Halle-Wittenberg, (2013) The chapter 'Biochar-Fungi Interactions in Soils' included in Soil Biogeochemistry, von-Seckendorff-Platz 3, Halle, Germany.

Wiener, N. (1954a). The human use of human beings: Cybernetics and society (2nd ed.). New York: Doubleday.

Wiener, N. (1954b). The human use of human beings: Cybernetics and society (2nd ed.). New York: Doubleday.

Wiener, N. (1961). Cybernetics: Or control and communication in the animal and the machine (2nd ed.). Cambridge, Mass.: M.I.T. Press.

Wiener, N. (1964). God and Golem, inc: A comment on certain points where cybernetics impinges on religion. Cambridge: M.I.T. Press.

Williamson, J (ed.) (1995). The Political Economy of Policy Reform. Institute for International Economics.

Wilson, David Sloan. (2019) This View of Life . Knopf Doubleday Publishing Group. Kindle Edition.

Wilson, D.S., et al., 'Generalizing the core design principles for the efficacy of groups.' Journal of Economic Behavior Organization (2013), http://dx.doi.org/10.1016/j.jebo.2012.12.010

Wilson, C. (1963). The outsider. London: Pan.

Wishart, I. (2011) The Paradise Conspiracy. Howling at the Moon. New Zealand.

Winsbury, R. Lane, M. (May, 1979). 'Prestel is the first to Start'. Intermedia – special survey: videotex. London. IIC. Vol.7. No.3, p.11

Wolfe, T. (2016). The kingdom of speech (First edition). New York: Little, Brown and Company.

Wolff, M. (2018). Fire and fury: Inside the Trump White House (First edition). New York: Henry Holt and Company.

Wolfmeyer, M. (2014). Math education for America?: Policy networks, big business, and pedagogy wars. New York: Routledge, 2014.

Wolfram, S. (2002). A new kind of science. Wolfram Media.

Woodward, B. (2018). Fear: Trump in the White House. London, England: Simon Schuster.

www.ingramcontent.com/pod-product-compliance
Lightning Source LLC
Chambersburg PA
CBHW072039020426
42334CB00017B/1327